64 95

National Building Codes Handbook

Other McGraw-Hill Books of Interest

BIANCHINA • *Forms and Documents for the Builder*
BOLT • *Roofing the Right Way, Third Edition*
BYNUM, WOODWARD, RUBINO • *Handbook of Alternative Materials in Residential Construction*
CORINCHOCK • *Technician's Guide to Refrigeration Systems*
DOMEL • *Basic Engineering Calculations for Contractors*
FRECHETTE • *Accessible Housing*
FRECHETTE • *Bathroom Remodeling*
GERHART • *Everyday Math for the Building Trades*
HARRIS • *Noise Control Manual for Residential Buildings*
HUTCHINGS • *CPM Construction Scheduler's Manual*
JAHN, DETTENMAIER • *Offsite Construction*
MILLER, BAKER • *Carpentry and Construction, Second Edition*
PETRUCCI, DOMEL • *Residential Contracting: Hands on Project Management for the Builder*
PHILBIN • *The Illustrated Dictionary of Building Terms*
PHILBIN • *Painting, Staining and Finishing*
POWERS • *Kitchens: Professional's Illustrated Design and Remodeling Guide*
SCHARFF AND THE STAFF OF ROOFER MAGAZINE • *Roofing Handbook*
SCHARFF AND THE STAFF OF WALLS & CEILINGS MAGAZINE • *Drywall Construction Handbook*
SHUSTER • *Structural Steel Fabrication Practices*
SKIMIN • *The Technician's Guide to HVAC Systems*
WOODSON • *Professional Modeler's Manual*
WOODSON • *Be a Successful Building Contractor, Second Edition*

Dodge Cost Books from McGraw-Hill

MARSHALL & SWIFT • *Dodge Unit Cost Book*
MARSHALL & SWIFT • *Dodge Repair & Remodel Cost Book*
MARSHALL & SWIFT • *Dodge Heavy Construction Unit Cost Book*
MARSHALL & SWIFT • *Dodge Electrical Construction Unit Cost Book*
MARSHALL & SWIFT • *Home Repair and Remodel Cost Guide*

National Building Codes Handbook

Jonathan F. Hutchings, BCM

McGraw-Hill
New York San Francisco Washington, D.C. Auckland Bogotá
Caracas Lisbon London Madrid Mexico City Milan
Montreal New Delhi San Juan Singapore
Sydney Tokyo Toronto

Library of Congress Cataloging-in-Publication Data

National building codes handbook / Jonathan F. Hutchings.
   p.   cm.
Includes index.
1. Building laws—United States. I. Title.
KF5701.H87   1997
343.73'07869—dc21   97-23520
                                                                      CIP

McGraw-Hill

A Division of The McGraw·Hill Companies

Copyright © 1998 by Jonathan F. Hutchings. All rights reserved. Printed in the United States of America. Except as permitted under the United States Copyright Act of 1976, no part of this publication may be reproduced or distributed in any form or by any means, or stored in a data base or retrieval system, without the prior written permission of the publisher.

1 2 3 4 5 6 7 8 9 0   FGR/FGR   9 0 2 1 0 9 8 7

ISBN 0-07-031819-0

The sponsoring editor for this book was Zoe G. Foundotos, the editing supervisor was Caroline R. Levine, and the production supervisor was Pamela A. Pelton.

Printed and bound by Quebecor/Fairfield.

This book is printed on recycled, acid-free paper containing a minimum of 50% recycled, de-inked fiber.

Information contained in this work has been obtained by The McGraw-Hill Companies, Inc. ("McGraw-Hill") from sources believed to be reliable. However, neither McGraw-Hill nor its authors guarantee the accuracy or completeness of any information published herein, and neither McGraw-Hill nor its authors shall be responsible for any errors, omissions, or damages arising out of use of this information. This work is published with the understanding that McGraw-Hill and its authors are supplying information, but are not attempting to render engineering or other professional services. If such services are required, the assistance of an appropriate professional should be sought.

This book is dedicated to the cast of the construction process: the designers, builders, engineers, and inspectors of the American Construction Industry.

There's a great old traditional saying that an *architect* is a person who knows a very little about a great deal and keeps knowing less and less about more and more until he knows practically nothing about everything. An *engineer* is a person who knows a great deal about very little and who goes along knowing more and more about less and less until he knows practically everything about nothing. A *contractor* is a person who starts out knowing practically everything about all things, but winds up knowing nothing about anything, due to his association with architects and engineers.

Inspectors, of course, are there to referee.

Entered words or products which we have reason to believe constitute trademarks have been designated as such. However, neither the presence nor the absence of such a designation should be regarded as affecting the legal status of any trademark.

"This publication is designed to provide accurate authoritative information in regard to the subject matter covered. It is sold with the understanding that neither the publisher nor author is engaged in rendering legal, accounting, or other professional service. If legal or expert assistance is required, the services of a competent professional person should be sought."

From a declaration of principles jointly adopted by a committee of the American Bar Association and committee of the Publishers Association.

Copyright acknowledgments to the Standard Building Code for the use of reprinted sections of the 1994 edition. Written permission to reproduce this material was sought from and granted by the copyright holder, Southern Building Code Congress International, Inc., 900 Montclair Road, Birmingham, Alabama 35213-1206. Copyright acknowledgments also to *Builder* magazine, *Automated Builder* magazine, and Simpson Strong-Tie Company for information on next-step construction technologies, new materials, and methodologies.

Contents

Preface xiii
Acknowledgments xxi

Chapter 1. Administration 1

101.0	General	2
102.0	Building Department	5
103.0	Powers and Duties of the Building Official	8
104.0	Permits	10
105.0	Required Inspections	20
106.0	Certificates	25
107.0	Tests	27
108.0	Construction Board of Adjustment and Appeals	27
110.0	Violations and Penalties	31

Chapter 2. Terms and Definitions 33

201.0	Definitions	33

Chapter 3. Occupancy 59

301.0	Occupancy Classification	59
302.0	Definitions	61
303.0	Mixed Occupancies	61
304.0	Assembly Occupancy–Group A	63
305.0	Business Occupancy–Group B	65
306.0	Educational Occupancy–Group E	67
307.0	Factory-Industrial Occupancy–Group F	68
308.0	Hazardous Occupancy–Group H	69
309.0	Institutional Occupancy–Group I	72
310.0	Mercantile Occupancy–Group M	73
311.0	Residential Occupancy–Group R	73
312.0	Storage Occupancy–Group S	75

Chapter 4. General Building Limitations — 77

- 500.0 General Limitations — 77
- 503.0 Height and Area — 77
- 504.0 Buildings Located on the Same Lot — 85
- 505.0 Buildings Located within a Fire District — 87

Chapter 5. Types of Code Construction — 89

- 601.0 Construction Types — 89
- 602.0 Definitions — 91
- 603.0 Type I Construction — 91
- 604.0 Type II Construction — 92
- 605.0 Type III Construction — 93
- 606.0 Type IV Construction — 96
- 607.0 Type V Construction — 96
- 608.0 Type VI Construction — 97
- 609.0 Partitions — 97
- 610.0 Mixed Types of Construction — 98

Chapter 6. Fire-resistant Materials and Construction — 99

- 700.0 Fire-resistant Materials and Construction — 99
- 702.0 Definitions — 103
- 703.0 Materials for Fire Resistance — 105
- 704.0 Fire-resistant Separations and Ratings — 108
- 705.0 Protection of Openings — 118
- 706.0 Combustibles in Fire-Rated Assemblies — 133
- 707.0 Combustibles in Concealed Spaces — 134
- 708.0 Thermal Insulating Materials — 134
- 709.0 Calculated Fire Resistance — 136

Chapter 7. Interior Finishes — 145

- 801.0 Interior — 145
- 803.0 Restrictions on Interior Finishes — 148
- 804.0 Acoustical Ceiling Systems — 153

Chapter 8. Fire Protection Systems — 155

- 901.0 Fire Protection — 155
- 902.0 Definitions — 156
- 903.0 Sprinklers — 157
- 904.0 Standpipes — 161
- 905.0 Fire Alarms — 164

Chapter 9. Means of Egress — 169

- 1001.0 Egress — 169

1002.0	Definitions	172
1003.0	Occupant Load and Means of Egress Capacity	173
1004.0	Arrangement and Number of Exits	176
1005.0	Special Exit Requirements	181
1006.0	Stairway Protection	188
1007.0	Stairway Construction	190
1008.0	Access to Roof	195
1009.0	Horizontal Exits	195
1010.0	Exit Discharge	196
1011.0	Fire Escapes	198
1012.0	Doors	199
1013.0	Ramps	204
1014.0	Balconies, Porches, and Galleries	206
1015.0	Guardrails	206
1016.0	Exit Illumination and Signs	207
1017.0	Exit Obstructions	210
1018.0	Special Egress Requirements by Occupancy	210
1019.0	Assembly	210
1020.0	Business	225
1021.0	Educational	226
1022.0	Factory-Industrial	227
1023.0	Hazardous	227
1024.0	Institutional	228
1025.0	Mercantile	234
1026.0	Residential	235
1027.0	Storage	236

Chapter 10. Exterior Walls 239

1400.0	Exterior Walls	239
1402.0	Definitions	239
1403.0	Veneered Walls	239
1404.0	Architectural Trim, Balconies, and Bay Windows	255
1405.0	Fire Department Access in Exterior Walls	255

Chapter 11. Roof Structures 259

1501.0	General	259
1502.0	Definitions	259
1503.0	Penthouse-Type Roof Structures	261
1504.0	Tanks	263
1505.0	Cooling Towers	264
1506.0	Other Roof Structures	264
1507.0	Parapet Walls	265
1508.0	Gutters and Leaders	266
1509.0	Roof Coverings	266

Chapter 12. Structural Loads 293

1600.0	Structural	293
1602.0	Definitions	296
1603.0	Dead Loads	296
1604.0	Live Loads	297
1605.0	Snow Loads	304
1606.0	Wind Loads	304
1607.0	Earthquake Loads	313

Chapter 13. Foundations and Retaining Walls   339

1800.0	Foundations	339
1802.0	Definitions	340
1803.0	Excavations	340
1804.0	Footings and Foundations	342
1805.0	Piles	363

Chapter 14. Concrete   373

1901.0	Concrete	373
1903.0	Materials	376
1904.0	Durability Requirements	378
1905.0	Concrete Quality	384
1906.0	Mixing and Placing of Concrete	391
1907.0	Formwork, Embedded Pipes, and Construction Joints	393
1908.0	Details of Reinforcement	397
1909.0	Slab on Ground	401
1910.0	GFRC Exterior Wall Panels	401
1911.0	Parapet Walls	402
1912.0	Seismic Provisions	402

Chapter 15. Electrical   411

2700.0	General	412
2702.0	Definitions	412

Chapter 16. Wood and Framing   427

2300.0	General	427
2301.0	Scope	429
2302.0	Definitions	436
2303.0	Construction Practices	436
2304.0	Protection Against Decay and Termites	437
2305.0	Fire Protection	441
2306.0	Fastenings	443
2307.0	Floor Framing	444
2308.0	Vertical Framing	450
2309.0	Roof and Ceiling Framing	457
2310.0	Wood Structural Panel Diaphragms	460

2311.0	Particleboard Diaphragms	464
2312.0	Seismic Provisions	466

Chapter 17. Gypsum Board and Plaster — 479

2501.0	General	479
2502.0	Definitions	479
2503.0	Materials	479
2504.0	Application	479
2505.0	Allowable Partition Heights	481
2506.0	Vertical Gypsum Board Diaphragms	481

Appendix. Code Specification Tables — 487

Index 531

Preface

In retrospect of my career in construction management, I have had the extraordinary good fortune of meeting a wide variety of great people in the industry from designers, contractors, engineers, inspectors, and architects to commercial developers, financiers, and city planners. Though all were very different individuals, those who were really successful in any field had two common traits: They loved their work, and they truly cared about the unseen public who would inhabit their buildings long after the building was certified for occupancy. That is really what the model national building codes are all about: The fire-, life-, and safety-protection of the public. Any builder worth his or her salt comes to learn that there is more to construction than just making a profit. Knowledge of the *intent* as well as the letter of the law of the codes is vital in modern code compliance construction and in the business survival of your company.

The construction industry has changed in the last decade dramatically from companies who once dominated their markets by size, bidding, and bonding power to the modern, lean, downsized builder/designer/contractor companies of today who use strategic personnel in definitive operations management. In the trades, these contractors are called *paper contractors,* because all they handle is paper. We also call them smart. No tools, no trucks, few employees. Their success formula is simple: Tight bid + lower overhead = we get the job. Other job markets outside your local region are the new markets that can profit through knowledge of what is being built and where, as well as the relative national building codes to plan and estimate with. This book provides the necessary cross-link between all the national codes and can be used effectively at work tomorrow.

In addition to new sophistication in modern construction codes, the new age of technololgy has also brought in new materials, new technology, and powerful computer systems using dedicated software programs, opening the way for further changes in the American construction industry. I have earned my living in the construction industry for more than 20 years now, from a high school laborer to apprentice, to journeyman, to foreman, to contractor; in management roles from estimator to project manager, to project scheduler, to state teaching credentials in construction technologies. In those two decades I have seen many changes that foreshadow even bigger changes in the future for the American builder as the new century accelerates toward us. Of all the things we build with today, 75% have been invented within the last 50 years.

Of those, 80% have been invented in the last 10 years. It is estimated that, of all the things we will be building with by the year 2025, *90% have not even been invented yet.* Graphite composite plastics that are stronger than steel, airentrained lightweight concrete, and ultralight metal alloys are all space-age technology being used in modern construction today. And even as you read this, the clock on their obsolescence is ticking. *Paradigm* is a word that means the currently accepted parameters of any specific thing, such as an industry. And the construction industry's paradigm is changing as fast as the clouds above your head. More important, as Alan Toffler so correctly stated in his prescient book, *Future Shock,* the *rate* of that change is accelerating exponentially.

Worldwide, in residential construction, American Western Platform Framing is the accepted standard of excellence in homebuilding. After serving as an estimator and consultant for the insurance industry's reconstruction efforts in both the 7.6 Loma Prieta Earthquake of 1989 and the 7.8 Northridge–Los Angeles firsthand, I now understand why the rest of the world sends their building officials and industry representatives to America to learn our methods of construction. When one stops and reflects, the reason the world looks to us for leadership in construction is quite clear.

We build the best. That is not just my opinion; that is documented fact. Here are the statistics. Sixty-one casualties in Northridge–Los Angeles compared to 5000 in Kobe, Japan. Both earthquakes rated 7.8 on the Richter scale. The difference is due to our highly evolved standards of construction building codes. In commercial and industrial construction, our methods of reinforced concrete and steel construction are unequaled in history. This is the last proud American industry still retaining global leadership and justifiably so. The construction industry is the largest single business enterprise in America. This industry is comprised of a diverse spectrum of highly trained and skilled professionals, but this book was geared for a specific foursome: the architect, the engineer, the contractor, and the inspector. The construction process begins with the design professionals. The two major professions within the design disciplines are the architect and engineer. Architecture is, in essence, the pursuit of the art and science of habitat for humans in the environment. However, architectural services usually extend beyond the phase of building design and the production of plans and drawings. An equally important task of the architect begins after the building design, cost estimates, and time schedules have been approved by the owner. The architect's knowledge of the national building codes relative to his or her clients' construction types in different cities and states is a distinct business advantage. But the job is not finished there.

Upon approval of the plans and specifications, the architect incorporates his or her design into the package of articles termed *contract documents*. These documents are a collection of related drawings, plans, specifications, legal notices, and contracts used by contractors to bid on a project. The drawings, plans, and specifications must be in compliance with the national building code having jurisdiction where the project is located. Multiproject jobs in different states, such as franchises, anchor tenants, and mall construction point out the need for a book like this that will provide the design professional a path to the required reference in any of the national building codes. The architect normally also presides over the bidding, assists in selection of a bidder, and aids in establishing contracts between the owner and the contractor. The architect, in essence, becomes the agent of the owner. The final and equally critical service the architect provides is overseeing the construction of the project. This includes the interpretation of the contract documents when necessary, assisting the contractor in dealings with the owner, ensuring that all contracts are being expedited in accordance with the plans and specifications, and that all contractual obligations have been satisfied by all parties to the contract.

During the construction process, an architect may retain consultants to supplement his or her own skills and to deliver a more complete, well-coordinated service to the client. The architect is trained in the integration of all building systems (structural, mechanical, electrical, etc.) and has basic knowledge and design skill in each of these areas. However, on large projects he or she will rely on the design capabilities of specialists in these engineering areas. In some architect offices, the engineers are placed on retainer or under contract for services. In large offices the staff usually includes one or more of these design professionals in a variety of disciplines. The major distinction between architects and engineers runs along generalist and specialist lines. Each is a true professional, certified and licensed through rigorous examinations. The engineer, however, is focused on an exhaustive study of a single technical area such as structural, civil, geotechnical and soils, or mechanical engineering. Both these design professionals must have as thorough an understanding of the letter and intent of the national building codes as do the builders and inspectors.

The national building codes have become standardized, and the differences in regulations between them have become less varied over the past years. Standardized section numbers and regulations have been done in a conscious effort by the national building codes publishers to standardize the many regulations into a cohesive nationwide format. The national building codes are revised and republished every 3 years in legal form for ready adoption by each local city and county jurisdictional authority nationwide. Most local communities typically adopt the relative code in routine entirety, but they are free to make changes in some sections, add material to others, or delete sections as they see fit. Therefore, the exact local code verbiage may differ slightly or significantly from the regional national building code. Always check with your local building department or official for specific local regulations and specifications. This handbook is unique and the first of its kind to cross-reference the national building codes. It does not replace the national building codes; it complements the national building code adopted by your local jurisdiction and provides commentary and analysis of those areas that have common historical points of contention.

The three national model building codes used in America are the *Standard Building Code* (SBC), *Uniform Building Code* (UBC), and the *National Building Code* (BOCA). All modern construction materials and practices are covered within these national building codes. These are the main focus of this book. Noted, but not covered herein, is the *1995 ASTM Standards in Building Codes*, 32d edition, the current edition at the time of this writing. It is a huge work, encompassing four volumes, some 7080 pages in all. This massive reference contains 1300 ASTM specifications as adopted by all model building codes including BOCA, UBC, CABO, and standard codes. The covered national building codes herein are

- The *Standard Building Code*. As do the other two model codes, this code adopts by reference the nationally recognized industry standards of materials and installations. The 1994 edition, current at the time of this writing, is referenced herein. The SBC was first adopted by the Southern Building Code Congress International in November of 1945 at the Annual Research Conference. Revised editions have been published since that time at approximately 3-year intervals. The 1994 Standard Building Code has been reformatted into the common building code format that was developed cooperatively by SBCCI, BOCA, ICBO, AIA, and SFPE under the auspices of the Council of American Building Officials (CABO).

- The *Uniform Building Code*. This code was first published in 1927 by the International Conference of Building Officials. The 1994 edition, current at the time of this writing, is cross-referenced by section number herein. Like all the model national building codes, the UBC has had revised editions published at 3-year intervals, and the new editions incorporate changes approved since the previous edition.

- The *National Building Code*. This code is the major model building code used in the eastern United States. The 1996 edition, current at the time of this writing, is cross-referenced by section number herein. The current edition includes changes approved through 1993 and changes made to standardize all 1996 BOCA codes and to update the references to standards. The oldest of the national building codes, BOCA was founded in 1915. It is a nonprofit, member-serviced organization dedicated to professional code administration and enforcement for the protection of public health, safety, and welfare.

Accordingly, the purpose of this book is not to reprint volumes of these many different code specifications, but to serve as a reference handbook of a specific topic section with discussion, commentary, and analysis of the most commonly encountered sections and regulations. It is based on the *Standard Building Code* with the related section numbers in UBC and BOCA codes bracketed for easy cross-reference. This book is also updated to contain both English units and metric equivalents to meet the new federal requirements of metric standardization on all federally funded projects. Analysis and explanations are included to give the reader a practical application understanding of the national building codes as they are meant to be adopted as law and are therefore worded in necessarily terse "legalese." Tables not shown in text appear in the Appendix.

The procedures under which the national building codes are prepared are designed to provide for an orderly introduction of new developments as construction technologies continue to improve in the industry (with particular emphasis on safety from the standpoint of end use). The committee panels who develop the different national building codes include parties of interest having expertise or technical competence in varying fields of construction management, building safety, or related industries working together with the objective of safeguarding the life-, safety-, and fire-protection of the public. This system of national building codes evolution has historically proven itself to be above the reach of special interests and has produced a worldwide standard for residential and light commercial buildings.

The designated purpose of the national building codes is to provide *minimum* standards to guard the life and safety of the public by regulating and controlling the design, construction, and quality of materials used in modern construction. The codes have also come to govern the use and occupancy, location of a type of building, and the ongoing maintenance of all buildings and facilities. Once adopted by a local jurisdiction, these national building code standards then become law. The provisions of the national building codes apply to the construction, alteration, moving, demolition, repair, structural maintenance, and use of any building or structure within the local jurisdiction, except public works projects located primarily in a deeded public way, utility towers and poles, mechanical equipment not specifically regulated in the codes, or hydraulic flood control structures.

The national building codes are the legal instruments that enforce public safety in modern construction of human habitation and assembly structures. They are used as such not only in the construction industry but also by the insurance industry for compensation appraisals and claims adjustments and by the legal industry for court litigation. Accordingly, the theme of the work is to show the intent as well as the letter of the regulations through analysis of the specific sections and regulations therein. As stated previously, the reader will note in this work that the section is preceded by the *Standard Building Code* (SBC) section number relative to that chapter, followed by the UBC- and BOCA-related section numbers respectively. These appear in brackets following the SBC section number throughout the book for ease in cross-referencing the national building codes. Brackets also indicate analysis to keep those passages separate from actual section wordage.

This work is focused on providing the construction designer, builder, property developer, project manager, or construction management professional a broader practical application knowledge and readily accessible understanding of the scope, meaning, and application of the their relative national building code regulations and specifications. Electrical, plumbing, and HVAC codes are not covered in this book as they each have their own respective national code. Not all the national building code chapters are covered herein as page limitation allows coverage of only the most commonly encountered or problematic of the code sections. Areas of special interest must be researched from within the national building codes themselves. The reader will note a compressed text with no traditional spacings between subheaders and text. To my previous readers, I apologize for the change in format continuity. I do this to get the maximum amount of wordage per page without going to a smaller (and harder to read) typeset point size. This allows me to retain the standard typeset that you are now reading. The style of presentation in this book is priority divisional referencing, reflecting the systems of the national building codes.

The standardization over the last decade of the national building codes has also brought into the world a distinct advantage in fire safety of public assembly buildings. Chapter 6 in each of the national building codes is "Types of Construction." When you see reference to the fire protection codes regulations (NFiPA), that fire code regulation will correlate to the fire wall ratings that are cross-referenced to chapter 6 of the national code you are using. The same is true for chapter 7 of all the national building codes, Fire Resistant Construction.

One of the critical factors that emphasize the importance of the national building codes is the highly competitive nature of the construction industry today. As in all businesses, safety and quality have historically been sacrificed for profit. But in this business if quality is sacrificed, people die. The national building codes are the public's only effective barrier against any compromise with basic structural safety. This book's analysis of the current national building codes is most effectively used by having your local jurisdictional approved copy of the relative code book at hand and referring to each section as it is discussed. The commentary given here is intended to supplement and clarify the actual wording of the covered code regulations as given in the respective code book itself.

For the most effective use of cross-referencing code regulations, the following SBC procedure for checking code compliance should be used by the reader. Sections referred to appear in their respective chapters.

1. Determine the occupancy classification of the structure. Select the occupancy classification which most accurately fits the use of the building.
2. Determine the actual physical properties of the building:
   (a) Determine the building area of each floor
   (b) Determine the grade elevation for the building
   (c) Determine the building height in feet above grade
   (d) Determine the building height in stories
   (e) Determine the separation distance from exterior walls to assumed and common property lines
   (f) Determine percent of exterior openings per floor
3. Determine the minimum type of construction necessary to accommodate the proposed structure:
   (a) Determine the maximum allowable heights and floor areas for types of construction and occupancy classification

(b) Check allowable height and area increases permitted
4. Check detailed occupancy requirements.
5. Check detailed construction requirements:
    (a) Fire protection of structural members
    (b) Fire protection requirements
    (c) Means of egress requirements
    (d) Special restrictions if in fire district
6. Review design as related to standards.
7. Check other requirements as necessary:
    (a) Construction projecting into public property
    (b) Elevators and conveying systems
    (c) Sprinklers, standpipes, and alarm systems
    (d) Use of combustible materials on the interior
    (e) Roofs and roof structures
    (f) Light, ventilation, and sanitation
    (g) Other

These steps are naturally varied in sequence by individual preferences; however, the first three are standard steps which should be followed in proper order to assist in design or review of buildings. Computing occupancy loads is tied to minimum egress requirements. Computing the size of the building is tied to allowable height and area. Classification of occupancy will determine type of construction and fire-resistance ratings.

Remember, in matters of the real-world applications of construction contracts, always research the contract specifications for code compliance thoroughly, make sure your cost estimates are as updated as possible, and consult an attorney in all contractual obligation matters. In the wordage of the national building codes, the word *may* is permissive and the word *shall* is mandatory, meaning that the code regulation is mandated by law. Further, please be informed that the facts, subjects, tables, fire ratings, and occupancy codes contained herein are as accurate as the model codes allow at this time of writing. This work is for general information purposes only and is not to be construed as legal advice. Both the publisher and the author assume no liability, either expressed or implied, for usage of this book. The contents of this book are protected from any form of physical or electronic copying, duplication, or storage retrieval, worldwide, by United States federal copyright law.

Acknowledgments

The following individuals and organizations contributed to the making of this Work. I would like to take this brief moment to thank them and give them their deserved moment of attention in the spotlight. These are professionals in the industry who are competent and caring people:

John Battles, P.E., and William B. Ware, General Counsel, Standard Building Code Council, Birmingham, AL

Sharon Goldstein, National Fire Protection Association, Quincy, MA

Robert Allen, International Fire Code Institute, Washington, DC

Barbara Reynolds, National Safety Council, Itasca, IL

The American National Standards Institute, New York, NY

Building Officials and Code Administrators International, Country Club Hills, IL

International Building Code Council, Inc., Whittier, CA

American National Standards Institute, New York, NY

Linde Altman, Builder magazine, Washington DC

Don O. Carlson, Automated Builder magazine, Carpinteria, CA

Richard R. Chapman, Simpson Strong-Tie Co., Pleasanton, CA

Thanks and a tip of the hat to my long-time editor and friend April D. Nolan, who helped conceive and develop the early stages of this book. Special thanks also to Zoe Foundotos, Senior Editor, for continued production assistance in the latter stages of the this book, Carol Levine, Editing Supervisor, and Lucy Mullins, Copy Editor. These exceptional individuals provided support and professional expertise that were pivotal factors in making this reference book available in the extensive McGraw-Hill Professional Book catalog. Without them, this book would have never made the long journey to your hands, your mind, and your business.

I would also like to thank all the fine folks I have dealt with over the past years in the many departments of the McGraw-Hill publishing house who have all been outstanding individuals and extraordinary literary professionals.

National Building Codes Handbook

Chapter 1

Administration

SECTION 100.0 TITLE and SCOPE [100.0; 100.0]

The national building codes are intended to be adopted by reference through statute or ordinance and enforced by state and local governments, governmental agencies, or other authorities having local jurisdiction. In preparation for adopting the code, the authority having local jurisdiction has the power to include such items as inspection fees, permit fees, and penalties or fines for noncompliance in the adopting legislation.

Enforcement is one of the basic and most important of code rules because it establishes the necessary conditions for the legal use of the respective code. The local building official's authority may be exercised either by enforcement of that code rule or by waiver of the code rule when the official is satisfied that a specific non-code-conforming method or technique satisfies the safety intent of the code. The code further stipulates that, when conflicts arise about the meaning or intent of any code rule as it applies to any particular installation, the building inspector's superior, the building official, is the only one authorized by the code to make final interpretations of the regulations.

It should be carefully noted that any code reference to special waiver permission as a basis for accepting any design or installation technique requires that such permission waiver be in written form. Whenever the inspection authority gives special permission for a condition that is at variance with code rules or not covered fully by the regulations, the authorization must be a written document and not simply verbal permission. All the official interpretations issued on a specific code edition are reviewed by the appropriate code-making panel during the period when the specific code edition is being revised. In reviewing an interpretation, a code panel may agree with the interpretation findings and clarify the code text to avoid misunderstanding of intent, or the panel may reject the findings of the interpretation and alter the code text to clarify the code panel's intent. On the other hand, the code panel may not recommend any change in the code text because of the special conditions described in the official interpretation.

Section 103 of the national building codes also establishes the right of the building official to render interpretations of the code for unique local circumstances and to determine code <u>intent</u> compliance of substitution materials or installations. This is done to provide each jurisdiction with necessary discretionary latitude in dealing with their local construction requirements. The section further gives the building official or authorized deputy (building inspector) the right of access on the job site to inspect all work. If denied access, this section further authorizes these individuals to use all legal means of entry including returning with the police and a warrant to enforce access to the project.

With the wide adoption of the national building codes throughout the country, the authority having jurisdiction has the prime responsibility of interpreting code rule in its area and disagreements on the intent of particular code regulations in its area. It is the intent of the national building codes that disagreements on the intent of particular code rules should be resolved at the local level. The provisions of this first chapter in all the national building codes govern the administration and lawful enforcement of the national building codes.

101.0 ADMINISTRATION [101.0; 101.0]
101.0 General
101.1. The provisions of this chapter govern the jurisdiction, administration, and enforcement of the standard building, gas, mechanical, and plumbing codes, and the National Electrical Code, hereinafter referred to as the <u>technical codes</u>, as may be adopted by the state or local jurisdiction.

[This section sets forth the outline for the authority of the building code. Both the UBC and BOCA have their own similar gas, mechanical, plumbing, and electrical codes referred to under this and other section numbers. Once adopted by the local jurisdiction, this section becomes law. From here to subsection 101.4, this section sets the code's legal parameters.]

101.3 Code Remedial

101.3.1. This code is hereby declared to be remedial and shall be construed to secure the beneficial interests and purposes thereof, (which are public safety, health, and general welfare) through structural strength, stability, sanitation, adequate light and ventilation, and safety to life and property from fire and other hazards attributed to the built environment including alteration, repair, removal, demolition, use and occupancy of buildings, structures, or premises, and through regulation of the installation and maintenance of all electrical, gas, mechanical, and plumbing systems, which may be referred to as service systems.

[This subsection, by defining remedial, allows a jurisdictional agency to adopt the code as enforceable law.]

101.3.2 Quality Control

Quality control of materials and workmanship is not within the purview of code except as it relates to the purposes stated herein.

101.3.3 Permitting and Inspection

The inspection or permitting of any building, system, or plan by any jurisdiction, under the requirements of code, shall not be construed in any court as a warranty of the physical condition or adequacy of such building, system, or plan. No jurisdiction nor any employee thereof shall be liable in tort for damages for any defect or hazardous or illegal condition or inadequacy in such building, system, or plan nor for any failure of any component of such, which may occur subsequent to such inspection or permitting.

[This subsection correlates to UBC 108.5 and BOCA 113.1. Basically it is a hold-harmless liability waiver clause each code has to prevent a building inspector or the local jurisdiction from being sued. Even if the building inspector makes a mistake or misses something on the plans, the court has held that the building inspector, building official, and local jurisdictional authority (the building department) are immune from liability unless malice can be proved.]

101.4 APPLICABILITY [106.0; 112.0]

101.4.1 General

Where, in any specific case, different sections of code specify different materials, methods of construction, or other requirements, the most restrictive shall govern. Where there is a conflict between a general requirement and a specific requirement, the specific requirement shall be applicable.

101.4.2 Building
The provisions of code shall apply to the construction, alteration, repair, equipment, use and occupancy, location, maintenance, removal, and demolition of every building or structure or any appurtenances connected or attached to such buildings or structures.

101.4.3 Electrical
The provisions of the National Electrical Code (NEC) shall apply to the installation of electrical systems, including alterations, repairs, replacement, equipment, appliances, fixtures, fittings and appurtenances thereto. [Both the UBC and BOCA also refer to the NEC for their electrical specifications.]

101.4.4 Gas
The provisions of the Standard Gas Code shall apply to the installation of consumers' gas piping, gas appliances, and related accessories as covered in code. These requirements apply to gas piping systems extending from the point of delivery to the inlet connections of appliances and to the installation and operation of installed residential and commercial gas appliances and related accessories. [UBC and BOCA have their own gas codes under this same section.]

101.4.5 Mechanical
The provisions of the Standard Mechanical Code shall apply to the installation of mechanical systems, including alterations, repairs, replacement, equipment, appliances, fixtures, fittings, and/or appurtenances. The mechanical systems include ventilating, heating, cooling, air-conditioning, and refrigeration systems; incinerators; and other energy-related systems. [UBC and BOCA have their own mechanical codes under this same section.]

101.4.6 Plumbing
The provisions of the Standard Plumbing Code shall apply to every plumbing installation, including alterations, repairs, replacement, equipment, appliances, fixtures, fittings, and appurtenances, and when connected to a water or sewerage system. [UBC and BOCA have their own plumbing codes under this same section.]

101.4.7 Federal and State Authority

The provisions of code shall not be held to deprive any federal or state agency, or any applicable governing authority having jurisdiction, of any power or authority which it had on the effective date of the adoption of code or of any remedy then existing for the enforcement of its orders, nor shall it deprive any individual or corporation of its legal rights as provided by law. [This follows legal precedent that federal law supersedes state law.]

101.4.8 Appendix

To be enforceable, the Appendix included in the technical codes must be referenced in the code text or specifically included in the adopting ordinance.

101.4.9 Referenced Standards

Standards referenced in the technical codes shall be considered an integral part of the codes without separate adoption. If specific portions of a standard are denoted by code text, only those portions of the standard shall be enforced. Where code provisions conflict with a standard, the code provisions shall be enforced. Permissive and advisory provisions in a standard shall not be construed as mandatory. [This is a touchy point for designers of franchise outlets. Often in this type of multiregional construction, the referenced standard is the current edition of the <u>ASTM Standards in Building Codes</u>. Check with the building official having local jurisdiction where the project is being built for advisory provisions on design materials substitution.]

102.0 BUILDING DEPARTMENT [104.0; 104.0]

This section creates the authority for the local jurisdiction to establish and maintain a building department to oversee the enforcement of the national building code. It further sets forth the criteria for the people who should run the office and how they should be hired and fired. It also delineates a standard of ethics by requiring that the building officials and their deputy inspectors not be financially interested in the construction of any building unless they are the direct owner. Officers or employees of the building department cannot engage in any work which is inconsistent with duties of the department.

102.1 Establishment

There is hereby established a department to be called the building department, and the person in charge shall be known as the building official. [The building official is typically the senior inspector in the local building department office. The building inspectors that check your job are subordinate to this building official who has final call on code compliance installations or materials.]

102.2 Employee Qualifications

102.2.1 Building Official Qualifications

The building official shall have at least 10 years experience or equivalent as an architect, engineer, inspector, contractor, or superintendent of construction, or any combination of these, 5 years of which shall have been in responsible charge of work. The building official should be certified through a recognized certification program.

The building official shall be appointed or hired by the applicable governing authority and shall not be removed from office except for cause after full opportunity has been given to be heard on specific charges before such applicable governing authority. [Both BOCA and UBC paraphrase this information.]

102.2.2 Chief Inspector Qualifications

The building official, with the approval of the applicable governing authority, may designate chief inspectors to administer the provisions of the building, electrical, gas, mechanical, and plumbing codes. Each chief inspector shall have at least 10 years experience or equivalent as an architect, engineer, inspector, contractor, or superintendent of construction, or any combination of these, 5 years of which shall have been in responsible charge of the work. They should be certified through a recognized certification program for the appropriate trade. They shall not be removed from office except for cause after full opportunity has been given to be heard on specific charges before each applicable governing authority.

102.2.3 Inspector Qualifications

The building official, with the approval of the applicable governing authority, may appoint or hire such number of officers, inspectors, assistants, and other employees as shall be authorized from time to time. A person shall not be appointed or hired as inspector of construction who has not had at least 5 years experience as a building inspector, engineer, architect, or superintendent, foreman, or competent mechanic in charge of construction. The inspector should be certified through a recognized certification program for the appropriate trade.

102.2.4 Deputy Building Official Qualifications
The building official may designate as his or her deputy an employee in the department who shall, during the absence or disability of the building official, exercise all the powers of the building official. The deputy building official should have the same qualifications listed in Section 102.2.2.

102.3 Restrictions on Employees
An officer or employee connected with the department, except one whose only connection is as a member of the board established by code, shall not be financially interested in the furnishing of labor, material, or appliances for the construction, alteration, or maintenance of a building, structure, service, system, or in the making of plans or of specifications thereof, unless he or she is the owner of such. This officer or employee shall not engage in any other work which is inconsistent with his or her duties or conflicts with the interests of the department.

102.4 Records
The building official shall keep, or cause to be kept, a record of the business of the department. The records of the department shall be open to public inspection. [Both BOCA and UBC paraphrase this information regarding public right-to-know access of building departments' records.]

102.5 Liability
Any officer, employee, or member of the board of adjustments and appeals charged with the enforcement of code, acting for the applicable governing authority in the discharge of his or her duties, shall not thereby render himself or herself personally liable and is hereby relieved from all personal liability, for any damage that may accrue to persons or property as a result of any act required or permitted in the discharge of his or her duties.

Any suit brought against any officer, employee, or member because of performing such an act in the enforcement of any provision of this code shall be defended by the department of law until the final termination of the proceedings. [The district attorney's office.]

102.6 Reports
The building official shall submit annually a report covering the work of the building department during the preceding year. A summary of the decisions of the board of adjustments and appeals during said year may be incorporated in said report.

103.0 POWERS AND DUTIES OF THE BUILDING OFFICIAL [103.0; 103.0]
103.1 General
The building official is hereby authorized and directed to enforce the provisions of code. The building official is further authorized to render interpretations of code, which are consistent with its spirit and purpose.
103.2 Right of Entry
Whenever necessary to make an inspection to enforce any of the provisions of code, or whenever the building official has reasonable cause to believe that there exists in any building or upon any premises any condition or code violation which makes such building; structure; premises; electrical, gas, mechanical, or plumbing systems unsafe, dangerous, or hazardous, the building official may enter such building, structure, or premises at all reasonable times to inspect the same or to perform any duty imposed upon the building official by code.

If such building or premises is occupied, the building official shall first present proper credentials and request entry. If such building, structure, or premises is unoccupied, the building official shall first make a reasonable effort to locate the owner or other persons having charge or control of such and request entry. If entry is refused, the building official shall have recourse to every remedy provided by law to secure entry. [This section is basically the same in BOCA and UBC.]

103.2.2. When the building official shall have first obtained a proper inspection warrant or other remedy provided by law to secure entry, no owner or occupant or any other persons having charge, care, or control of any building, structure, or premises shall fail or neglect, after proper request is made as herein provided, to promptly permit entry therein by the building official for the purpose of inspection and examination pursuant to code. [This establishes the building department's right, if denied access to construction inspection, to return with a court order and the local law enforcement to serve it.]

103.3 Stop Work Orders

Upon notice from the building official, work on any building; structure; or electrical, gas, mechanical, or plumbing system that is being done contrary to the provisions of code or in a dangerous or unsafe manner shall immediately cease. Such notice shall be in writing and shall be given to the owner of the property, to his or her agent, or to the person doing the work and shall state the conditions under which work may be resumed. Where an emergency exists, the building official shall not be required to give a written notice prior to stopping the work. [Note that the stop work order must be in writing. All orders or waivers from either the building inspector or building official must be in document form to be legally enforceable.]

103.4 Revocation of Permits

103.4.1 Misrepresentation of Application

The building official may revoke a permit or approval, issued under the provisions of code, in case there has been any false statement or misrepresentation as to the material fact in the application or plans on which the permit or approval was based.

103.4.2 Violation of Code Provisions

The building official may revoke a permit upon determination by the building official that the construction, erection, alteration, repair, moving, demolition, installation, or replacement of the building; structure; or electrical, gas, mechanical, or plumbing systems for which the permit was issued is in violation of, or not in conformity with, the provisions of code.

103.5 Unsafe Buildings or Systems

All buildings; structures; and electrical, gas, mechanical, or plumbing systems which are unsafe, unsanitary, do not provide adequate egress; constitute a fire hazard or are otherwise dangerous to human life; or which in relation to existing use constitute a hazard to safety or health are considered unsafe.

All such unsafe buildings, structures, or service systems are hereby declared illegal and shall be abated by repair and rehabilitation or by demolition in accordance with the provisions of the Standard Unsafe Building Abatement Code. [UBC and BOCA each also have a similar unsafe building abatement code.]

103.6 Requirements Not Covered by Code
Any requirements necessary for the strength, stability, or proper operation of an existing or proposed building; structure; or electrical, gas, mechanical, or plumbing system, or for the public safety, health, and general welfare, not specifically covered by this or the other technical codes, shall be determined by the building official.

[It is important to note here that the determining official is the building official and not the building inspector. This again is indicative of the chain of command within the building department having jurisdictional authority. Successful designers and builders understand this subtle difference in the national building codes authority and use it to their advantage in dealing with the bureaucratic hierarchy.]

103.7 Alternate Materials and Methods
The provisions of the technical codes are not intended to prevent the use of any material or method of construction not specifically prescribed by them, provided any such alternate has been reviewed by the building official. The building official shall approve any such alternate, provided the building official finds that the alternate for the purpose intended is at least the equivalent of that prescribed in the technical codes, in quality, strength, durability, effectiveness, fire-resistance, and safety. The building official shall require that sufficient evidence or proof be submitted to substantiate any claim made regarding the alternate.

[Here again is an example of the building official being the only individual with authority to approve alternative construction in variance of code compliance.]

104.0 PERMITS [106.0; 108.0]
The national building codes allow for permit fees to be set by the independent jurisdictional authorities that adopt the codes. With the exception of small storage outbuildings, fences, and painting, permits are required for all new construction, repairs, or alterations. Permit fees serve two purposes for the local jurisdiction. First, they are calculated to underwrite the expenses of the building department in general and administrative costs. This avoids putting the burden of the cost of the building department on the taxpayers

of the community and puts the cost burden upon the project owners and developers of the construction projects within the jurisdiction.

The second purpose of permits is to initiate the timetable and scheduling of the required inspections. Additionally, a copy of the permit is forwarded to the county or city tax assessor who will reassess the property valuation for an increased tax rate. In most jurisdictions, permits are required to be posted on the job site in a conspicuous place for accessibility of the building inspector.

104.1 PERMIT APPLICATION [108.5; 106.1]
104.1.1 When Required
Any owner, authorized agent, or contractor who desires to construct, enlarge, alter, repair, move, demolish, or change the occupancy of a building or structure or to erect, install, enlarge, alter, repair, remove, convert, or replace any electrical, gas, mechanical, or plumbing system, the installation of which is regulated by the technical codes, or to cause any such work to be done, shall first make application to the building official and obtain the required permit for the work. [Both UBC and BOCA paraphrase this information in the above correlated section.]

> **EXCEPTION**: Permits shall not be required for the following mechanical work:
> 1. Any portable heating appliance
> 2. Any portable ventilation equipment
> 3. Any portable cooling unit
> 4. Any steam, hot, or chilled water piping within any heating or cooling equipment regulated by code
> 5. Replacement of any part which does not alter its approval or make it unsafe
> 6. Any portable evaporative cooler
> 7. Any self-contained refrigeration system containing 10 pounds (lb) [4.54 kilograms (kg)] or less of refrigerant and actuated by motors of one horsepower [746 watts (W)] or less.

104.1.2 Temporary Structures
A special building permit for a limited time shall be obtained before the erection of temporary structures such as construction sheds, seats, canopies, tents, and fences used in construction work or for temporary purposes such as reviewing stands. Such structures shall

be completely removed upon the expiration of the time limit stated in the permit.

104.1.3 Work Authorized

A building, electrical, gas, mechanical, or plumbing permit shall carry with it the right to construct or install the work, provided the same are shown on the drawings and set forth in the specifications filed with the application for the permit. Where these are not shown on the drawings and covered by the specifications submitted with the application, separate permits shall be required.

104.1.4 Minor Repairs

Ordinary minor repairs may be made with the approval of the building official without a permit, provided that such repairs shall not violate any of the provisions of the technical codes.

104.1.5 Information Required

Each application for a permit, with the required fee, shall be filed with the building official on a form furnished for that purpose and shall contain a general description of the proposed work and its location. The application shall be signed by the owner, or an authorized agent. The building permit application shall indicate the proposed occupancy of all parts of the building and of that portion of the site or lot, if any, not covered by the building or structure and shall contain such other information as may be required by the building official.

104.1.6 Time Limitations

An application for a permit for any proposed work shall be deemed to have been abandoned 6 months after the date of filing for the permit, unless a permit has been issued before then. One or more extensions of time for periods of not more than 90 days each may be allowed by the building official for the application, provided the extension is requested in writing and justifiable cause is demonstrated.

104.2 Drawings and Specifications

Some difficulty in the terms of the national building codes has evolved concerning the difference between plans and specifications. Problems sometimes arise in conflicts between plans and specifications; therefore, the builder must understand the relationship between the two. In construction law, approved plans, specifications, and general conditions have equal authority and none takes precedence (sequence or logical order of authority) over the others.

The working blueprints of a project are known as the plans. The only set of plans that are legal to build with are the stamped and approved set. These can be (and often are) revised prior to or during the construction process. Plans represent the proposed description of the construction process including installation and materials to be used. Officially stamped plans represent the approved description of the project's construction process. Specifications, on the other hand, provide greater detail on facts regarding equipment, materials, and installation than are shown on the plans.

For normal residential construction, plans are broken down into six categories: plot plan, foundation plan, floor plan, roof plan, elevations, and detail cross sectionals. Larger commercial and industrial projects will call for many additional architectural, structural, mechanical, plumbing, and electrical plans. [The commercial plans typically encountered on today's projects run about 200 pages.]

104.2.1 Requirements

When required by the building official, two or more copies of specifications, and of drawings drawn to scale with sufficient clarity and detail to indicate the nature and character of the work, shall accompany the application for a permit. Such drawings and specifications shall contain information, in the form of notes or otherwise, as to the quality of materials, where quality is essential to conformity with the technical codes.

Such information shall be specific, and the technical codes shall not be cited as a whole or in part, nor shall the term legal or its equivalent be used as a substitute for specific information. All information, drawings, specifications and accompanying data shall bear the name and signature of the person responsible for the design. [Basically, this means design professionals must show their work and specs on the construction drawings and not simply refer to a referenced standard.]

104.2.2 Additional Data

The building official may require details, computations, stress diagrams, and other data necessary to describe the construction or installation and the basis of calculations. All drawings, specifications, and accompanying data required by the building official to be prepared by an architect or engineer shall be affixed with the architect's or engineer's official seal.

[Once stamped, these then become the official blueprints of record and the only official job working drawings. Any revisions must also carry this stamp.]

104.2.3 Design Professional

The design professional shall be an architect or engineer legally registered under the laws of this state regulating the practice of architecture or engineering and shall affix his or her official seal to said drawings, specifications, and accompanying data for the following:

1. All Group A, E, and I Occupancies
2. Buildings and structures three stories or more high
3. Buildings and structures 5000 sq ft (465 m^2) or more in area

For all other buildings and structures, the submittal shall bear the certification of the applicant that some specific state law exception permits its preparation by a person not so registered.

> **EXCEPTION**: Group R3 buildings, regardless of size, shall require neither a registered architect or engineer, nor a certification that an architect or engineer is not required.

[UBC and BOCA have different requirements on Occupancy Group R3. Check UBC section 310.1, and BOCA section 302.1. Straightforward and self-explanatory, this section requires the individuals who drew the prints to be registered, unless exempted, and affix their seals which include the individuals' registration number.]

104.2.4 Structural and Fire-Resistance Integrity

Plans for all buildings shall indicate how required structural and fire-resistance integrity will be maintained where a penetration of a required fire-resistant wall, floor, or partition will be made for electrical, gas, mechanical, plumbing, and communication conduits, pipes, and systems and also indicate in sufficient detail how the fire integrity will be maintained where required fire-resistant floors intersect the exterior walls. [UBC and BOCA have similar requirements for plans meeting these same following subsection criteria.]

Administration 15

104.2.5 Site Drawings
Drawings shall show the location of the proposed building or structure and of every existing building or structure on the site or lot. The building official may require a boundary line survey prepared by a qualified surveyor. [Registered property corners, known as hubs, must be located and filed as accurate with the Assessor's Parcel Number (APN) on file at the city or county's tax assessor's office. Invariably, unless the hubs are visible, registered surveyors will have to be hired to locate these property hubs.]

104.2.6 Hazardous Occupancies
The building official may require the following:

> **General site plan.** A general site plan drawn at a legible scale which shall include, but not be limited to, the location of all buildings, exterior storage facilities, permanent access ways, evacuation routes, parking lots, internal roads, chemical loading areas, equipment cleaning areas, storm and sanitary sewer accesses, emergency equipment, and adjacent property uses. The exterior storage areas shall be identified with the hazard classes and the maximum quantities per hazard class of hazardous materials stored.
>
> **Building floor plan.** A building floor plan drawn to a legible scale which shall include, but not be limited to, all hazardous materials storage facilities within the building and shall indicate rooms, doorways, corridors, exits, fire-rated assemblies with their hourly rating, location of liquid-tight rooms, and evacuation routes. Each hazardous materials storage facility shall be identified on the plan with the hazard classes and quantity range per hazard class of the hazardous materials stored.

[No may about it, the building department will want these plans and specifications in blueprint form. And they typically want three to five complete sets of the prints. Commercial prints can average 200 pages each.]

104.3 Examination of Documents
104.3.1 Plan Review
The building official shall examine or cause to be examined each application for a permit and the related accompanying documents, consisting of drawings, specifications, computations, and additional data, and shall ascertain by such examinations whether the

construction indicated and described is in accordance with the requirements of the technical codes and all other pertinent laws or ordinances.

[This job is usually handled in the building department by the entry-level plans checkers. They are usually very accurate, but people make mistakes, and plans checkers have been known to miss things. If problems are not brought to the attention of the building official until later in the job, the project's schedule will take a hit if the building inspector discovers them during construction installation. Always double check the plans specifications at this point.]

104.3.2 Affidavits
The building official may accept a sworn affidavit from a registered architect or engineer stating that the plans submitted conform to the technical codes. For buildings and structures, the affidavit shall state that the plans conform to the laws as to egress, type of construction, and general arrangement and, if accompanied by drawings, show the structural design and that the plans and design conform to the requirements of the technical codes as to strength, stresses, strains, loads, and stability.

The building official may, without any examination or inspection, accept such affidavit, provided the architect or engineer who made such affidavit agrees to submit to the building official copies of inspection reports as inspections are performed and certification upon completion of the structure or electrical; gas, mechanical, or plumbing systems that the systems have been erected in accordance with the requirements of the technical codes. Where the building official relies upon such affidavit, the architect or engineer shall assume full responsibility for the compliance with all provisions of the technical codes and other pertinent laws or ordinances. [Any such affidavit binds the signee to responsibility and liability for the construction.]

104.4 Issuing Permits
104.4.1 Action on Permits
The building official shall act upon an application for a permit without unreasonable or unnecessary delay. If the building official is satisfied that the work described in an application for a permit and the contract documents filed therewith conform to the requirements

of the technical codes and other pertinent laws and ordinances, he or she shall issue a permit to the applicant.

104.4.2 Refusal to Issue Permit
If the application for a permit and the accompanying contract documents describing the work do not conform to the requirements of the technical codes or other pertinent laws or ordinances, the building official shall not issue a permit but shall return the contract documents to the applicant indicating that the permit has been refused. Such refusal shall, when requested, be in writing and shall contain the reason for refusal.

104.4.3 Special Foundation Permit
When application for a permit to erect or enlarge a building has been filed and pending issuance of such a permit, the building official may, at his or her discretion, issue a special permit for the foundation only. The holder of such a special permit is proceeding at his or her own risk and without assurance that a permit for the remainder of the work will be granted or that corrections will not be required in order to meet provisions of the technical codes.

104.4.4 Public Right of Way
A permit shall not be given by the building official for the construction of any building or for the alteration of any building where said building is to be changed and such change will affect the exterior walls, bays, balconies, or other appendages or projections fronting on any street, alley, or public lane, or for the placing on any lot or premises of any building or structure removed from another lot or premises, unless application has been made at the office of the director of public works for the lines of the public street on which said building is proposed to be built, erected, or located. It shall be the duty of the building official to see that the street lines are not encroached upon except as provided for in SBC chapter 32. [Public right of way is also defined as such in both BOCA and UBC.]

104.5 Contractor's Responsibilities
It shall be the duty of every contractor who shall make contracts for the installation or repairs of buildings; structures; or electrical, gas, mechanical, or plumbing systems, for which a permit is required, to comply with state or local rules and regulations concerning licensing which the applicable governing authority may have adopted. [This subsection requires that all contractors building the

approved construction be licensed in their respective states and/or the state the project is being constructed in, pursuant to the local governing authority's regulations.]

104.6 Conditions of the Permit

104.6.1 Permit Intent

A permit issued shall be construed to be a license to proceed with the work and not as authority to violate, cancel, alter, or set aside any of the provisions of the technical codes, nor shall issuance of a permit prevent the building official from thereafter requiring a correction of errors in plans, construction, or violations of code.

Every permit issued shall become invalid unless the work authorized by such permit is commenced within 6 months after its issuance or if the work authorized by such permit is suspended or abandoned for a period of 6 months after the time the work is commenced. One or more extensions of time, for periods not more than 90 days each, may be allowed for the permit. The extension shall be requested in writing and justifiable cause demonstrated. Extensions shall be made in writing by the building official.

104.6.2 Permit Issued on Basis of an Affidavit

Whenever a permit is issued in reliance upon an affidavit or whenever the work to be covered by a permit involves installation under conditions which, in the opinion of the building official, are hazardous or complex, the building official shall require that the architect or engineer who signed the affidavit or prepared the drawings or computations shall supervise such work.

In addition, the architect or engineer shall be responsible for conformity with the permit, provide copies of inspection reports as inspections are performed, and upon completion make and file with the building official a written affidavit that the work has been done in conformity with the reviewed plans and with the structural provisions of the technical codes. In the event such architect or engineer is not available, the owner shall employ a competent person or agency whose qualifications are reviewed by the building official.

104.6.3 Plans

When the building official issues a permit, he or she shall endorse, in writing or by stamp, both sets of plans with the words "Reviewed for code compliance." One set of drawings so reviewed shall be retained by the building official, and the other set shall be returned to the applicant. The permit drawings shall be kept at the site of work and shall be open to inspection by the building official or his or her authorized representative.

[Plans should always include a shop drawings log to show equipment meeting design specifications and delivery and installation dates. In design planning, this is the stage equipment and materials code compliance research is begun. A shop drawings log should be started so that when the shop drawings start arriving from subcontractors and suppliers, each drawing can be properly checked for building codes compliance. Not only will a log of shop drawings activity allow a project designer to keep track of what drawings received are in compliance but, if the shop drawings log is computerized, it can show where the drawings have been sent and how long they have been there. The project designer needs to prod each subcontractor and supplier to submit their drawings promptly so this stage of researching code compliance can be completed. At the first project meeting, the major subcontractors and/or material suppliers should be requested to submit a preliminary shop drawing submission schedule.

The log should be cross-referenced with this handbook and the national code having authority at the project's location to determine code compliance of all the major pieces of equipment for which shop drawings are required. The subcontractor and/or supplier should also include the approximate delivery date of equipment after the approved shop drawings have been returned to them. Once shop drawings have been received, the next hurdle is getting them approved in a timely manner. The project designer should always review the incoming shop drawings before logging them into the computer, to determine whether they conform to the national building codes specifications and requirements. If compliance is questionable, contact the party who made the submission to reexamine the shop drawings to verify contract specifications. It is at this point that the architect and engineers should show that the shop drawings are being reviewed for compliance with the plans and specifications and are not merely being passed through without any scrutiny whatsoever.]

104.7 Fees
104.7.1 Prescribed Fees

A permit shall not be issued until the fees prescribed in Section 104.7 have been paid. Nor shall an amendment to a permit be released until the additional fee, if any, due to an increase in the estimated cost of the building; structure; or electrical, plumbing, mechanical, or gas system has been paid. [Both UBC and BOCA paraphrase this section's information.]

104.7.2 Work Commencing Before Permit Issuance

Any person who commences any work on a building; structure; or electrical, gas, mechanical, or plumbing system before obtaining the necessary permits shall be subject to a penalty of 100% of the usual permit fee in addition to the required permit fees. [This penalty fee for work without a permit is the same in UBC and BOCA.]

104.7.3 Accounting

The building official shall keep a permanent and accurate accounting of all permit fees and other monies collected and the names of all persons upon whose account the same was paid, along with the date and amount thereof.

104.7.4 Schedule of Permit Fees

On all buildings; structures; or electrical, plumbing, mechanical, and gas systems; or alterations requiring a permit, a fee for each permit shall be paid as required at the time of filing application, in accordance with the schedule as established by the applicable governing authority.

104.7.5 Building Permit Valuations

If, in the opinion of the building official, the valuation of building; alteration; structure; or electrical, gas, mechanical, or plumbing system appears to be underestimated on the application, the permit shall be denied, unless the applicant can show detailed estimates to meet the approval of the building official. Permit valuations shall include total cost, such as electrical, gas, mechanical, plumbing equipment, and other systems, including materials and labor.

105.0 REQUIRED INSPECTIONS [108.0; 113.0]

[Both BOCA and UBC paraphrase this section information. This section lists and defines the inspections required during the various construction phases before a final inspection can be called for. For most residential projects, there are five inspections required by each of the national building codes:

- **Foundation inspection.** This initial inspection is done after footings are excavated and foundation forms and reinforcement steel (rebar) are installed. This inspection checks for footing depth and width; conditions of footing base and sidewalls; and proper sizing, amount, and condition of rebar.
- **Concrete slab and under-floor inspection.** This inspection is done after all under-floor and in-slab plumbing, conduit, service, and ancillary equipment is in place but before the concrete is poured. This area must remain open for this inspection which includes no subfloor, and floor sheathing may be installed prior to this inspection.
- **Frame inspection.** This inspection is done after the exterior and interior walls are framed; fire stops, freeze blocks, and lateral bracing are installed; electrical and plumbing systems are roughed-in; and all vents are installed.
- **Gypsum board, lath, and plaster inspection.** This inspection is done after all gypsum board (Sheetrock) or lath work is installed but before it taped and textured or plaster is overlaid upon the lath. Here the inspector is checking for proper fastener schedule (placement) and condition of installation.
- **Final inspection.** When the construction is complete and the finish grading is done, the final inspection clears the project for issuance of a certificate of occupancy. The building is now ready to occupy.

Should any of these required inspections fail, the inspector will issue a correction notice or notice of non-compliance calling for correction of the specific work that failed inspection before scheduling a follow-up reinspection. Typically, reinspections are an extra charge to the owner or developer. Depending on the project's complexity, there may be other special inspections called for by the building inspector. Commercial and industrial projects typically require a full-time inspector on site during working hours. This extra cost of the inspector's salary is charged to the owner or developer as part of the project cost.]

105.1 Existing Building Inspections
Before issuing a permit, the building official may examine or cause to be examined any building or electrical, gas, mechanical, or plumbing system for which an application has been received for a

permit to enlarge, alter, repair, move, demolish, install, or change the occupancy.

The building official or designated deputy inspector shall inspect all buildings; structures; and electrical, gas, mechanical, and plumbing systems from time to time during and upon completion of the work for which a permit was issued. The building official shall make a record of every such examination and inspection and of all violations of the technical codes. [As stated earlier, these records are public-domain information accessible by the local citizenry.]

105.2 Manufacturers and Fabricators

When deemed necessary, the building official shall make, or cause to be made, an inspection of materials or assemblies at the point of manufacture or fabrication. A record shall be made of every such examination and inspection and of all violations of the technical codes.

105.3 Inspection Service

The building official may make, or cause to be made, the inspections required by Section 105. He or she may accept reports of inspectors of recognized inspection services, provided that after investigation he or she is satisfied as to their qualifications and reliability. A certificate called for by any provision of the technical codes shall not be based on such reports unless the same are in writing and certified by a responsible officer of such service.

105.4 Inspections Prior to Issuance of Certificate

The building official shall inspect, or cause to be inspected, at various intervals all construction or work for which a permit is required, and a final inspection shall be made of every building; structure; and electrical, gas, mechanical, and plumbing system upon completion, prior to the issuance of the certificate of occupancy or completion. [Both UBC and BOCA paraphrase this section information.]

105.5 Posting of Permit

Work requiring a permit shall not commence until the permit holder or his or her agent posts the permit card in a conspicuous place on the premises. The permit shall be protected from the weather and located in such a position as to permit the building official or representative to conveniently make the required entries thereon. This permit card shall be maintained in such position by the permit holder until the certificate of occupancy or completion is issued by the building official.

105.6 Required Inspections

The building official upon notification from the permit holder or his or her agent shall make the following inspections and such other inspections as necessary and shall either release that portion of the construction or notify the permit holder or his or her agent of any violations which must be corrected in order to comply with the technical codes:

Building
- Foundation inspection. To be made after trenches are excavated and forms erected.
- Frame inspection. To be made after the roof and all framing, fireblocking, and bracing are in place and after all concealed wiring, pipes, chimneys, ducts, and vents are complete.
- Final inspection. To be made after the building is completed and ready for occupancy.

Electrical
- Underground inspection. To be made after trenches or ditches are excavated, conduit or cable is installed, and before any backfill is put in place.
- Rough-in inspection. To be made after the roof, framing, fireblocking, and bracing are in place and prior to the installation of wall or ceiling membranes.
- Final inspection. To be made after the building is complete, all required electrical fixtures are in place and properly connected or protected, and the structure is ready for occupancy.

Plumbing
- Underground inspection. To be made after trenches or ditches are excavated, piping is installed, and before any backfill is put in place.
- Rough-in inspection. To be made after the roof, framing, fireblocking, and bracing are in place; after all soil, waste, and vent piping is complete; and prior to installation of wall or ceiling membranes.
- Final inspection. To be made after the building is complete, all plumbing fixtures are in place and properly connected, and the structure is ready for occupancy.

Mechanical
- <u>Underground inspection</u>. To be made after trenches or ditches are excavated, underground duct and fuel piping are installed, and before any backfill is put in place.
- <u>Rough-in inspection</u>. To be made after the roof, framing, fireblocking, and bracing are in place; after all ducting and other concealed components are complete; and prior to the installation of wall or ceiling membranes.
- <u>Final inspection</u>. To be made after the building is complete, the mechanical system is in place and properly connected, and the structure is ready for occupancy.

Gas
- <u>Rough piping inspection</u>. To be made after all new piping authorized by the permit has been installed and before any such piping has been covered or concealed or any fixtures or gas appliances have been connected.
- <u>Final piping inspection</u>. To be made after all piping authorized by the permit has been installed, and after all portions which are to be concealed by plastering or otherwise have been so concealed, and before any fixtures or gas appliances have been connected. This inspection shall include a pressure test.
- <u>Final inspection</u>. To be made on all new gas work authorized by the permit and such portions of existing systems as may be affected by new work or any changes, to ensure compliance with all the requirements of code and to assure that the installation and construction of the gas system is in accordance with reviewed plans.

105.7 Written Release
Work shall not be done on any part of a building; structure; or electrical, gas, mechanical, or plumbing system beyond the point indicated in each successive inspection without first obtaining a written release from the building official. Such written release shall be given only after an inspection has been made of each successive step in the construction or installation as indicated by each of the foregoing three inspections. [Both UBC and BOCA set up similar sequential inspections which parallel this section's information.]

105.8 Reinforcing Steel and Structural Frames
Reinforcing steel or structural framework of any part of any building or structure shall not be covered or concealed without first obtaining a release from the building official.

105.9 Plaster Fire Protection
In all buildings where plaster is used for fire protection purposes, the permit holder or his or her agent shall notify the building official after all lathing and backing are in place. Plaster shall not be applied until the release from the building official has been received.

106.0 CERTIFICATES [109.0; 118.0]
[A certificate of occupancy is mandated by all the national building codes. It is a document that brings the legal record of construction of a building to a finish and allows occupancy to begin. If this certificate is not issued, technically the building is unoccupiable as it is not finished being built. Utility companies will not service the building until the certificate is issued. Contractors who do not follow through to this stage in the construction recordation phase can face abandonment contractual breach and loss of license from the registrar of contractors.]

106.1 Certificate of Occupancy

106.1.1 Building Occupancy
A new building shall not be occupied or a change made in the occupancy, nature, or use of a building or part of a building until after the building official has issued a certificate of occupancy. Said certificate shall not be issued until all required electrical, gas, mechanical, plumbing, and fire-protection systems have been inspected for compliance with the technical codes and other applicable laws and ordinances and released by the building official.

106.1.2 Issuing Certificate of Occupancy
Upon satisfactory completion of construction of a building or structure and installation of electrical, gas, mechanical, and plumbing systems in accordance with the technical codes; after review of plans and specifications; and after the final inspection, the building official shall issue a certificate of occupancy stating the nature of the occupancy permitted, the number of persons for each floor when limited by law and the allowable load per square foot for each floor in accordance with the provisions of code.

106.1.3 Temporary/Partial Occupancy

A temporary/partial certificate of occupancy may be issued for a portion or portions of a building which may safely be occupied prior to final completion of the building.

106.2 Certificate of Completion

Upon satisfactory completion of a building, structure, electrical, gas, mechanical or plumbing system, a certificate of completion may be issued. This certificate is proof that a structure or system is complete and for certain types of permits is released for use and may be connected to a utility system. This certificate does not grant authority to occupy or connect a building, such as a shell building, prior to the issuance of a certificate of occupancy.

106.3 Service Utilities

106.3.1 Connection of Service Utilities

No person shall make connections from a utility, source of energy, fuel, or power to any building or system which is regulated by the technical codes for which a permit is required, until released by the building official and a certificate of occupancy or completion is issued.

106.3.2 Temporary Connection

The building official may authorize the temporary connection of the building or system to the utility source of energy, fuel, or power for the purpose of testing building service systems or for use under a temporary certificate of occupancy.

106.3.3 Authority to Disconnect Service Utilities

The building official shall have the authority to authorize disconnection of utility service to the building, structure, or system regulated by the technical codes, in case of emergency where necessary to eliminate an immediate hazard to life or property. The building official shall notify the serving utility, and whenever possible the owner and occupant of the building, structure, or service system of the decision to disconnect prior to taking such action. If not notified prior to disconnecting, the owner or occupant of the building, structure, or service system shall be notified in writing, as soon as practical thereafter. [This gives the local authority the ability to shut the power off for any building or structure and later condemn any building they feel unsafe to the public under an unsafe building abatement ordinance.]

106.4 Posting Floor Loads
106.4.1 Occupancy
An existing or new building shall not be occupied for any purpose which will cause the floors thereof to be loaded beyond their safe capacity. Occupancy of a building for mercantile, commercial, or industrial purposes, by a specific business, may be permitted when the building official is satisfied that such capacity will not thereby be exceeded.

106.4.2 Storage and Factory-Industrial Occupancies
It shall be the responsibility of the owner, agent, proprietor, or occupant of Group S and Group F Occupancies, or any occupancy where excessive floor loading is likely to occur, to employ a competent architect or engineer in computing the safe load capacity. All such computations shall be accompanied by an affidavit from the architect or engineer stating the safe allowable floor load on each floor in pounds per square foot uniformly distributed. The computations and affidavit shall be filed as a permanent record of the building department.

106.4.3 Signs Required
In every building or part of a building used for storage, industrial, or hazardous purposes, the safe floor loads, as reviewed by the building official on the plan, shall be marked on plates of approved design which shall be supplied and securely affixed by the owner of the building in a conspicuous place in each story to which they relate. Such plates shall not be removed or defaced; if they are lost, removed, or defaced, they must be replaced by the owner of the building.

107.0 TESTS [108.5; 107.1]
The building official may require tests or test reports as proof of compliance. Required tests are to be made at the expense of the owner, or his or her agent, by an approved testing laboratory or other approved agency.

108.0 CONSTRUCTION BOARD OF ADJUSTMENT AND APPEALS [105.0; 121.0]
[If a builder does not agree with the decision of the building official, this is the venue for appealing that decision. This section provides for an appellate court to be established and maintained to oversee

the building department's decisions and act as a board of adjustment. It further delineates the criteria for those individuals, their term of appointment, and the parameters of their power.]

108.1 Appointment

There is hereby established a board to be called the Construction Board of Adjustment and Appeals, which shall consist of seven members and two alternates. The board shall be appointed by the applicable governing body.

108.2 Membership and Terms

108.2.1 Membership

The Construction Board of Adjustment and Appeals should consist of seven members. Such board members should be composed of individuals with knowledge and experience in the technical codes, such as design professionals, contractors, or building industry representatives. In addition to the regular members, there should be two alternate members, one member at large from the building industry and one member at large from the public. Board members shall not act in a case in which they have a personal or financial interest.

108.2.2 Terms

The terms of office of the board members shall be staggered so no more than one-third of the board is fully appointed or replaced in any 12-month period. The two alternates, if appointed, shall serve only 1-year terms. Vacancies shall be filled for an unexpired term in the manner in which original appointments are required to be made. Continued absence of any member from required meetings of the board shall, at the discretion of the applicable governing body, render any such member subject to immediate removal from office.

108.2.3 Quorum and Voting

A simple majority of the board shall constitute a quorum. In varying any provision of code, the affirmative votes of the majority present, but not less than three affirmative votes, shall be required. In modifying a decision of the building official, not less than four affirmative votes, but not less than a majority of the board, shall be required. In the event that regular members are unable to attend a meeting, the alternate members, if appointed, shall vote.

108.2.4 Secretary of Board

The building official shall act as secretary of the board and shall make a detailed record of all its proceedings, which shall set forth the reasons for its decision, the vote of each member, the absence of a member, and any failure of a member to vote.

108.3 Powers
The Construction Board of Adjustments and Appeals shall have the power, as further defined in Section 108.4, to hear appeals of decisions and interpretations of the building official and consider variances of the technical codes.

108.4 Appeals
108.4.1 Decision of the Building Official
The owner of a building, structure, or service system, or a duly authorized agent, may appeal a decision of the building official to the Construction Board of Adjustment and Appeals whenever any one of the following conditions is claimed to exist:
- The building official rejected or refused to approve the mode or manner of construction proposed to be followed or the materials to be used in the installation or alteration of a building, structure, or service system.
- The provisions of code do not apply to this specific case.
- An equally good or more desirable form of installation can be employed in any specific case.
- The true intent and meaning of code or any of the regulations thereunder have been misconstrued or incorrectly interpreted.

108.4.2 Variances
The Construction Board of Adjustments and Appeals, when so appealed to and after a hearing, may vary the application of any provision of code to any particular case when, in its opinion, the enforcement thereof would do manifest injustice and would be contrary to the spirit and purpose of this or the technical codes or public interest and also finds all the following:
- That special conditions and circumstances exist which are peculiar to the building, structure, or service system involved and which are not applicable to others
- That the special conditions and circumstances do not result from the action or inaction of the applicant
- That granting the variance requested will not confer on the applicant any special privilege that is denied by code to other buildings, structures, or service system

- That the variance granted is the minimum variance that will make possible the reasonable use of the building, structure, or service system
- That the grant of the variance will be in harmony with the general intent and purpose of code and will not be detrimental to the public health, safety, and general welfare

108.4.2.1 Conditions of the Variance

In granting the variance, the board may prescribe a reasonable time limit within which the action for which the variance is required shall be commenced or completed or both. In addition, the board may prescribe appropriate conditions and safeguards in conformity with code. Violation of the conditions of a variance shall be deemed a violation of code.

108.4.3 Notice of Appeal

Notice of appeal shall be in writing and filed within 30 calendar days after the decision is rendered by the building official. Appeals shall be in a form acceptable to the building official.

108.4.4 Unsafe or Dangerous Buildings or Service Systems

In the case of a building, structure, or service system which, in the opinion of the building official, is unsafe, unsanitary, or dangerous, the building official may, in the order, limit the time for such appeals to a shorter period.

108.5 Procedures of the Board

108.5.1 Rules and Regulations

The board shall establish rules and regulations for its own procedure not inconsistent with the provisions of code. The board shall meet when called by the chairperson. The board shall meet within 30 calendar days after notice of appeal has been received.

108.5.2 Decisions

The Construction Board of Adjustment and Appeals shall, in every case, reach a decision without unreasonable or unnecessary delay. Each decision of the board shall also include the reasons for the decision. If a decision of the board reverses or modifies a refusal, order, or disallowance of the building official or varies the application of any provision of code, the building official shall immediately take action in accordance with such decision.

Every decision shall be promptly filed in writing in the office of the building official and shall be open to public inspection. A certified copy of the decision shall be sent by mail or otherwise to the appellant, and a copy shall be kept publicly posted in the office of the building official for 2 weeks after filing. Every decision of the board shall be final, subject however to such remedy as any aggrieved party might have at law or in equity.

110.0 VIOLATIONS AND PENALTIES [103.0; 116.0]

110.1. Any person, firm, corporation, or agent who shall violate a provision of code; shall fail to comply therewith or with any of the requirements thereof; or shall erect, construct, alter, install, demolish, or move any structure or service system, or has erected, constructed, altered, repaired, moved, or demolished a building, structure, or service system, in violation of a detailed statement or drawing submitted and permitted thereunder; shall be guilty of a misdemeanor.

Each such person shall be considered guilty of a separate offense for each and every day or portion thereof during which any violation of any of the provisions of code is committed or continued, and upon conviction of any such violation such person shall be punished within the limits and as provided by state laws. [Penalty for work done without a permit is payment in full of the original total fees plus a 50% penalty fine.]

Chapter 2

Terms and Definitions

SECTION 201.0 DEFINITIONS [201.0; 201.0]

Many of the national building codes' regulations are couched in "legalese" as they are meant for ready adoption into law. Terms and definitions are therefore crucial to understand in both construction design and project execution. Because these words are critically important to applications of structural integrity and material usage, their definitions must be carefully studied and cross-referenced with one another, as well as related to specific code rules using these words. Any installation that is closed in by any structural surfaces is considered to be <u>concealed</u>. Also observe that there are definitions for words that apply to installation methods and other definitions that apply to materials.

The term <u>approved</u> has proven to be a source of contention between builders and inspectors. To clarify, the building official having local jurisdiction on any specific installation is the only person who can legally decide what materials and/or equipment are approved. The label of a nationally recognized testing laboratory on a piece of equipment is a sure and ready way to be assured that the equipment is properly made and will function safely when used in accordance with the application data and limitations established by the testing organization. Each label used on a product gives the exact name of the type of equipment as it appears in the listing book of the testing organization.

It should be noted that the definitions for <u>labeled</u> and <u>listed</u> do not always require that the listing laboratory be <u>nationally recognized</u>. The code definitions acknowledge that a local inspector may accept the label or listing of a product by a testing organization that is qualified and capable even though it operates in a small area or section of the country and is not nationally recognized.

201.1 Scope
For the purpose of code, certain abbreviations, terms, phrases, words, and their derivatives shall be construed as set forth in this chapter or the chapter to which they are unique.

201.3 Words Not Defined
Words not defined herein shall have the meanings stated in the Standard Mechanical Code, Standard Plumbing Code, Standard Gas Code, or Standard Fire Prevention Code. Words not defined in the standard codes shall have the meanings in Webster's Ninth New Collegiate Dictionary, as revised. [UBC and BOCA use their own respective gas, mechanical, and plumbing codes; however, all three national building codes use Webster's Ninth New Collegiate Dictionary, as revised, for legal meaning of words and terms.]

202.0 Terms and Definitions
ACCESS FLOOR SYSTEM - An assembly consisting of panels mounted on pedestals to provide an under-floor space for the installation of mechanical, electrical, communication, or similar systems or to serve as an air-supply or return-air plenum.

ACCESSIBLE - Having access to but which first may require the removal of a panel, door, or similar covering of the item described. [This is interpreted to mean the item may be covered with a removable access panel or exposed, but not concealed.]

ACCESSORY STRUCTURE - A building, the occupancy of which is incidental to that of the main building, that is located on the same lot as the main building.

ACI - The American Concrete Institute, Box 19150, Redford Station, Detroit, Michigan 48219.

ADDITION - An extension or increase in floor area or height of a building or structure.

AGRICULTURAL BUILDING - A structure utilized to store farm tools, implements, hay, feed, grain, or other agricultural or horticultural products or to house poultry, livestock, or other farm animals. Such structure cannot include habitable or occupiable spaces or spaces in which agricultural products are processed, treated, or packaged.

ALLEY - Any public space or thoroughfare 20 ft (6096 mm) or less wide which has been dedicated or deeded for public use.

ALTER or ALTERATION - Any change or modification in construction or occupancy.

AND/OR - In a choice of two code provisions, signifies that use of both provisions will satisfy the code requirement and use of either provision is acceptable also.

APARTMENT - See Dwelling Unit.

APARTMENT HOUSE - Any building or portion thereof used as a multiple dwelling for the purpose of providing three or more separate dwelling units which may share means of egress and other essential facilities.

APPLICABLE GOVERNING BODY - A city, county, state, state agency, or other political government subdivision or entity authorized to administer and enforce the provisions of code, as adopted or amended.

APPROVED - Approved by the building official or other authority having jurisdiction.

APPROVED AGENCY - An established and recognized agency regularly engaged in conducting tests or furnishing inspection services, when such agency has been approved.

APPROVED FABRICATOR - An established and qualified person, firm, business, or corporation approved by the building official.

APPROVED MATERIAL - Material or equipment that have been tested and approved.

APPURTENANT STRUCTURE - A structure attached to the exterior or erected on the roof of a building designed to support or provide connection with service equipment, or be used for advertising or display purposes.

ARCHITECT - A duly registered and licensed architect.

ARCHITECTURAL TRIM - The ornamental or protective framing or edging around openings or at corners or eaves and other architectural elements attached to the exterior walls of buildings, usually of a color and material different from that of the adjacent wall surface and serving no structural purpose.

AREA, BUILDING - The area included within surrounding exterior walls, or exterior walls and fire walls, exclusive of courts. The area of a building or portion of a building without surrounding walls shall be the usable area under the horizontal projection of the roof or floor above.

AREA, GROSS FLOOR - The area within the inside perimeter of the exterior walls with no deduction for corridors, stairs, closets, thickness of walls, columns, or other features, exclusive of areas open and unobstructed to the sky.

AREA, NET FLOOR - The area actually occupied not including accessory unoccupied areas such as corridors, stairs, closets, thickness of walls, columns, toilet room, mechanical area, or other features.

ASSEMBLY BUILDING - A building or portion of a building used for the gathering together of 50 or more persons for such purposes as meetings, education, sports, instruction, worship, entertainment, amusement, taverns, restaurants, facilities for awaiting transportation, etc.

ASSEMBLY OCCUPANCY - Different code classifications of occupancy for public assembly. Specific definitions in Section 304.

ASTM - The American Society for Testing and Materials, 1916 Race Street, Philadelphia, Pennsylvania 19103.

ATRIUM - A space, intended for occupancy within a building, extending vertically through the building and enclosed at the top.

AUTOMATIC - As applied to fire-protection devices, is a device or system providing an emergency function without the necessity for human intervention and activated as a result of a predetermined temperature rise, rate of temperature rise, or combustion products, such as incorporated in an automatic sprinkler system, automatic fire door, automatic fire shutter, or automatic fire vent.

AUTOMATIC FIRE-EXTINGUISHING SYSTEM - An approved system of devices and equipment which automatically detects a fire and discharges an approved fire-extinguishing agent onto or in the area of a fire.

AWNING - An architectural projection that provides weather protection, identity, and/or decoration and is wholly supported by the building to which it is attached. An awning is comprised of a lightweight, rigid, and/or retractable skeleton structure over which an approved cover is attached.

BALCONY, ASSEMBLY ROOM - That portion of the seating space of an assembly room, the lowest part of which is raised 4 ft (1219 mm) or more above the level of the main floor.

BASEMENT - Any building story having a floor below grade.

BLEACHERS - Tiered or stepped seating facilities without backrests.

BOILER - A heating appliance intended to supply hot water or steam.

BOILER, HIGH PRESSURE - A boiler furnishing steam at pressures in excess of 15 pounds per square inch (psi) [103.3 kilopascals (kPa)] or hot water at temperatures in excess of 250 degrees Fahrenheit (250^0 F) [121 degrees Celsius (121^0 C)], or at pressures in excess of 160 psi (1002.4 kPa).

BOILER ROOM - Any room containing a steam or hot-water boiler.

BUILDING - Any structure that encloses a space used for sheltering any occupancy. Each portion of a building separated from other portions by a fire wall shall be considered as a separate building.

BUILDING, EXISTING - Any structure occupied prior to the date of adoption of the appropriate code, or one for which a legal building permit has been issued.

BUILDING LINE - The line, established by law, beyond which the building shall not extend, except as specifically provided by law.

BUILDING OFFICIAL - The officer or other designated authority, or their duly authorized representative, charged with the administration and enforcement of code.

BUILDING SERVICE EQUIPMENT - The mechanical, electrical, and elevator equipment including piping, wiring, fixtures, and other accessories, which provides sanitation, lighting, heating, ventilation, fire protection, and transportation facilities essential for the habitable occupancy of the building or structure for its designated occupancy.

BUILDING SITE - The area occupied by a building or structure, including the yards and courts required for light and ventilation, and such areas that are prescribed for access to the street.

BULK HANDLING - The transferring of flammable or combustible liquids from tanks or drums into smaller containers for distribution.

BUSINESS OCCUPANCY - Different code classifications of occupancy for business purposes. Specific definitions in Section 305.

CAST STONE - A precast building stone manufactured from portland cement concrete and used as a trim, veneer, or facing on or in buildings or structures.

CENTRAL HEATING PLANT - Environmental heating equipment which directly utilizes fuel to generate heat in a medium for distribution by means of ducts or pipes to areas other than the room or space in which the equipment is located.

CFR - The Code of Federal Regulations, a regulation of the United States of America available from the Superintendent of Documents, United States Government Printing Office, Washington, D.C. 20402.

CHIEF OF THE FIRE DEPARTMENT - The head of the fire department or a regularly authorized deputy.

CITY - See Applicable Governing Body.

CODE OFFICIAL - The building official or other designated authority charged with the administration and enforcement of code, or a duly authorized representative.

COMBUSTIBLE LIQUID - See Fire Code.

COMBUSTIBLE MATERIAL - A material which cannot be classified as noncombustible in accordance with that definition.

CONDOMINIUM, RESIDENTIAL - See apartment house.

CONGREGATE RESIDENCE - Any building, or portion thereof for occupancy by other than a family, which contains facilities for living, sleeping, and sanitation, as required by code, and may include facilities for eating and cooking. A congregate residence may be a shelter, convent, monastery, dormitory, or fraternity or sorority house but does not include jails, hospitals, nursing homes, hotels, or lodging houses.

CONSTRUCTION DOCUMENTS - All of the written, graphic and pictorial plans and documents prepared or assembled for describing the design, location, and physical characteristics of the elements of the project necessary for obtaining a building permit. The construction drawings shall be drawn to an appropriate scale.

CONSTRUCTION TYPES - See noted section number in each related national code:

SBC
  Type I - section 603.0
  Type II - section 604.0
  Type III - section 605.0
  Type IV - section 606.0
  Type V - section 607.0

Type VI - section 608.0

UBC
Type I - section 602.1
Type II - section 603.1
Type III - section 604.1
Type IV - section 605.1
Type V - section 606.1

BOCA
Type I - section 603.0
Type II - section 603.0
Type III - section 604.0
Type IV - section 605.0
Type V - section 607.0

CONTROL AREA - A building or portion of a building within which the exempted amount of hazardous materials may be stored, dispensed, handled, or used.

CORRIDOR - A passageway into which compartments or rooms open and which is enclosed by partitions, other than partial partitions, and/or walls and a ceiling or a floor or roof deck above.

CORROSIVE - A chemical that causes visible destruction of, or irreversible alterations in, living tissue by chemical action at the site of contact. A chemical is considered to be corrosive if, when tested on the intact skin of albino rabbits by the method described in the United States Department of Transportation in Appendix to page 49, CFR 173, it destroys or changes irreversibly the structure of the tissue at the site of contact following an exposure period of 4 hours. This term shall not refer to action on inanimate surfaces.

COURT - A space, open and unobstructed to the sky, located at or above grade level on a lot and bounded on at least three sides by walls of a building.

DEAD END - A hallway, corridor, or space open to a corridor so arranged that it can be entered from an exit access corridor without passage through a door but does not lead to an exit.

DEAD LOAD - The weight of all permanent construction, including walls, floors, roofs, ceilings, stairways, and fixed service equipment, plus the net effect of prestressing.

DISPENSING - The pouring or transferring of any material from a container, tank, or similar vessel whereby vapors, dusts, fumes, mists, or gases may be liberated to the atmosphere.

DORMITORY - A space in a unit where group sleeping accommodations are provided with or without meals for persons not members of the same family group. The space may be one room or a series of closely associated rooms under joint occupancy and single management, as in college dormitories, fraternity houses, military barracks, and ski lodges.

DRAFT STOP -A material, device, or construction installed to restrict the movement of air within open spaces of concealed areas of building components such as crawl spaces, floor-ceiling assemblies, roof-ceiling assemblies, and attics.

DWELLING - A building occupied exclusively for residential purposes by not more than two families, unless qualified otherwise in code text.

DWELLING UNIT - A single unit providing complete, independent living facilities for one or more persons including permanent provisions for living, sleeping, eating, cooking, and sanitation.

EDUCATIONAL OCCUPANCY - Different code classifications for educational assembly ratings. See Section 306.0 and section 300.0 Occupancy Codes in all national codes.

EFFICIENCY DWELLING UNIT - A dwelling unit containing only one habitable room.

ELECTRICAL CODE - The National Electrical Code, as adopted by the local jurisdiction.

ELEVATOR CODE - The safety code for elevators, dumbwaiters, escalators, and moving walks as adopted by this jurisdiction.

ENGINEER - A duly registered and licensed engineer.

EQUIPMENT, EXISTING - Any equipment regulated by code which was legally installed prior to the effective date of code or for which a permit to install has been issued.

EVALUATION REPORT - A report indicating compliance with the provisions of the national building codes as analyzed by an evaluation committee appointed by the local building official. This is a report document that verifies that the equipment or installation complies with or exceeds the codes' minimum acceptance criteria. The term <u>approved</u> means the equipment or installation has an evaluation report on file that is approved by the local code jurisdictional authority. Both UBC and BOCA have similar evaluation reports.

EXIT - That portion of the means of egress which is separated from all other spaces of a building or structure by construction and opening protectives, as required for exits, to provide a protected way of travel to the exit discharge. Exits include exterior exit doors, separated exit stairs, exit passageways, and horizontal exits.

EXIT ACCESS - That portion of a means of egress which leads to an entrance to an exit.

EXIT COURT - An outside space with building walls on three or more sides and open to the sky.

EXIT DISCHARGE - That portion of a means of egress between the termination of an exit and a public way.

FAMILY - One or more persons living together, whether related to each other by birth or not, and having common housekeeping facilities.

FARM BUILDINGS - Structures, other than residences and structures appurtenant thereto, for on-farm use (barns, sheds, poultry houses, etc.).

FIRE CODE - The National Fire Protection Code produced by the National Fire Protection Association.

FIRE DOOR - A door and its assembly, so constructed and assembled in place as to give the specified protection against the passage of fire.

FIRE HAZARD - The potential degree of fire severity based on the occupancy of a structure, classified as high, moderate, or low.

> **High** - All occupancies which involve the use, storage, sale, manufacture, or processing of highly combustible, volatile, flammable, or any explosive products which are capable of burning with extreme rapidity and producing explosions or large volumes of smoke, poisonous fumes, or release gases in the event of fire.
>
> **Moderate** - All occupancies which involve the use, storage, sale, manufacture, or processing of any materials which are capable of burning with moderate rapidity and producing a considerable volume of smoke, but which do not produce either poisonous fumes, or an explosion, in the event of fire.

Low - All occupancies which involve the use, storage, sale, or manufacture of any materials that do not ordinarily burn rapidly, or produce excessive smoke, poisonous fumes, or explosions in the event of fire.

FIRE PROTECTION - The provision of construction safeguards and exit facilities and the installation of fire alarm, fire-detecting, and fire-extinguishing service equipment to reduce the fire risk, including the risk involved in the spread of fire by exterior exposure to and from adjoining buildings and structures.

FIRE RESISTANCE or FIRE-RESISTANCE RATING - The period of time a building or building component maintains the ability to confine a fire or continues to perform a given structural function or both, as determined by tests prescribed in Section 701.2.

FIRE-RETARDANT-TREATED WOOD - Any wood product which, when impregnated with chemicals by pressure process or other means during manufacture, shall have, when tested in accordance with ASTM E 84, a flamespread index of 25 or less and show no evidence of significant progressive combustion when the test is continued for an additional 20-min. In addition, the flame front shall not progress more than 10' 6" (3200 mm) beyond the centerline of the burners at any time during the test. See Section 2301.8 for acceptance criteria for fire-retardant-treated wood.

FIRE WALL - A 4-hour fire-resistant wall, having protective openings, which restricts the spread of fire and extends continuously from the foundation to or through the roof, with sufficient structural stability under fire conditions to allow collapse of construction on either side without collapse of the wall.

FIREBLOCKING - Barriers installed to resist the movement of flame and gases to other areas of a building through small concealed passages in building components such as floors, walls, and stairs.

FLAMESPREAD - The propagation of flame over a surface.

FLAMESPREAD RATING - That numerical value assigned to a material tested in accordance with ASTM E 84.

FLAMMABLE LIQUID - See Fire Code.

FLOOR AREA - The area included within the surrounding exterior walls of a building or portion thereof, exclusive of vent shafts and courts. The floor area of a building, or a portion thereof, not provided with surrounding exterior walls is the usable area under the horizontal projection of the roof or floor above.

FOAM PLASTIC INSULATION - A plastic which is intentionally expanded by the use of a foaming agent to produce a reduced-density plastic containing voids consisting of hollow spheres or interconnected cells distributed throughout the plastic for thermal insulating or acoustical purposes and which has a density less than 20 lb/cu ft (320 kg/m^3).

FOOTBOARDS - That part of a raised seating facility other than an aisle or cross aisle upon which the occupant walks to reach a seat. Applies to reviewing stands, grandstands, and bleachers.

FOOTING - That portion of the foundation of a structure which spreads and transmits loads directly to the soil or the piles.

FRONT OF LOT - The front boundary line of a lot bordering on the street and, in the case of a corner lot, may be either frontage.

GALLERY - That portion of the seating space of an assembly room having a seating capacity of more than 10 located above a balcony.

GARAGE - A building or portion thereof in which a motor vehicle containing flammable or combustible liquids or gas in its tank is stored, repaired, or kept.

GARAGE, PRIVATE - A building or any portion of a building, which is not more than 1000 sq ft (93 m^2) in area, in which only motor vehicles used by the tenants of the building or buildings on the premises are stored or kept.

GARAGE, PUBLIC - Any garage other than a private garage.

GRADE - A reference plane representing the average of finished ground level adjoining the building at all exterior walls. When the finished ground level slopes away from the exterior walls, the reference lane shall be established by the lowest points within the area between the building and the lot line or between the building and a point 6 ft (1829 mm) from the building, whichever is closer to the building.

GRADE, LUMBER - The division of sawn lumber into quality classes with respect to its physical and mechanical properties as defined in published lumber manufacturers' standard grading rules.

GRANDSTANDS - Tiered or stepped seating facilities.

GROUT - Mixture of cementitious materials and aggregate to which sufficient water is added to produce pouring consistency without segregation of the constituents.

GUARDRAIL SYSTEM - A system of building components located near the open sides of elevated walking surfaces for the purposes of minimizing the possibility of an accidental fall from the walking surface to the lower level.

GUEST - Any person hiring or occupying a room for living or sleeping purposes.

GUEST ROOM - Any room or rooms used by a guest for sleeping purposes. Every 100 sq ft (9.3 m^2) of superficial floor area in a dormitory shall be considered to be a guest room.

GYPSUM SHEATHING - A gypsum board used as a backing for exterior surface materials, manufactured with water-repellent paper and which may be manufactured with a water-resistant core, in accordance with ASTM C 79.

GYPSUM WALLBOARD - A gypsum board manufactured in accordance with ASTM C 36 used primarily as an interior surfacing for building structures.

GYPSUM WALLBOARD, TYPE X - A gypsum board specially manufactured to provide specific fire-resistant characteristics.

HABITABLE SPACE - A space in a structure for living, sleeping, eating, or cooking. Bathrooms, toilet compartments, closets, halls, storage or utility space, and similar areas are not considered habitable space.

HANDLING - The deliberate transport of materials by any means to a point of storage or use.

HANDRAIL - A horizontal or sloping rail grasped by hand for guidance or support.

HAZARD CONTENTS, HIGH - Contents which are liable to burn with extreme rapidity or from which poisonous fumes or explosions are to be feared in case of fire.

HAZARD CONTENTS, LOW - Contents of such low combustibility that no self-propagating fire therein can occur. The only probable danger requiring the use of emergency exits will be from panic, fumes, smoke, or fire from some external source.

HAZARD CONTENTS, ORDINARY - Contents which are liable to burn with moderate rapidity or to generate a considerable volume of smoke but from which neither poisonous fumes nor explosions are to be feared in case of fire.

HAZARDOUS OCCUPANCY - See Section 308.

HAZARDOUS PRODUCTION MATERIAL (HPM) - A solid, liquid, or gas that has a degree of hazardous rating in health, flammability, or reactivity of 3 or 4 and which is used directly in research, laboratory, or production processes which have, as their end product, materials which are not hazardous.

HEALTH HAZARD - A classification of a chemical for which there is statistically significant evidence based on at least one study conducted in accordance with established scientific principles that acute or chronic health effects may occur in exposed persons. The term health hazard includes chemicals which are carcinogens, toxic or highly toxic agents, reproductive toxins, irritants, corrosives, sensitizers, hepatotoxins, nephrotoxins, neurotoxins, agents which act on the hematopoietic system, and agents which damage the lungs, skin, eyes, or mucous membranes.

HEATING - See SBC chapter 28 and the Standard Mechanical Code. [BOCA and UBC both have similar heating, ventilation, and air-conditioning sections and related mechanical codes, the National Mechanical Code (NMC) and Uniform Mechanical Code (UMC), respectively.]

HEIGHT, BUILDING - The vertical distance from grade to the highest finished roof surface in the case of flat roofs or to a point at the average height of the highest roof having a pitch. Height of a building in stories includes basements, except as specifically provided for in Section 503.2.4. [Here the codes differ as to what is permitted in building height, which is also affected by any local ordinance. It is advisable to check with local building officials for regional height definition and requirements.]

HEIGHT, STORY - The vertical distance from top to top of two successive finished floor surfaces.

HEIGHT, WALL - The vertical distance to the top measured from the foundation wall, or from a girder or other intermediate support of such wall.

HELIPORT - An area of land or water or a structural surface which is used, or intended to be used, for the landing and takeoff of helicopters, and any appurtenant areas which are used, or intended to be used, for heliport buildings and other heliport facilities.

HELISTOP - The same as a heliport, except that no refueling, maintenance, repairs, or storage of helicopters is permitted.

HIGHLY TOXIC MATERIAL - A material which produces a lethal dose or a lethal concentration which falls within any of the following categories:

1. A chemical that has a median lethal dose (LD_{50}) of 50 mg or less per kilogram of body weight when administered orally to albino rats weighing between 200 and 300 g each.
2. A chemical that has a median lethal dose (LD_{50}) of 200 mg or less per kilogram of body weight when administered by continuous contact for 24 hours (or less if death occurs within 24 hours) with the bare skin of albino rabbits weighing between 2 and 3 kg each.
3. A chemical that has a median lethal dose (LD_{50}) in air of 200 parts per million by volume or less of gas or vapor, or 2 mg/L or less of mist, fume, or dust, when administered by continuous inhalation for 1 hour (or less if death occurs within 1 hour) to albino rats weighing between 200 and 300g each.

Mixtures of these materials with ordinary materials, such as water, may not warrant a classification of highly toxic. While this system is basically simple in application, any hazard evaluation which is required for the precise categorization of this type of material shall be performed by experienced, technically competent persons. [This toxic materials criteria is virtually the same in UBC and BOCA.]

HORIZONTAL EXIT - Way of passage from one building to an area of refuge in another building on approximately the same level, or a way of passage through or around a wall or partition to an area of refuge on approximately the same level in the same building, which affords safety from fire or smoke from an area of incidence and areas communicating therewith.

HORIZONTAL SEPARATION - The distance in feet measured from the building face to the closest interior lot line; to the centerline of a street, alley, or public way; or to an imaginary line between two buildings on the same property.

HOTEL - Any building containing six or more guest rooms intended or designed to be used, or which are used, rented, or hired out to be occupied or which are occupied for sleeping purposes by guests.

HOT-WATER-HEATING BOILER - A boiler having a volume exceeding 120 gal [454.2 liters (L)], or a heat input exceeding 210 degrees Fahrenheit (210^0 F) [99 degrees Celsius (99^0 C)], that provides hot water to be used externally to itself.

INDUSTRIAL OCCUPANCY - See Section 306.0 [306.0; 306.1].

INSTITUTIONAL OCCUPANCY - See Section 308.0 [308.0; 308.1].

INTERIOR LOT LINE - See Property Line, Common.

IRRITANT - A chemical which is not a corrosive but which causes a reversible inflammatory effect on living tissue by chemical action at the site of contact. A chemical is a skin irritant if, when tested on the intact skin of albino rabbits by the methods of 16 CFR section 1500.41 for 4 hr exposure or other appropriate techniques, it results in an empirical score of 5 or more. A chemical is an eye irritant if so determined under the procedure listed in 16 CFR section 1500.42 or other appropriate techniques. [UBC and BOCA use the same regulations and refer to the same sections in the CFR.]

JURISDICTION - The government unit which has adopted code under due legislative authority. [At the municipal or county level.]

LABELED - Devices, equipment, or materials to which have been affixed a label, seal, symbol, or other identifying mark of a nationally recognized testing laboratory, inspection agency, or other organization concerned with product evaluation that maintains periodic inspection of the production of the above labeled items and by whose label the manufacturer attests to compliance with applicable nationally recognized standards. [This does not necessarily mean approved.]

LINTEL - The member placed over an opening in a wall which supports the wall construction above. [Header, of the header assembly. Lintel implies load bearing and therefore requires structural specifications as per code regulations.]

LISTED - Describes equipment or materials included in a list published by a nationally recognized testing laboratory, inspection agency, or other organization concerned with product evaluation that maintains periodic inspection of production of listed equipment or materials and whose listing states either that the equipment or material meets nationally recognized standards or has been tested and found suitable for use in a specified manner.

The means for identifying listed equipment may vary for each testing laboratory, inspection agency, or other organization concerned with product evaluation, some of which do not recognize equipment as listed unless it is also labeled. The building official should utilize the system employed by the listing organization to identify a listed product. [Again, listing does not necessarily equate to <u>approved</u>. For equipment to be approved, it must be both listed and labeled.]

LIVE LOAD - The weight superimposed by the use and occupancy of the building, not including crane load, dead load, earthquake load, snow load, or wind load.

LOAD DURATION - The period of continuous application of a given load or the aggregate of periods of intermittent applications of the same load.

LODGING HOUSE - Any building or portion thereof containing not more than five guest rooms where rent is paid in money, goods, labor, or otherwise.

LOT - A parcel of land considered as a unit, with one or more of the following features of that unit:

 Lot, corner - A lot with two adjacent sides abutting upon streets or other public spaces

 Lot, interior - A lot which faces on one street or with opposite sides on two streets

 Lot line - A line dividing one lot from another, or from a street or any public place

 Lot line, interior - Any lot line other than one adjoining a street or public space

 Lot line, street - The lot line dividing a lot from a street or other public way

LOW-PRESSURE HOT-WATER-HEATING BOILER - A boiler furnishing hot water at pressures not exceeding 160 psi (1102.4 kPa) and at temperatures not exceeding 250 degrees Fahrenheit (210^0 F) [121 degrees Celsius (121^0 C)].

LOW-PRESSURE STEAM-HEATING BOILER - A boiler furnishing steam at pressures not exceeding 15 psi (103.4 kPa).

MARQUEE - A permanent roofed structure attached to and supported by the building and projecting over public property.

MEANS OF EGRESS - A continuous and unobstructed way of exit travel from any point in a building or structure to a public way, consisting of three separate and distinct parts: (1) the way of exit access, (2) the exit, and (3) the way of exit discharge. A means of egress comprises the vertical and horizontal ways of travel and shall include the intervening room space, doors, corridors, passageways, balconies, stairs, ramps, enclosures, lobbies, horizontal exits, courts, and yards. See Exit and Exit Access.

MEMBRANE PENETRATION FIRE STOP - A material, device, or construction installed to resist, for a prescribed time period, the passage of flame, heat, and hot gases through openings in a protective membrane in order to accommodate cables, cable trays, conduit, tubing, pipes, or similar items.

MEZZANINE - One or more intermediate levels between the floor and ceiling of a story, meeting the requirements of Section 503.2.3 [500.1; 500.0].

MOTEL - See Hotel.

MULTIPLE DWELLING - See Apartment House.

NONCOMBUSTIBLE BUILDING MATERIAL - A material which meets either of the following requirements:
1. Materials which pass the test procedure set forth in ASTM E 136.
2. Materials having a structural base of noncombustible materials as defined in part 1, with a surfacing not more than 1/8 in (3.17 mm) thick which has a flamespread rating not greater than 50 when tested in accordance with ASTM E 84.

The term noncombustible does not apply to the flamespread characteristics of interior finish or trim materials. Any material shall not be classed as noncombustible which is subject to increase in combustibility or flamespread rating beyond the limits herein established through the effects of age, moisture, or other atmospheric conditions.

OCCUPANCY - The purpose for which a building, or part thereof, is used or intended to be used.

OCCUPANCY, MIXED - A building used for two or more occupancies classified in different occupancy groups.

OCCUPANT CONTENT - The actual number of total occupants permitted to occupy a floor area in accordance with the maximum capacity of the exits serving that floor area.

OCCUPANT LOAD - The calculated minimum number of persons for which the means of egress of a building or portion thereof is designed, based on Section 1003.1 [1003.0; 1003.1].

OCCUPIABLE ROOM - A room or enclosed space designed for human occupancy in which individuals congregate for amusement, educational or similar purposes, or in which occupants are engaged at labor; and which is equipped with means of egress, light, and ventilation facilities meeting the requirements of code.

OPEN-AIR GRANDSTANDS and BLEACHERS - Seating facilities which are located so that the side toward which the audience faces is unroofed and without an enclosing wall.

ORIEL WINDOW - A window which projects from the main line of an enclosing wall of a building and is carried on brackets or corbels.

OWNER - Any person, agent, firm, or corporation having a legal or equitable interest in the property.

PARTITION - An interior wall, other than folding or portable, that subdivides spaces within any story, attic, or basement of a building.

PARTITION, PARTIAL - A partition with a maximum height of 72 in (1829 mm).

PEDESTRIAN WALKWAY - A walkway used exclusively for pedestrian traffic.

PENETRATION FIRE STOP - A through-penetration fire stop or a membrane-penetration fire stop.

PERMANENT SEATING - Seating facilities which remain at a location for more than 90 days. Applies to reviewing stands, grandstands, and bleachers.

PERMIT - An official document or certificate issued by the building official authorizing performance of a specified activity.

PERSON - A natural person, his or her heirs, executors, administrator, or assigns; a firm, partnership, or corporation and its successors or assigns; or the agent of any of the aforesaid; or any other group acting as a unit as well as individuals. It shall also include an executor, administrator, trustee, receiver, or other representative appointed according to law.

Whenever the word <u>person</u> is used in any section of code that is prescribing a penalty or fine, as to partnerships or associations, the word shall include the partners or members thereof, and as to corporations, shall include the officer, agents, or members thereof who are responsible for any violation of such section.

PHOTOLUMINESCENT - The property of emitting light as the result of absorption of visible or invisible light, which continues for a length of time after excitation.

PLASTIC, APPROVED - A thermoplastic, thermosetting, or reinforced plastic material which has self-ignition temperature 650 degrees F (343 degrees C) or greater when tested in accordance with ASTM D 1929, a smoke density rating no greater than 450 when tested in accordance with ASTM E 84 in the way intended for use, or a smoke density rating no greater than 75 when tested in the thickness intended for use by ASTM D 2843 and which meets one of the combustibility classifications listed below:

- CC 1 - Plastic materials which have a burning extent of 1 in (25.4 mm) or less when tested in nominal 0.060 in (1.52 mm) thickness by ASTM D 635, or
- CC 2 - Plastic materials which have a burning rate of 2-1/2 in (64 mm) per minute or less when tested in nominal 0.60 in (1.52 mm) thickness by ASTM D 635 or in the thickness intended for use.

PLASTIC, GLASS-FIBER-REINFORCED - Plastic reinforced with glass fiber and having not less than 20% of glass fibers by weight.

PLASTIC, GLAZING - Plastic materials which are glazed or set in a frame or sash and not held by mechanical fasteners which pass through the glazing material.

PLATFORM, PERMANENT - A platform used within an area for more than 30 days.

PLATFORM, TEMPORARY - A platform used within an area for 30 days or less.

PLENUM - An air compartment or chamber to which one or more ducts are connected and which forms part of an air distribution system. This definition is intended to clarify use of this word, which is referred to in HVAC sections: A <u>plenum</u> is a compartment or chamber to which one or more ducts are connected and which forms part of an air distribution system. As now noted in one of the national building codes and soon to be adopted in all, a plenum is an enclosure "specially fabricated to transport environmental air." See Figure 2-1.

The definition further clarifies that an air-handling space above a suspended ceiling or under a raised floor (such as in a computer room) is not a plenum, but is "other spaces" used for environmental air.

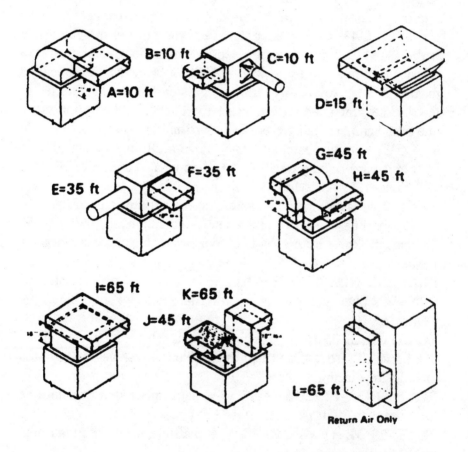

Figure 2-1
Plenums

PROPERTY LINE, ASSUMED - The centerline of a street where an exterior building wall faces a street, or an imaginary line between the exterior walls of two buildings on the same lot.

PROPERTY LINE, COMMON - A line dividing one lot from another.

PUBLIC SPACE - A legal open space on the premises, accessible to a public way or street, such as yards, courts, or open spaces permanently devoted to public use, which abuts the premises and is permanently maintained accessible to the fire department and free of all encumbrances that might interfere with its use by the fire department.

PUBLIC WAY - Any street, alley, or other parcel of land open to the outside air, deeded, dedicated, or otherwise permanently appropriated for public use and having a clear and unobstructed width and height of not less than 10 ft (3048 mm).

READILY ACCESSIBLE - Having direct access without the need of removing any panel, door or similar covering of the item described, and without requiring the use of portable ladders, chairs, etc. See Accessible.

REGISTERED DESIGN PROFESSIONAL - An architect or engineer who is registered or licensed to practice professional architecture or engineering, as defined by the statutory requirements of the professional registration laws of the state in which the project is to be constructed.

REPAIR - The reconstruction or renewal of any part of an existing building for the purpose of its maintenance.

REQUIRED - Shall be construed to be mandatory by provisions of code.

RESIDENTIAL OCCUPANCY - See Section 311.0 [310.0; 310.1].

REVIEWING STANDS - Elevated platforms accommodating not more than 50 persons. Seating facilities, if provided, are normally in the nature of loose chairs. Reviewing stands accommodating more than 50 persons are grandstands.

SAFE DISPERSAL AREA - An area which will accommodate a number of persons equal to the total capacity of the stand and building which it serves in such a manner that no person within the area need be closer than 50 ft (15.2 m) from the stand or building. Dispersal areas are based on an area of not less than 3 sq ft

(0.28m²) per person. Applies to reviewing stands, grandstands, and bleachers.

SEISMIC LOAD - The forces superimposed on a building or structure by an earthquake.

SELF-CLOSING - As applied to a fire door or other opening, means normally closed and equipped with an approved device which will ensure closing after having been opened for use.

SELF-LUMINOUS - Powered continuously by a self-contained power source other than a battery or batteries, such as radioactive tritium gas. A self-luminous sign is independent of external power supplies or other energy for its operation.

SENSITIZER - A chemical that causes a substantial proportion of exposed people or animals to develop an allergic reaction in normal tissue after repeated exposure to the chemical.

SERVICE CORRIDOR - A fully enclosed passage used for transporting hazardous production materials and for purposes other than required exiting.

SHAFT - A vertical opening extending through one or more stories of a building. An interior space, enclosed by walls or construction, extending through one or more stories or basements which connects openings in successive floors and roof, to accommodate elevators, dumbwaiters, mechanical equipment, or similar devices or to transmit light or ventilation air.

SHAFT ENCLOSURE - The walls or construction forming the boundaries of a shaft.

SHALL - As used in code, the term means "mandatory."

SMOKE DETECTOR - An approved, listed device that senses visible or invisible particles of combustion.

SMOKEPROOF ENCLOSURE - An exit consisting of a vestibule and continuous stairway enclosed from the highest point to the lowest point and designed so that the movement of products of combustion produced by a fire occurring in any part of the building into the smokeproof tower is limited.

SNOW LOAD - The forces superimposed on a building or structure resulting from the accumulation of snow.

SPRINKLERED - Equipped with an approved automatic sprinkler system properly maintained.

STAIRWAY - One or more flights of stairs, either exterior or interior, with the necessary landings and platforms connecting them, to form a continuous and uninterrupted passage from one level to another in a building or structure.

STORY - That portion of a building included between the upper surface of a floor and upper surface of the floor or roof next above. [UBC and BOCA specify that the topmost story is that portion of the building between the finished floor of the uppermost floor and the structure's ceiling or roof above. "If the finished floor level directly above a usable or unused under-floor space is more than 6 feet (1829 mm) above grade as defined within each code for more than 50 percent of the total perimeter or is more than 12 feet (3658 mm) above grade at any point, such usable or unused under-floor space shall be considered as a story."]

STORY, FIRST - The lowest story in a building which qualifies as a story, as defined herein, except that a floor level in a building having only one floor level shall be classified as a first story, provided such floor level is not more than 4 ft (1219 mm) below grade, as defined herein, for more than 50 % of the total perimeter, or not more than 8 ft (2438 mm) below grade, as defined herein, at any point.

STREET - Any public thoroughfare, street, avenue, boulevard, park, or space more than 20 ft (6096 mm) wide which has been dedicated or deeded to the public for public use. [UBC and BOCA: "Any public thoroughfare, street, avenue, boulevard, park or space more than 16 feet (4877 mm) wide which has been dedicated or deeded to the public for public use."

STREET LINE - A lot line dividing a lot from a street.

STRUCTURAL ALTERATION or WORK - The installation or assembly of any new structural components, or any change to existing structural components, in a system, building, or structure.

STRUCTURAL OBSERVATION - The visual observation of the structural system, for general conformance to the approved plans and specifications, at significant construction stages and at completion of the structural system. Structural observation does not include or waive the responsibility for the inspections required by Section 108 or other sections of code.

STRUCTURE - That which is built or constructed or a portion thereof. An edifice or building of any kind, or any piece of work artificially built up or composed of parts joined together in some definite manner.

SURGICAL AREA - The preoperating, operating, recovery, and similar rooms within an outpatient health-care center.

TENANT - Any person, agent, firm, corporation, or division who uses or occupies land, a building, or portion of a building by title, under a lease, by payment of rent, or who exercises limited control over the space.

TENANT SEPARATION - A partition or floor-ceiling assembly, or both, between tenants.

THEATER - A building, or part thereof, which contains an assembly hall with or without a stage which may be equipped with curtains and permanent stage scenery or mechanical equipment adaptable to the showing of plays, operas, motion pictures, performances, spectacles, and similar forms of entertainment.

THROUGH-PENETRATION FIRE STOP - A material, device, or construction installed to resist, for a prescribed time period, the passage of flame, heat, and hot gases through openings which penetrate the entire fire-resistive assembly in order to accommodate cables, cable trays, conduit, tubing, pipes, or similar items.

TOWNHOUSE - A single family dwelling constructed in a series or group of attached units with property lines separating each unit.

UL - The Underwriters Laboratories, Inc., 333 Pfingsten Road, Northbrook, Illinois 60062.

USABLE CRAWL SPACE - A crawl space designed to be used for equipment or storage.

USE - With reference to hazardous materials other than flammable or combustible liquids, the placing in action or making available for service by opening or connecting any container utilized for confinement of material whether a solid, liquid, or gas.

USE, CLOSED SYSTEM - The use of a solid or liquid hazardous material in a closed vessel or system that remains closed during normal operations where vapors emitted by the product are not liberated outside the vessel or system and the product is not exposed to the atmosphere during normal operations; and all uses of compressed gases. Examples of closed systems for solids and liquids include the product conveyed through a piping system into a closed vessel, system, or piece of equipment and the reaction process operations.

USE, OPEN SYSTEM - Use of a solid or liquid hazardous material in a vessel or system that is continuously open to the atmosphere during normal operations. Examples of open systems for solids and liquids include dispensing from or into open breakers or containers and dip tank and plating tank operations.

VALUATION or VALUE - When applied to a building, means the estimated cost to replace the building in like kind.

VENEER - A facing attached to a wall for the purpose of providing ornamentation, protection, or insulation but not counted as adding strength to the wall.

VERTICAL OPENING - An opening through a floor or roof.

WALL, BEARING - A wall supporting any vertical load in addition to its own weight.

WALL, CURTAIN - A nonbearing wall between columns or piers which is not supported by girders or beams but is supported on the ground.

WALL, EXTERIOR - A wall, bearing or nonbearing, which is used as an enclosing wall for a building, other than a party wall or fire wall.

WALL, FOUNDATION - A wall below the first floor extending below the adjacent ground level and serving as a support for a wall, pier, column, or other structural part of a building.

WALL, NONBEARING - A wall which supports no vertical load other than its own weight.

WALL, PANEL - A nonbearing wall in skeleton or framed construction, built between columns or piers and wholly supported at each story.

WALL, PARAPET - That part of any wall entirely above the roof line.

WALL, PARTY - A fire wall on an interior lot line, used or adapted for joint service between two buildings.

WALL, RETAINING - A wall designed to prevent the lateral displacement of soil or other material.

WEATHER-EXPOSED SURFACES - All surfaces of walls, ceilings, floors, roofs, soffits, and similar surfaces exposed to the weather, except the following:
- Ceilings and roof soffits enclosed by walls or by beams which extend a minimum of 12 in (305 mm) below such ceilings or roof soffits.

- Walls or portions of walls within an enclosed roof area, when located a horizontal distance from an exterior opening equal to twice the height of the opening.
- Ceiling and roof soffits beyond a horizontal distance of 10 ft (3048 mm) from the outer edge of the ceiling or roof soffits.

WIND LOAD - The forces superimposed on a building or structure by the movement of an air mass at a specified velocity.

WINDOW WELL - A soil-retaining structure at a window having a sill height lower than the adjacent ground elevation.

WRITING - Includes printing and typewriting.

WRITTEN NOTICE - A notification in writing delivered in person to the individual or parties intended or delivered at or sent by certified or registered mail to the last residential or business address of legal record.

YARD - An unoccupied open space other than a court which is unobstructed from the ground to the sky, except where specifically provided by code, on the lot on which the building is situated.

ZONING - The reservation of certain specified areas within a community or city for buildings and structures, or use of land for certain purposes with other limitations such as height, lot coverage, and other stipulated requirements.

Chapter

3

Occupancy

SECTION 301.0 OCCUPANCY CLASSIFICATION [301.0; 301.0]

Once adopted by a jurisdiction, a national building code mandates that every existing and proposed building or structure be classified according to its use and occupancy purposes. The national building codes do this for a very careful and correct reason: to protect human life. These occupancy classification separations allow for the consideration of highest hazard to sleeping or assembled people. This further allows for the critical analysis of construction design to factor in fire-resistant construction and materials and fire-protection systems. Fire-protection systems are cost-effective in construction bids if the designer will cross check specs with these occupancy codes and the Fire Code.

The national building codes look at it from a worst-case scenario of hazard to assembled people. The more people assembled in an area of smaller square footage, the higher the degree of hazard. People congregate in different buildings for different purposes and divisions are necessary for the different types of construction and fire resistance. People assemble in large numbers for transportation centers; hospitals, business, schools, sports, and entertainment. Stadiums and reviewing stands are also covered under this section.

Occupancy is defined as the purpose for which a building or any portion thereof is used. Occupied is defined, as applied to a building, as though followed by the words "or intended, arranged, or designed to be occupied." A change of occupancy is defined as any change in the purpose or level of activity within a structure that involves a change in application of the requirements of the code. To compute occupant load, see SBC section 1002. To compute occupiable living space, see SBC section 1202, UBC section 1201.0, and NBC section 1200.1.

301.1 Scope

Provisions of this chapter shall govern the classification of building occupancies.

301.2 Occupancy or Use Categories

Every new and existing building, structure, or part thereof shall, for the purpose of code, be classified as one of the following occupancy groups according to its use or occupancy as a building or structure:

SBC:
- Group A - Assembly (see section 304)
- Group B - Business (see section 305)
- Group E - Educational (see section 306)
- Group F - Factory-Industrial (see section 307)
- Group H - Hazardous (see section 308)
- Group I - Institutional (see section 309)
- Group M - Mercantile (see section 310)
- Group R - Residential (see section 311)
- Group S - Storage (see section 312)

UBC:
- Group A - Assembly (see section 303)
- Group B - Business (see section 304)
- Group E - Educational (see section 305)
- Group F - Factory-Industrial (see section 306)
- Group H - Hazardous (see section 307)
- Group I - Institutional (see section 308)
- Group M - Mercantile (see section 309)
- Group R - Residential (see section 310)
- Group S - Storage (see section 311)
- Group U - Utility (see section 312)

NBC:
- Group A - Assembly (see section 303)
- Group B - Business (see section 304)
- Group E - Educational (see section 305)
- Group F - Factory-Industrial (see section 306)
- Group H - Hazardous (see section 307)

Group I - Institutional (see section 308)
Group M - Mercantile (see section 309)
Group R - Residential (see section 310)
Group S - Storage (see section 311)
Group U - Utility (see section 312)

301.3 Uncertain Classification
Each occupancy group is intended to include buildings as hereinafter defined and those of similar character or use. Wherever there is any uncertainty as to the classification of a building, the building official shall determine the classification within which it falls, according to the life, safety, and relative fire hazard involved.

302.0 DEFINITIONS [302.0; 302.1]
The following words and terms shall, for the purposes of this chapter and as stated elsewhere in code, have the meanings shown herein. Refer to Chapter 2 for general definitions.

ROOMING HOUSE, Transient - Any building or portion thereof containing not more than five guest rooms, occupied by not more than five guests, where rent is paid and guests are transient.
ROOMING HOUSE, Not Transient - Any building or portion thereof containing guest rooms where rent is paid and guests are not transient.

303.0 MIXED OCCUPANCIES [302.0; 302.1]
There is a major difference between determining occupancy and computing occupancy load for mixed occupancies. Occupancy is determined by the proposed usage of the building or structure, and occupancy load is factored by the amount of square footage available to the people inside the building.

Area separations are required by the codes to reduce hazard to occupants by providing fire breaks that retard flamespread between areas of congregation. Accordingly, area separation walls are factored in terms of one or more hours fire protection. A 1-hr fire rating on partition walls will protect adjacent rooms for 1-hr from flamespread and fire. Area separations are used by construction designers to mix different occupancies in the same building or structure.

Without area separations offering fire breaks, the building will always be classified as the highest hazard occupancy for public safety. Access between occupancies can be provided by a common hallway which will also serve as an area separation.

303.1 Multiple Occupancies
A building that is used for two or more occupancies, classified within different occupancy groups, shall be considered a mixed occupancy building.

> **EXCEPTION**: A building containing two or more occupancies, none of them Group H, may be considered a single occupancy when
> 1. The required type of construction for the building is determined by applying the height and area limitations for each of the applicable occupancy groups to the entire building with the most restrictive type of construction requirements being applied.
> 2. The entire building conforms with the most code restrictive occupancy group fire-protection requirements, as determined by SBC chapters 7 and 8.
> 3. The entire building conforms with the most restrictive occupancy group sprinkler, standpipe, and alarm system requirements, as determined by SBC chapter 9.
> 4. All other requirements of code are applied to each portion of the building based on the use of that space.

303.2 Height and Area
A mixed occupancy building shall be governed by the height and area limitations applying to the principal intended use. However, each portion of the building shall conform to all other requirements of code for the occupancy contained therein. Accessory occupancies shall not exceed the area limitation nor be located at a height greater than that permitted for such occupancy group in the type of construction being used.

304.0 ASSEMBLY OCCUPANCY - GROUP A [303.0; 303.0]

Each occupancy group requires a different type of construction. Unless dealing with a mixed-use occupancy building, all occupancy groups classify the building by construction type or a portion thereof within area separation walls. The building must conform to the type of construction specified to meet the minimum requirements for that classification. The designer can provide a higher classification of construction, but the code regulations are minimum requirements for that type of construction.

The national building codes use Group A Occupancy for assembly of large numbers of people, thus constituting the highest degree of hazard to the public. In large buildings or structures where occupant load will increase with people assembling in large numbers for sports, entertainment, transportation, or lodging, the hazard to public safety is considered to be the highest.

304.1 Scope

Group A Occupancy is the use of a building or structure, or any portion thereof, for the gathering together of persons for purposes such as civic, social, or religious functions; recreation; or for food or drink consumption; or awaiting transportation. These buildings include

SBC - Group A Occupancy shall include, among others, the following:

Amusement park buildings	Passenger depots
Auditoriums	Public assembly halls
Churches	Recreation halls
Dance halls	Restaurants
Gymnasiums	Stadiums and grandstands
Motion picture theaters	Tents for assembly
Museums	Theaters (stage production)

UBC - UBC assembly occupancies are divided into four divisions and one subdivision. Division 1 is for buildings having an assembly room with an occupant load of 1000 or more and a legitimate stage. Division 2 is for buildings having an assembly room with an occupant load of less than 1,000 and a legitimate stage. Division 2.1 is for buildings having an assembly room with an occupant load of 300 or more without a legitimate stage, including such buildings used for educational purposes and not classified as Group B or E Occupancies.

Division 3 is for buildings having an assembly room with an occupant load of less than 300 without a legitimate stage, including such buildings used for educational purposes and not classified as Group B or E Occupancies. Division 4 includes stadiums, reviewing stands, and amusement park structures not included within other Group A Occupancies.

UBC Group A Occupancies also include the use of a building or structure for a gathering of 50 or more persons for civic, social, or religious functions; recreation; education or instruction; food or drink consumption; or awaiting transportation. A room or space used for assembly purposes by less than 50 persons and accessory to another occupancy is included as part of that major occupancy.

NBC - BOCA assembly occupancies are divided into four groups. Use Group A-1 includes all buildings and structures intended for the production and viewing of performing arts or motion pictures and which are usually provided with fixed seats (including theaters, motion picture theaters, and television and radio studios admitting an audience). Use Group A-3 includes all buildings with or without an auditorium where people assemble for amusement, entertainment, or recreation purposes, as well as incidental motion picture, dramatic or theatrical presentations, lectures, or other similar purposes without a theatrical stage other than a raised platform; and which are principally occupied without permanent seating facilities (including art galleries, exhibition halls, museums, lecture halls, libraries, restaurants other than nightclubs, recreation centers; and buildings designed for similar assembly purposes, including passenger terminals).

Use Group A-4 includes all buildings and structures which are occupied exclusively for the purpose of worship or other religious services. Use Group A-5 includes structures utilized for outdoor assembly intended for participation in or reviewing activities, including grandstands, bleachers, coliseums, stadiums, amusement park structures, and fair or carnival structures.

304.2 Subclassifications
304.2.1. Assembly occupancies shall be divided into two subclassifications as set forth in this section, both of which shall comply with the requirements for Group A Occupancy unless otherwise specified:
1. A-1. Large assembly shall include theaters and other places of assembly without a legitimate stage and with an occupant load of 1000 or more persons. Large assembly shall also include theaters and other places of assembly with a stage requiring proscenium opening protection and with an occupant load of 700 or more persons.
2. A-2. Small assembly shall include theaters and other places of assembly with or without a stage requiring proscenium opening protection and with an occupant load of 100 or more persons, but with an occupant load less than designated for large assembly.

304.2.2. Assembly occupancies with an occupant load less than 100 persons shall be classified as Group B.

305.0 BUSINESS OCCUPANCY - GROUP B [304.0; 304.1]
The most commonly used occupancy code regulations of the national building codes are Group B (business and commercial) and Group R (residential). Group B Occupancy covers small businesses and offices. This classification is the usage of a building or structure, or any portion thereof, for office, professional, or service-type transactions including normal accessory storage and the keeping of records and accounts.

SBC - Group B Occupancy includes, among others, the following:

Animal hospitals, kennels
Automobile, motor vehicle
   showrooms
Automobile or other vehicle
   service stations
Banks
Barbershops
Beauty shops
Bowling alleys

Electronic processing areas
Florist and nurseries
General post offices
Greenhouses
Laboratories - testing and
   research (nonhazardous)
Laundries: pickup and delivery
   stations and self-service
Libraries (other than school)

Chapter Three

Car washes
Civic administration areas
Clinics - outpatient
Dry cleaning: self-service and pickup and delivery
Educational occupancies above the twelfth grade

Office buildings
Police stations
Print shops
Professionals: attorney, dentist, physician, engineer, etc.
Radio and television stations
Telephone exchanges

UBC - UBC Group B Occupancies include buildings for office, professional, or service-type transactions which are not classified as Group H Occupancies. This includes occupancies for the storage of records and accounts and for eating and drinking establishments with an occupant load of less than 50. These business occupancies include, but are not limited to, the following:

Animal hospitals, kennels, pounds
Banks
Barbershops
Car washes
Dry cleaning: pickup and delivery stations and self-service
Electronic data processing
Florists and nurseries
Laundry: pickup and delivery stations and self-service
Outpatient and medical clinics
Print shops
Radio and television stations
Telephone exchanges

Automobile and other motor vehicle showrooms
Beauty shops
Civic administration
Educational occupancies above the twelfth grade
Fire stations
Laboratories: testing and research
Police stations
Post offices
Professional services such as attorney, dentist, physician, or engineer

NBC - BOCA's Group B Occupancy includes all buildings and structures which are occupied for the transaction of business; for the rendering of professional services; or for other services that involve stocks of goods, wares, or merchandise in limited quantities which are incidental to office occupancies or sample purposes. These include, but are not limited to,

Air traffic control towers
Animal hospitals, kennels, pounds

Fire stations
Florists and nurseries

Automobile and other motor
  vehicle showrooms
Banks
Barbershops
Police stations
Beauty shops
Car washes
Civic administration
Clinic, outpatient
Dry cleaning: pickup and delivery
  stations and self-service
Electronic data processing

Laboratories; testing
  and research
Laundry: pickup and delivery
  stations and self-service
Post offices
Print shops
Professional services;
  attorney, physician, etc.
Radio and television stations
Telecommunication
  equipment building

305.1.3. Assembly occupancies with an occupant load less than 100 persons shall be classified as Group B.
> **EXCEPTION**: Provisions of SBC sections 403.1.3.1, 403.2, 403.3, 1019.10, 1019.11, and 3103.0 shall apply to buildings used for assembly purposes, regardless of occupant load.

305.1.4. Dry cleaning establishments using solvents which are nonflammable or nonflammable at ordinary temperatures and only moderately flammable at higher temperatures (Class IV system), shall be classified as Group B Occupancy.

306.0 EDUCATIONAL OCCUPANCY - GROUP E [305.0; 305.1]

The national building codes use this occupancy class to designate educational use. Occupancy load is determined by maximum occupants during specific hours. The subsections are factored by student occupant loads during peak usage. This occupancy group is not designed for sleeping and lodging facilities. These types of postsecondary schools with living quarter dormitories would fall under the heading of <u>Mixed Occupancies</u>.

306.1 Scope

Group E Occupancy is the use of a building or structure, or any portion thereof, by six or more persons at any one time for educational purposes through the twelfth grade.

306.1.2. Child care facilities which accommodate six or more children of any age who stay less than 24 hr per day shall be classified as Group E.

306.1.3. Parts of buildings used for the congregating or gathering of 100 or more persons in one room shall be classified as Group A Occupancy, regardless of whether or not such gathering is of an educational or instructional nature.

306.1.4. Schools for business or vocational training shall be classified in the same occupancies and conform to the same requirements as the trade, vocation, or business taught, provided the concentration of persons will not exceed that listed in Section 1003 for the occupancy classification used.

307.0 FACTORY-INDUSTRIAL OCCUPANCY - GROUP F [306.0; 306.1]

307.1 Scope

Group F Occupancy is use of a building or structure, or any portion thereof, for assembling, disassembling, repairing, fabricating, finishing, manufacturing, packaging, or processing operations that are not otherwise classified in code.

307.1.2. Group F Occupancy shall include, among others, the occupancies listed in this section but does not include buildings used principally for any purpose involving highly combustible, flammable, or explosive products or materials. (See Section 108.)

> Assembly plant   Mill
> Factory   Processing plant
> Manufacturing plant

307.1.3. Portions of Group F Occupancy involving highly combustible, flammable, or explosive products or materials shall be properly ventilated, protected, and separated from the remainder of the building in accordance with the appropriate National Fire Protection Association (NFiPA) standard or the entire building will be classified as Hazardous Occupancy. (See Section 308.)

308.0 HAZARDOUS OCCUPANCY - GROUP H [307.0; 307.1]

Group H Occupancy is the principal use of a building or structure, or any portion thereof, that involves the manufacturing, processing, generation, storage, or other use of hazardous materials in excess of the exempt quantities listed in this section.

Group H Occupancies include all hazardous group occupancies where highly combustible materials are manufactured or highly flammable materials are stored, used, or processed. Height limitations of the building are factored by hourly increments of fire-resistance of wall construction. Buildings approaching 100 ft in height must be of 4-hr fire-resistant construction.

308.2 Subclassification

308.2.1. Group H hazardous occupancies shall be divided into H1 through H4 according to the hazards presented by each material as described below:

H1. Buildings or parts thereof used for the manufacturing, processing, generation, or storage of materials which present a detonation hazard. Detonation hazards include any and all explosives, blasting agents, and fireworks; Class 4 liquid and solid oxidizers; any unclassified but detonatable organic peroxides; and Class 3 and 4 detonatable unstable (reactive) materials in excess of the amounts given in Table 308.2A (see Appendix). [UBC and BOCA, see section 307.]

H2. Buildings or parts thereof used for the manufacturing, processing, generation, or storage of materials which present a deflagration hazard or a hazard from accelerated burning. Defined deflagration hazards include all Class I, II, and III-A flammable or combustible liquids in open containers or in containers pressurized at more than 15 psi (103 kPa); all combustible dusts stored in piles or within open containers; Class 3 liquid and solid oxidizers; any and all Class I organic peroxides, solid, liquid, and gaseous pyrophorics. Class I nondetonatable unstable (reactive) materials and flammable cryogenic fluids in excess of the amounts given in Table 308.2B (See Appendix). [UBC and BOCA, see section 307.]

H3. Buildings or parts thereof used for the manufacturing, processing, generation, or storage of materials which readily support combustion or present a physical hazard. Physical hazards include Class I, II, and III flammable and combustible liquids in closed containers pressurized at 15 psi (103 kPa) or less. Level 2 and 3 aerosol products, flammable solids, Class 1 and 2 liquid and solid oxidizers, Class II, III, or IV organic peroxides, Class 1 and 2 unstable (reactive) materials, water reactive materials, and oxidizing cryogenic fluids in excess of the amounts given in Table 308.2C. [UBC and BOCA, see section 307.]

H4: Buildings or parts thereof used for the manufacturing, processing, generation, or storage of materials which are health hazards. Health hazards include toxic compressed gases; highly toxic and toxic solids and liquids; corrosives, irritants, sensitizers, and other health hazard solids, liquids and gases in excess of the amounts given in Table 308.2D (see Appendix). [UBC and BOCA, see section 307.]

EXCEPTIONS:
1. The quantities of alcoholic beverages used in retail sales are unlimited provided the liquids are packaged in individual containers not exceeding 1 gal (3.79 L).
2. The quantities of medicines, foodstuffs, and cosmetics containing not more than 50% by volume of water-miscible flammable liquids and with the remainder of the solution not being flammable are unlimited when packaged in containers not exceeding 1 gal (3.79 L).
3. For mercantile occupancies, the storage of all flammable and combustible liquids shall be limited to quantities needed for display and normal merchandising purposes but shall not exceed the quantities permitted in the Fire Code.
4. Explosives, blasting agents, and ammunition preempted by the requirements of chapters 19 and 20 of the Fire Code.

Occupancy 71

5. Refrigeration systems.
6. Storage, use, and handling of pesticides and agricultural materials used for weed abatement, pest control, erosion control, soil amendments, or similar application for use on the premises.
7. Materials contained within fuel tanks or batteries on automobiles.
8. Stationary batteries used for facility emergency power, uninterrupted power supply, or telecommunication facilities, provided that the batteries are provided with safety venting caps and sufficient ventilation to produce a vapor-air mixture which is less than 25% of the lower explosive limit.
9. Control areas in accordance with Section 308.2.2.
10. Corrosives, irritants and sensitizers shall not include commonly used building materials.
11. Corrosives, irritants, and sensitizers shall not include personal or household products in their original packaging for retail display.
12. Level 1 aerosol products shall be considered equivalent to a Class III commodity, as defined by NFiPA 30B and the Fire Code.

308.2.2. Control areas containing hazardous material not exceeding the exempt quantities shall be permitted in all occupancies except assembly occupancies. The control area shall be separated by 1-hr fire-resistant construction with 3/4-hr opening protection with a self-closing or automatic closing device in accordance with Section 705.1.3.2.3. The maximum number of control areas per floor in multistory buildings shall be limited to four.

The maximum number of control areas in any building shall be limited to 10. When control areas are on different floors but adjacent to each other, the floor and ceiling between these control areas shall have not less than a 2-hr fire-resistant construction. When the floor and ceiling assembly forms part of the separation, the separation walls of the control area shall extend from the floor below to the floor deck above. [UBC and BOCA, see Section 307.]

308.2.3. When the stored amount of any hazardous material listed in Tables 308.2A through 308.2D (see Appendix) is exceeded in any one control area, such storage shall be within a room or building conforming to the code requirements for H1, H2, H3, or H4 Occupancies. [UBC and BOCA, see section 307.]

308.2.4 Multiple Hazards
Materials representing hazards that are classified in one or more of the hazardous occupancy subgroups shall conform to the code requirements for each of the use groups represented. [UBC and BOCA, see section 307.]

309.0 INSTITUTIONAL OCCUPANCY - GROUP I [308.0; 308.1]

Group I Occupancies are the institutional buildings that are occupied by people who are incarcerated, committed, or otherwise not able to freely leave the building. Because Group I Occupancies present a higher hazard to the people who occupy them, much more restrictive regulations are used in Group I to protect those occupants.

Requirements for height, area, and exit egress are more critical, and specifications are tighter for type of construction and fire-resistance. The codes call for both audible and visible alarms to be used in different areas. In nonpatient areas, audible alarms will suffice, but in patient areas, visible alarms are required. All smoke and heat detectors must be hard-wired to the building's primary electrical system and equipped with battery backup.

309.1 Group I Unrestrained Occupancy
Group I Unrestrained includes buildings or portions thereof used for medical, surgical, psychiatric, nursing, or custodial care on a 24 hr basis of six or more persons who are not capable of self-preservation and shall include among others:
- Detoxification facilities
- Hospitals
- Mental hospitals
- Nursing homes (both intermediate care facilities and skilled nursing facilities)

Facilities such as the above with five or less persons not ancillary to other uses shall be classified as a residential occupancy.

309.2 Group I Restrained Occupancy
Group I Restrained includes buildings or portions thereof which provide sleeping accommodations for six or more persons under

some degree of restraint or security who are generally incapable of self-preservation due to security measures not under the occupants' control and shall include among others:
- Correctional institutions
- Detention centers
- Jails
- Reformatories

> **EXCEPTION**: Group I Restrained qualifying for Use Condition 1 may be classified as a Group R Occupancy.

310.0 MERCANTILE OCCUPANCY - GROUP M [309.0; 309.1]
310.1 Scope
Group M Occupancy is the use of a building or structure or any portion thereof, for the display and sale of merchandise including stocks of goods, wares, or merchandise incidental to such purposes and accessible to the public and shall include, among others, the following:

Department stores	Sales rooms
Drug stores	Shopping centers
Markets	Wholesale stores
Retail stores	(other than warehouses)

311.0 RESIDENTIAL OCCUPANCY - GROUP R [310.0; 310.1]

In the national building codes, the least restricted of all occupied buildings is the Residential R3 Occupancy. Because all living, dining, and sleeping rooms are required to have windows that must open directly to the outside, this classification represents far less hazard to the public than other types of occupancies. The exception to this is that the windows may open onto a roofed porch if it has an inside ceiling height of at least 7 ft (2 m) and is 65% open on the longer side.

In design computations for number of windows required for each room, the codes call for each room's required windows to have a total area of at least 10% of the floor area of the room or at least 10 sq ft in area. Bathroom windows are required to have a window area of at least 3 sq ft. These must be manufactured with half of the window openable. Bathrooms without windows must be equipped with mechanical ventilation directly ducted to the outside.

R3 requires that every dwelling have at least one bathroom containing a water closet, lavatory, and either a shower or bathtub. This room must also be separated from food preparation areas or storage rooms by a tight-fitting door. The water closet compartments must be at least 30 in wide and have a unobstructed space in front of the water closet of at least 24 in. Kitchens must contain at least one sink. All these fixtures (with the exception of the water closet) must be serviced with hot and cold domestic water from the dwelling's plumbing service side.

Attached garages must have a 1-hr fire-resistive separation wall construction between the garage and the dwelling. Passage doors from the dwelling to the garage must be solid-core and equipped with self-closing mechanisms as additional fire-resistive separation. The minimum R3 floor space requirements are that one room must have at least 150 sq ft of area and bedrooms must have at least 70 sq ft. No room, with the exception of a kitchen, may be less than 7 ft at any dimension. Ceiling height for R3 is set at a minimum of 7 ft 6 in (2.2 m) for at least 50% of the total room area and no part of the ceiling can be less than 5 ft from finish floor elevation. Smoke detectors are required in living and sleeping areas; check with local code regulations for number and placement. R3 buildings that are three or more stories high must be equipped with an approved manual or automatic fire alarm system. Most jurisdictions are now requiring fire-protection systems for new residential construction, regardless of stories or building height.

311.1 Scope

Group R Occupancy is the use of a building or structure, or any portion thereof, for sleeping accommodations not classed as a Group I Occupancy.

311.2 Subclassifications

Group R Occupancies shall include, among others, the following:

> R1. Residential occupancies where the occupants are all primarily transient in nature including
>> Boarding houses (transient)
>> Hotels
>> Motels

R2. Multiple dwellings where the occupants are primarily permanent in nature, including
    Apartment houses
    Convents
    Dormitory facilities which accommodate six or more persons of more than 2-1/2 years of age who stay more than 24 hr.
    Fraternities and sororities
    Monasteries
    Rectories
    Rooming houses (not transient)

R3. Residential occupancies including the following:
    Child care facilities which accommodate five or less children of any age for any time period
    One- and two-family dwellings where the occupants are primarily permanent in nature and not classified as R1, R2, or I
    Rooming houses (transient)

[In the R1 classification, motels have become a point of contention as each room has an exterior door to the outside; therefore, there is immediate egress access unlike for hotels. However, the national building codes still classify them together. In code classification corridors serving 30 or more people must have 1-hr fire-resistive construction. Doors leading into this corridor must be exterior-grade solid core with no grilles or louvers unless protected by a fusible link. Corridors exceeding 20 ft in length must have an egress door and cannot be terminated in a dead end.

These R subclassifications can be found in section 310.1 of the UBC and section 310.0 of the NBC.]

312.0 STORAGE OCCUPANCY - GROUP S [311.0; 311.1]
312.1 Scope
Group S Occupancy is the principal use of a building or structure, or any portion thereof, for storage that is not classed as a Group H Occupancy, including buildings or structures used for the purpose

of sheltering animals. For buildings used for the storage of hazardous materials, see Section 308.

312.2 Subclassifications

312.2.1. S1 Moderate Hazard Storage shall include buildings used for the storage of combustible materials when not classified as S2 Low Hazard or Group H. [UBC and BOCA, see section 307.]

312.2.2. S2 Low Hazard Storage shall include buildings used for the storage of noncombustible materials such as products on wood pallets or in paper cartons without significant amounts of combustible wrappings. Such products may have a negligible amount of plastic trim such as knobs, handles, or film wrapping. S2 Low Hazard Storage shall include but not be limited to the following:

Beer or wine up to 12% alcohol in metal, glass, or ceramic containers
Cement in bags
Dairy products in nonwaxed containers
Dry cell batteries
Dry insecticides
Electrical insulators
Electric motors
Empty cans
Foods in noncombustible containers
Frozen foods
Foods in noncombustible containers
Gypsum board
Glass bottles, empty or filled with noncombustible liquid
Inert pigments
Oil-filled and other types of distribution transformers
Meats
Metal cabinets
Metal desks with plastic tops
Stoves
Washers and dryers

312.2.3. Portions of Group S Occupancy involving highly combustible, flammable, or explosive products or materials shall be properly ventilated, protected, and separated from the remainder of the building in accordance with the appropriate NFiPA standard or the entire building will be classified as Group H Occupancy.

Chapter 4

General Building Limitations

SECTION 500.0 GENERAL LIMITATIONS [500.0; 500.0]

This section sets forth the parameters of the building envelope, placement of the structure on the property, allowable height and floor space areas, and property setbacks. This section also includes areas of openings, fire-resistance of walls, allowable area increases, mixed occupancies allowable floor areas, fire-resistive substitution, and mezzanines.

In design work, there are three conventional ways to increase allowable floor area. The first is to install an automatic sprinkler system throughout which will increase allowable area in certain occupancies. The second way is to provide an increase in perimeter separation between the building and others adjacent to it on surrounding properties. Another is to build the structure using a higher type of construction. All these methods incur more cost in per-square-foot construction or in extra land purchase, so in computing for allowable area increases a balance must be struck between occupancy load and type of construction.

In determining allowable building height, remember that a roof structure designed to house anything other than mechanical service equipment will be counted as another additional building story.

501.1 Scope

Provisions of this chapter shall govern the height and area of buildings. This chapter contains no unique definitions. For general definitions, see Chapter 2.

503.0 HEIGHT and AREA

[In the UBC, section 503.0 also has to do with the building's location on the property. In that section, buildings are mandated to adjoin or have access to a public way or yard on not less than one side. Required open-space yards must be permanently maintained.

For the purpose of this UBC section, the centerline of an adjoining public way is considered an adjacent property line. In BOCA, this information appears under section 503.1 which governs the heights and areas of all buildings and structures between exterior walls, and between exterior walls and fire walls, governed by the type of construction and the use group classification as defined in BOCA chapters 3 and 6. Height and area limitations are fixed in all national building codes.

503.1.1. For the purpose of code, height and area, as applied to a building, have the meanings as designated in Chapter 2, Section 200.0, Definitions. [Same for UBC and BOCA.]

503.1.2. The height and area for buildings or structures of the different types of construction shall be governed by the intended occupancy or use of the building, as provided for in this chapter, and shall not exceed the limits set forth in Table 500 (see Appendix) except as modified in Sections 503.2 and 503.3 and the specific use provisions of this chapter. For the purpose of code, each part of a building or structure included within fire walls shall be considered a separate building.

> **EXCEPTION**: A building permitted to be unlimited in area by Section 503.4.1 shall be permitted to have interior fire walls.

503.1.3. A building heretofore erected shall not be extended to exceed the allowable floor area set forth in this chapter, governed by the occupancy and type of construction. A building heretofore lawfully erected, which exceeds such area, may be extended horizontally, provided such extension does not exceed the area prescribed and provided such extension is separated from the existing building by a fire wall as set forth in Section 503.1.2.

UBC requirements under section 503.2, Fire Resistance of Walls, state that exterior walls must have fire-resistance and opening protection as set forth in UBC Table 5-A and in accordance with such additional provisions as set forth in UBC chapter 6. In the UBC, distance is measured at right angles from the property line. However, the provision does not apply to walls at right angles to the property line. Projections beyond the exterior walls must comply with UBC section 705 and cannot exceed beyond:

1. A point one third the distance to the property line from an assumed vertical plane located where fire-resistive protection of openings is first required due to location on property
2. More than 12 in (305 mm) into areas where openings are prohibited

In BOCA, the area limitations are specified in BOCA Table 503 and apply to the maximum horizontally projected area of all buildings fronting on a street or a public space not less than 30 ft (9144 mm) in width with access from a public street. BOCA section 503.3 sets height limitations in feet and the number of stories above grade specified in BOCA Table 503 which apply to all buildings and to all separate parts of a building that are enclosed within fire walls complying with the provisions of BOCA chapter 7 and section 700.0.

A basement is considered as a story above grade where the finished surface of the floor above the basement is more than 6 ft (1829 mm) above grade plane; more than 6 ft (1829 mm) above the finished ground level for more than 50% of the total building perimeter; or more than 12 ft (3658 mm) above the finished ground level at any point.

503.2 Height Modifications

UBC section 504.6 concerns buildings of different heights. Where any area separation wall separates portions of buildings having different heights, it may terminate at a point 30 in (762 mm) above the lower roof level, provided the exterior wall for a height of 10 ft (3048 mm) above the lower roof is made of 1-hr fire-resistive construction with openings protected by assemblies having a 3/4-hr fire-protection rating.

Two-hr fire-resistive area separation walls may terminate at the underside of the roof sheathing, deck, or slab of the lower roof provided that

1. When the roof-ceiling framing elements are parallel to the wall, such framing and elements supporting such framing shall not be of less than 1-hr fire-resistive construction for a width of 10 ft (3048 mm) along the wall at the lower roof.
2. When the lower roof-ceiling framing elements are perpendicular to the wall, the entire span of such framing

and elements supporting such framing shall not be of less than 1-hr fire-resistive construction.

503.2.1 Rooftop Structures

Church spires, chimneys, tanks and supports, aerial supports, parapet walls not over 4 ft (1219 mm) high, bulkheads and penthouses used solely to enclose stairways, tanks, elevator machinery or shafts, or ventilation or air-conditioning apparatus need not be considered in determining the highest point of the building, provided that the highest point is determined to be taken to be the highest point of the roof of the highest penthouse when the aggregate area of all penthouses and other roof structures exceeds one-third of the area of the roof upon which they stand.

[UBC section 504.6.5 has additional specs covering rooftop parapet faces. "Parapets of area separation walls shall have noncombustible faces for the uppermost 18 inches (457 mm), including counterflashing and coping materials."]

503.2.2 Parking Under Group R

Where a one-story automobile parking garage, enclosed or open of Type I or II construction, or open of Type III construction, with grade entrance, is provided under a building of Group R Occupancy, the number of stories to be used in determining the minimum type of construction may be measured from the floor above such parking area. The floor and ceiling assembly between the parking garage and the Group R Occupancy above shall comply with the type of construction required for the parking garage and shall also provide a fire-resistance rating not less than the occupancy separation required in Section 704.1.1.

503.2.3 Mezzanines

A mezzanine shall not be counted as a story when it meets the following requirements of Sections 503.2.3.1 through 503.2.3.4.

503.2.3.1. The construction of a mezzanine shall be consistent with the type of materials and fire-resistance ratings required for the building in which it is constructed.

503.2.3.2. The total area of mezzanines within a room shall not exceed one-third that of the room or space in which they are located. The enclosed space under a mezzanine shall not be included in a determination of the size of the room or space in which the mezzanine is located.

503.2.3.3. All portions of a mezzanine shall be open and unobstructed to the room in which it is located except for walls not more than 42 in (1067 mm) high, columns, and posts.
EXCEPTIONS:
1. Mezzanines or portions thereof need not be open to the room in which they are located, provided the occupant load of the aggregate area of the enclosed space does not exceed 10.
2. In sprinklered buildings, a mezzanine having two or more means of egress need not open into the room in which it is located, if at least one of the means of egress provides direct access to an exit at the mezzanine level.

503.2.3.4. Means of Egress
Means of egress shall be in accordance with Section 1005.6. [Section 504.6 of the UBC carries further specs under Area Separation Walls. That section establishes that "any building separated by one or more area separation walls which comply with the provisions of that section may be considered a separate building. The extent and location of such area separation walls shall provide a complete separation. When an area separation wall also separates occupancies that are required to be separated by an occupancy separation, the most restrictive requirements of each separation shall apply."]

503.2.4 Basements
A basement of a building shall not count as a story, when applying Table 500 for allowable building height, if the upper surface of the first floor above such basement complies with all of the following:
- It is less than 7 ft (2134 mm) above grade.
- It is less than 7 ft (2134 mm) above finished ground level for more than 50% of the perimeter of a building.
- It is less than 12 ft (3658 mm) above finished ground level around the entire building perimeter.

503.2.5 Group A and E Basements [504.5; 803.4]
Group A and Group E basements used as classrooms or assembly rooms shall be counted as a story.

[Section 504.5 of the UBC has a different interpretation of basements, and it states "A basement need not be included in the total allowable area, provided such basement does not exceed the area permitted for a one-story building. In BOCA, see Section 803.4.1.]

503.2.6 Special Unlimited Height [504.6; 503.3]

503.2.6.1. The height of Group B, Group M, and Group R Occupancies of Type II construction shall not be limited, provided the fire-resistance of all columns is 1 hr and of the other structural members including floors is not less than that shown in SBC chapter 6, but in no case less than 2 hr except that roofs and their supporting beams, girders, trusses, and arches shall be not less than 1-1/2 hr.

503.2.6.2. For unlimited-height open air grandstands and bleachers, see SBC section 403.6.2.1 [1013.0; 303.6 Use Group A-5].

503.3 General Area Modifications [506.0; 506.0]

503.3.1. The exceptions and requirements of Sections 503.3 and 503.4 shall modify unsprinklered areas permitted by Table 500 (see Appendix) and the specific use provisions of this chapter.

503.3.2. Where streets, public spaces, horizontal separation between property lines of not less than 30 ft (9144 mm), or separations of buildings of 30 ft (9144 mm) on commonly owned property extend along the building perimeter, the areas permitted by Table 500 may be increased as follows (except for hazardous occupancies):

$$I = 4/3[100\ (F/P - 0.25)]$$

where I = percent increase of unsprinklered areas in Table 500
F = building perimeter which fronts on streets, public spaces, or horizontal separation not less than 30 ft (9144 mm) wide
P = total perimeter of building

503.3.3. For both an unsprinklered building and a sprinklered building, the percent increase is multiplied by the unsprinklered area permitted in Table 500 for the type of construction of the building, and the resulting area increase is added to either the sprinklered or unsprinklered areas in Table 500. When there are no unsprinklered areas permitted for the building in Table 500, an unsprinklered area can be computed for use in this section. The corresponding unsprinklered areas are computed as one-third of the sprinklered area for one story only and as one-half of the sprinklered area for multistories.

503.4 Occupancy Area Modifications [506.0; 506.0]

503.4.1. The area of a one-story building of Group B, Group F, Group M, or Group S Occupancy shall not be limited provided the building is equipped with an approved automatic sprinkler system throughout, in accordance with Section 903, or other automatic extinguishing systems as approved by the building official and provided it is surrounded on all sides by a permanent open space of not less than 60 ft (18.2 m). High-piled combustible storage shall be protected in accordance with chapter 36 of the Fire Code.

EXCEPTIONS:
1. Where water may cause or increase a fire, other fire-extinguishing systems shall be required in any rooms or buildings used for the manufacture or storage of hazardous materials including but not limited to, aluminum powder, calcium carbide, calcium phosphate, metallic sodium and potassium, quicklime, magnesium powder, and sodium peroxide.
2. In Group F and S occupancies where noncombustible products are manufactured or stored, such as metal processing and manufacturing plants, and where metal products are not stored in combustible wrappings, containers, or palletized, the sprinkler system may be omitted upon approval of the building official.

503.4.2. The area of a one-story building of Type IV construction used for Group E Occupancy shall not be limited provided the building is equipped throughout with an approved automatic sprinkler system in accordance with Section 903, is surrounded on all sides by a permanent open space of not less than 60 ft (18.2 m), and is provided with 1-hr fire-resistant smokestop partitions dividing the building into areas not to exceed 30,000 sq ft (2787 m^2) in floor area.

503.4.3. One-story Group A buildings without a stage requiring proscenium opening protection of Type III, IV, or V 1-hr construction which are surrounded on all sides by a permanent open space of not less than 60 ft (18.2 m); which are provided with an approved automatic sprinkler system; which have the assembly floor located at, or within, 21 in (533 mm) of street or grade level;

and which have all exits meet the street or grade level by ramps having a slope not exceeding 1:10 shall not be limited in area.

503.4.4. Where there are no balconies or galleries in Group A - Large Assemblies without a stage requiring proscenium opening protection or in Group A - Small Assembly with or without a stage requiring proscenium opening protection, and the assembly floor is located at or within 21 in (533 mm) of street or grade level and all exits meet the street or grade level by ramps having a slope not exceeding 1:10, the maximum allowable areas of Type III, IV, and V construction may be increased 50% over the unsprinklered areas specified in Table 500. This increase may be added to the area increase permitted by Section 503.3.2 [1001.2; 1011.5].

503.4.5. One-story buildings used for participation sports such as tennis and skating and for similar activities, limited in occupant content to those participating in the sports activity, and with no spectator seating permitted, may be unlimited in area when of Types III, IV, and V construction and are surrounded on all sides by not less than 30 ft (9144 mm) of permanent open space.

503.4.6. When used as a place of worship, the allowable areas for Group A - Small Assembly without a stage requiring proscenium opening protection may be increased 33-1/3% over the unsprinklered areas specified in Table 500. This increase may be added to the area increase permitted by Section 503.3.2 [303.0 Assembly Use Groups; 303.5 Use Group A-4 structures].

503.4.7. The area of a one-story Group E Type III, IV, or V building may be increased 100% over the unsprinklered areas specified in Table 500 if the building is surrounded on all sides by a permanent open space of not less than 60 ft (18.2 m) and there are at least two exits provided from each classroom, one of which opens directly to the exterior of the building. This increase may be added to the area increase permitted by Section 503.3.2.

503.4.8. The permanent open space of 60 ft (18.2 m) required in Sections 503.4.1, 503.4.2, 503.4.3 and 503.4.7 shall be permitted to be reduced to not less than 40 ft (12.2 m) provided all the following requirements are met:
 1. The reduced open space shall not be allowed for more than 75% of the perimeter of the building.
 2. The exterior wall facing the reduced open space shall have a minimum fire-resistance rating of 3 hr.

3. All openings in the exterior wall, facing the reduced open space, shall have opening protectives with a fire-resistance rating of 3 hr.

503.4.9. Group A area modification: Open-air grandstands and bleachers; see SBC section 403.6.2.1.

503.4.10. Group M area modifications: Covered mall buildings; see SBC section 413.6.

503.4.11. Group S area modifications:
1. Aircraft hangars; see SBC section 411.7.2.
2. Automobile parking garages; see SBC section 411.3.

504.0 BUILDINGS LOCATED ON THE SAME LOT [503.1; 503.0]

Where the exterior walls of two or more buildings located on the same lot face one another, and one of the walls is not constructed as required for a fire wall, a property line shall be assumed between them. The fire-resistance requirements for such facing walls and for the protection of openings therein shall be the same as required by code for walls and openings facing an assumed property line, as provided in Table 600 (see Appendix).

> **EXCEPTION**: Fire-resistance separation shall not be required between a dwelling and its detached private garage.

[In UBC regulations, when a new building is to be erected on the same property as an existing building, the location of the assumed property line with relation to the existing building must be such that the exterior wall and opening protection of the existing building meet the criteria as set forth in Table 5-A and UBC chapter 6.

> **UBC EXCEPTION**: Two or more buildings on the same property may be considered as portions of one building if the aggregate area of such buildings is within the limits specified in UBC section 504 for a single building. When the buildings so considered house different occupancies or are of different types of construction, the area is that amount allowed for the most restricted occupancy or construction.]

504.6.2 Fire Resistance and Openings [706.0; 721.6.4]

Area separation walls shall not be of less than 4-hr fire-resistive construction in Types I, II-FR., III, and IV buildings and 2-hr fire-resistive construction in Types II 1-hr, II-N, or V buildings. The total width of all openings in such walls shall not exceed 25% of the length of the wall in each story. All openings shall be protected by a fire assembly having a 3-hr fire-protection rating in 4-hr fire-resistive walls and 1- and 1.5-hr fire-protection rating in 2-hr fire-resistive walls.

504.6.3 Extensions Beyond Exterior Walls [506.0; 504.0]
Area separation walls shall extend horizontally to the outer edges of horizontal projecting elements such as balconies, roof overhangs, canopies, marquees, or architectural projections extending beyond the floor area as defined in UBC section 207.

> **EXCEPTIONS**:
> 1. When horizontal projecting elements do not contain concealed spaces, the area separation wall may terminate at the exterior wall.
> 2. When the horizontal projecting elements contain concealed spaces, the area separation wall need only extend through the concealed space to the outer edges of the projecting elements.

In either exception 1 or 2, the exterior walls and the projecting elements above shall not be less than 1-hr fire-resistive construction for a distance not less than the depth of the projecting elements on both sides of the area separation wall. Openings within such widths shall be protected by fire assemblies having a fire-protection rating of not less than 3/4 hr.

504.6.4 Terminating
Area separation walls shall extend vertically from the foundation to a point at least 30 in (762 mm) above the roof.

> **EXCEPTIONS**:
> 1. Any area separation wall may terminate at the underside of the roof sheathing, deck, or slab provided the roof-ceiling assembly is of at least 2-hr fire-resistive construction.
> 2. Two-hr fire-resistive area separation walls may terminate at the underside of the roof sheathing, deck, or slab provided
> 2.1 When the roof-ceiling framing elements are parallel to the walls, such framing and elements supporting such framing shall not be less than 1-hr fire-resistive

construction for a width of not less than 5 ft (1524 mm) on each side of the wall.

2.2 When roof-ceiling framing elements are perpendicular to the wall, the entire span of such framing and elements supporting such framing shall not be of less than 1-hr fire-resistive construction.

2.3 Openings in the roof shall not be located within 5 ft (1524 mm) of the area separation wall.

2.4 The entire building shall be provided with not less than a Class B roof covering as specified in SBC Table 15-A.

3.0 Two-hr area separation walls terminate at underside of noncombustible roof sheathing, deck or slabs of roofs of noncombustible construction provided:

3.1 Openings in the roof are not located within 5 ft (1524 mm) of the area separation wall.

3.2 The entire building is provided with not less than a Class B roof covering as specified in UBC Table 15-A.

505.0 BUILDINGS LOCATED WITHIN A FIRE DISTRICT [503.0; 503.0]

Additional provisions for buildings located within a fire district are contained in SBC Appendix F. Those provisions are applicable only where specifically included in the adopting ordinance [UBC n/a; BOCA n/a].

Chapter

5

Types of Construction

SECTION 601.0 CONSTRUCTION TYPES [601.0; 601.0]

Because so many of the national building code rules involve the words <u>dwelling</u> and <u>residential</u>, there have been problems applying code regulations to the various types of dwellings: one-family houses, two-family houses, apartment houses, condominium units, dormitories, hotels, youth hostels, motels, etc. The present editions of the codes include terminology to eliminate such problems and use definitions of <u>dwelling</u> coordinated with the words used in specific code rules.

    A <u>dwelling unit</u> is now defined as "one or more rooms" used "as a house-keeping unit" and must contain space or areas specifically dedicated to "eating, living, and sleeping" and must have "permanent provisions for cooking and sanitation." A one-family house is a "dwelling unit." So is an apartment in an apartment house or a condominium unit. But a guest room in a hotel or motel or a dormitory room or unit is not a <u>dwelling unit</u> if it does not contain "permanent provisions for cooking"–which means a built-in range or counter-mounted unit (with or without an oven).

    Types of construction are classified from the highest fire-resistive to the least, Type I being the most fire-resistive construction and Type V being the least. The type of construction will also determine the allowable building height. Building height limitations are given as maximum height in stories. It is important to note that because of high-hazard situations some occupancies are not allowed in specific types of construction regardless of area separation, floor area, or building height.

601.2 Classification by Type of Construction

Provisions of this chapter shall govern the classification of construction type by materials and fire resistance of its elements and the use of more than one construction type in a building. Every building shall be classified by the building official into one of the types of construction as set forth in this section.

SBC

> Type I
> Type II
> Type III
> Type IV
>   1-hr protected
>   Unprotected
>
> Type V
>   1-hr protected
>   Unprotected
> Type VI
>   1-hr protected
>   Unprotected

UBC

> **600.0 Types of Construction**
> 601.0 General
> 602.0 Construction classification
> 603.0 Types 1 and 2 construction
> 604.0 Type 3 construction
> 605.0 Type 4 construction
> 606.0 Type 5 construction

BOCA

> **Construction, type of:** section 602.0
> **Type 1:** section 603.0
> **Type 2:** section 603.0
> **Type 3:** section 604.0
> **Type 4:** section 605.0
> **Type 5:** section 606.0

601.2.2. Materials for any one of the six types of construction may be used as specified in Table 600 (see Appendix), or as permitted in this chapter.

601.3 Fire-Resistance Requirements

601.3.1. All fire resistance requirements are expressed in terms of the number of hours of satisfactory performance in accordance with ASTM E 119. Construction required to have a fire resistance rating shall be supported by construction of equal or greater fire resistance.

> **EXCEPTION**: In construction Types IV Unprotected, V Unprotected, and VI Unprotected, structural elements supporting exit access corridor walls and tenant separation walls of not more than 1-hr fire resistance need not be rated provided a fire resistance rating is not required by other provisions of code.

601.4 Materials and Construction Approved for Fire Resistance

601.4.1. The degree of fire resistance and the materials, assemblies, and constructions providing such resistance are defined in Chapter 6 except that other materials, assemblies, and constructions shall be approved, provided test data of a recognized engineering or testing laboratory are submitted, establishing that they develop the required fire resistance rating under tests made in accordance with ASTM E 119 or based on calculations and accepted engineering practice as set forth in Section 709.

601.4.2. Where structural requirements necessitate assemblies providing greater fire resistance than specified in this chapter, such structural requirements shall govern.

602.0 DEFINITIONS [601.0; 602.0]

This chapter contains no unique definitions. For general definitions, see Chapter 2, Section 200.0.

603.0 TYPE I CONSTRUCTION [602.0; 603.1]

Type I construction is the most fire-resistant construction classification. Type I buildings are required to have 4-hr fire-resistive exterior walls. With the exception of Group H, all occupancies with Type I construction nonbearing walls are allowed to have a 1-hr fire rating where unprotected openings are permitted. Nonbearing walls that front streets that are 50 or more feet wide can be of noncombustible, unprotected construction. Structural frames of Type I must be made of reinforced concrete, reinforced masonry, steel, or iron and have a 3-hr fire rating. Any load-bearing members that are components of an exterior wall must have the same fire protection rating as the wall or meet a 4-hr fire rating. The height of Type I buildings is unlimited and the allowable area is limited only in Group A Occupancy.

The fire resistance of building components such as structural framing, interior and exterior walls, non-loadbearing partition walls, roofs, floors, exterior doors, and windows determines the type of construction. If any of these primary building components is reduced in type of construction, the entire building is automatically given a lower classification. If one of these elements were reduced, a Type I building with required 3-hr fire-resistive structural frame, 2-hr floors, and 4-hr exterior bearing walls would be lowered in classification to Type V due to the lack of full compliance with Type I construction requirements.

Mezzanines in Type I can be made of either unprotected metal or wood, but a mezzanine floor must not cover more than one-third of the entire floor area of any room. Code also restricts the number of allowable mezzanine floors in any room to two.

603.1. Type I is construction in which the structural members including exterior walls, interior bearing walls, columns, beams, girders, trusses, arches, floors, and roofs are of noncombustible materials and are protected so as to have a fire resistance not less than that specified for the structural elements as specified in Section 600. For interior nonbearing partition requirements, see Section 704.2. For provisions governing combustibles in concealed spaces, see Section 707.

604.0 TYPE II CONSTRUCTION [603.0; 603.2]

Type II is construction in which the structural members including exterior walls, interior bearing walls, columns, beams, girders, trusses, arches, floors, and roofs are of noncombustible materials and are protected so as to have a fire resistance not less than that specified for the structural elements as specified in Table 600. For interior nonbearing partition requirements, see Section 704.2. For provisions governing combustibles in concealed spaces, see Section 707.

Type II construction requires all construction to be of noncombustible components: concrete, iron, steel, or masonry.

Nonbearing partition walls that are permanent may be constructed of fire-retardant-treated wood. This is the most common type of construction for commercial and industrial buildings. In the UBC, a Type II 1-hr fire-resistive rating requires that all nonstructural components meet a 1-hr fire rating. Type II-N has no fire ratings on any components with the exception of the structural frame. These are typically all steel construction buildings. In Type II construction, the maximum allowable area and height is factored by the building's purposed occupancy.

605.0 TYPE III CONSTRUCTION [604.0; 604.1]
Type III is construction in which fire resistance is attained by the sizes of heavy timber members (sawn or glue-laminated) being not less than indicated in this section; or by providing fire resistance not less than 1-hr where materials other than wood of heavy timber sizes are used; by the avoidance of concealed spaces under floors and roofs; by the use of approved fastenings, construction details and adhesives for structural members; and by providing the required degree of fire resistance in exterior and interior walls.

Heavy timber construction has certain requirements to fall under this classification. Roofs must be a minimum of 2-in nominal thickness. Framed timber trusses and load-bearing columns must be a nominal thickness of 8 in by 8 in to classify as heavy timber construction. Glue-laminated support beams that serve as floor supports must be a minimum nominal thickness of 10 in by 8 in and glue-laminated roof supports must be a minimum nominal thickness of 6 in by 8 in.

In UBC and BOCA, Type III construction requires the building to have 4-hr fire-rated exterior walls. In most occupancy groups, Type III construction limits the number of stories to two. Internal components, including the structural frame, must meet only a 1-hr fire rating. In the 1-hr rated category, the building height is limited to 65 ft and for Type III-N is limited to 55 ft in height.

605.2 Columns
605.2.1. Wood columns may be sawn or glue-laminated and shall be not less than 8 in nominal in any dimension when supporting floor loads, and not less than 6 in nominal wide and 8 in nominal deep when supporting roof and ceiling loads only.

605.2.2. Columns shall be continuous or superimposed throughout all stories by means of reinforced concrete or metal caps with brackets or shall be connected by properly designed steel or iron caps, with pintles and base plates; or by timber splice plates affixed to the columns by means of metal connectors housed within the contact faces, or by other approved methods.

605.3 Floor Framing

605.3.1. Beams and girders of wood may be sawn or glue-laminated and shall be not less than 6 in nominal wide and not less than 10 in nominal deep.

605.3.2. Framed or glue-laminated arches which spring from the floor line and support floor loads shall be not less than 8 in nominal in any dimension.

605.3.3. Framed timber trusses supporting floor loads shall have members of not less than 8 in nominal in any dimension.

605.4 Roof Framing

605.4.1. Framed or glue-laminated arches for roof construction which spring from the floor line and do not support floor loads shall have members not less than 6 in nominal wide and 8 in nominal deep for the lower half of the height and not less than 6 in nominal in any dimension for the upper half of the height.

605.4.2. Framed or glue-laminated arches for roof construction which spring from the top of walls or wall abutments, framed timber trusses, and other roof framing which do not support floor loads shall have members not less than 4 in nominal wide and not less than 6 in nominal deep.

Spaced members may be composed of two or more pieces not less than 3 in nominal thick when blocked solidly throughout their intervening spaces or when such spaces are tightly closed by a continuous wood cover plate not less than 2 in nominal thickness, secured to the underside of the members. Splice places shall be not less than 3 in nominal thickness. When protected by approved automatic sprinklers under the roof deck, such framing members shall be not less than 3 in nominal width.

605.5 Construction Details

605.5.1. Wall plate boxes of self-releasing type, or approved hangers, shall be provided where beams and girders enter masonry. An air space of 1/2 in (12.7 mm) shall be provided at the top, ends, and sides of the member unless approved durable or treated wood is used.

605.5.2. Girders and beams shall be closely fitted around columns, and adjoining ends shall be cross-tied to each other, or intertied by caps or ties, to transfer horizontal loads across the joint. Wood bolsters may be placed on top of columns which support roof loads only.

605.5.3. Where intermediate beams are used to support floors, they shall rest on top of the girders or shall be supported by ledgers or blocks securely fastened to the sides of the girders, or they may be supported by approved metal hangers into which the ends of the beams shall be fitted closely.

605.5.4. Columns, beams, girders, arches, and trusses of material other than wood shall have a fire resistance rating of not less than 1 hr.

605.5.5. Wood beams and girders supported by walls required to have a fire resistance rating of 2 hr or more shall have not less than 4 in (102 mm) of solid masonry between their ends and the outside face of the wall and between adjacent beams.

605.5.6. Adequate roof anchorage shall be provided.

605.6 Floor Decks

605.6.1. Floors shall be without concealed spaces. They shall be of sawn or glue-laminated, splined, or tongue-and-grooved plank, not less than 3 in nominal thickness, or of planks not less than 4 in nominal width set on edge and well-spiked together. The planks shall be laid so that no continuous line of joints will occur except at points of support, and they shall not be spiked to supporting girders.

Planks shall be covered with 1-in nominal tongue-and-groove flooring laid crosswise or diagonally or with 15/32-in (11.9-mm) wood structural panels. Planks and flooring shall not extend closer than 1/2 in (12.7 mm) to walls to provide an expansion joint, and the joint shall be covered at top and bottom.

605.7 Roof Decks

Roofs shall be without concealed spaces, and roof decks shall be sawn or glue-laminated, splined, or tongue-and-grooved plank, not less than 2 in nominal thickness, or of planks not less than 3 in nominal width, set on edge and spiked together as required for floors, or of 1-1/8 in (29 mm) tongue-and-grooved wood structural panels bonded with exterior glue. Other types of decking may be used when approved by the building official.

605.8 Walls

605.8.1. Bearing portions of exterior and interior walls shall be of approved noncombustible materials and shall provide fire resistance ratings in accordance with Table 600 (see Appendix).

605.8.2. Nonbearing portions of exterior walls shall be of approved noncombustible materials and shall provide fire resistance ratings in accordance with Table 600.

> **EXCEPTION**: Where a horizontal separation of at least 20 ft (6096 mm) is provided, wood columns, arches, beams, and roof decks conforming to heavy timber sizes may be used externally.

606.0 TYPE IV CONSTRUCTION [605.0; 605.1]

Type IV is construction in which the structural members including exterior walls, interior bearing walls, columns, beams, girders, trusses, arches, floors, and roofs are of noncombustible materials. Type IV construction may be protected or unprotected. Fire-resistance requirements for structural elements of Type IV construction shall be as specified in Table 600 (see Appendix). For provisions governing combustibles in concealed spaces, see Section 707.

Type IV construction requires that wood floors be ventilated in the underfloor areas as this type of construction is usually not built on slabs as are the other types but rather on perimeter foundations with pier and posts supporting wooden girders.

607.0 TYPE V CONSTRUCTION [606.0; 606.0]

Type V is construction in which the exterior bearing and nonbearing walls are of noncombustible material and have a fire resistance not less than that specified in Table 600 (see Appendix); bearing portions of interior walls are of material permitted in Table 600 and have a fire resistance not less than that specified in Table 600; and beams, girders, trusses, arches, floors, roofs, and interior and framing are wholly or partly of wood or other approved materials and have a fire resistance not less than that specified in Table 600. Type V construction may be either protected or unprotected. Fire resistance requirements for structural elements of Type V construction shall be as specified in Table 600.

In the UBC, Type V construction allows a wood frame if the exterior walls are not required to be noncombustible. All Type V 1-hr requires a 1-hr fire rating for all construction components. Most R3 and multifamily residence buildings are required to be of Type V-N construction.

608.0 TYPE VI CONSTRUCTION [N/A; N/A]

Type VI is construction in which the exterior bearing and nonbearing walls and partitions, beams, girders, trusses, arches, floors, and roofs and their supports are wholly or partly of wood or other approved materials. Type VI construction may be either protected or unprotected. Fire-resistance requirements for the structural elements of Type VI construction shall be as specified in Table 600 (see Appendix).

609.0 PARTITIONS [602.1; 602.0]

609.1. Bearing walls shall comply with the provisions of SBC chapter 6, but shall provide not less than the degree of fire resistance specified in Table 600 (see Appendix).

609.2 Nonbearing partitions shall conform to Sections 609.2.1 through 609.2.4 and have the fire resistance specified in Table 700 (see SBC chapter 6) except as specified elsewhere in code.

609.2.1 Type I and Type II Construction

Partitions shall be constructed of noncombustible materials.
    EXCEPTIONS:
      1. Framing members may be of fire-retardant-treated wood.
      2. Pocket doors and their frames may be of wood.

609.2.2 Type III Construction

Partitions may be of any material permitted by code.

609.2.3 Type IV Construction

Partitions shall be constructed of noncombustible materials except that framing members of fire-retardant-treated wood may be used and pocket doors and their frames may be of wood. Partitions in one-story buildings only may be of any material permitted by code. Partitions in fully sprinklered buildings, regardless of height, may be of any material permitted by code.

609.2.4 Type V and Type VI Construction
Partitions may be of any material permitted by code.

610.0 MIXED TYPES OF CONSTRUCTION
[602.1; 602.0]
610.1 Area Limitations
Where two or more types of construction not separated by fire walls occur in the same building, the area of the entire building shall not exceed the least area permitted based on occupancy for the types of construction used in the building.
610.2 Height Limitations
Where two or more types of construction occur in the same building, the height of the entire building shall not exceed the least height permitted based on the occupancy for the types of construction used in the building.
610.3 Open Parking Structures
Open parking structures which comply with Section 411.3 may be constructed beneath other occupancies in buildings of mixed types of construction in accordance with the following:

1. The height and area of the open parking structure shall not exceed that permitted by SBC Table 411.3.1 for the type of construction of the open parking structure.
2. The total height of the structure shall not exceed that allowed for its primary occupancy in accordance with Section 303 and Table 500 (see Appendix).
3. The fire resistance of structural members within the open parking structure that support any part of the building above the open parking structure shall have the same or greater fire resistance as the supported type of construction. This provision applies to all columns; beams, girders, and trusses directly connected to the columns; and all other structural members which directly brace the columns.
4. The entire structure shall be of noncombustible construction.
5. Occupancy separations shall be maintained in accordance with Section 704.1.

Chapter

6

Fire-resistant Materials and Construction

SECTION 700.0 FIRE-RESISTANT MATERIALS AND CONSTRUCTION [700.0; 701.1]

It is important to understand that the national building codes are intended only to assure that construction is done with end-user safety in mind, i.e., to provide a system that is essentially free from hazard for the public. Occupancy and use are rated by potential hazard to people using the building or structure. Fire is the number one life-threatening hazard in buildings, and this chapter on fire resistance, is paramount to all modern construction design.

The fire resistance of building components such as structural framing, interior and exterior walls, non-loadbearing partition walls, roofs, floors, exterior doors, and windows determines the type of construction. If any of these primary building components is reduced in type of construction, the entire building is automatically given in a lower classification. If one of these fire-resistive elements were reduced, a Type I building with required 3-hr structural frame, 2-hr floors and 4-hr exterior bearing walls would be lowered in classification to Type V due to the lack of full compliance with Type I construction requirements.

Types of construction are classified by two factors that need to be comprehensively defined here: noncombustible and fire-resistive. Noncombustible materials will not burn, but they will transmit heat in a fire and contribute to flamespread which maintains combustion if the fire becomes hot enough. Noncombustible materials include steel, masonry, iron, and concrete; however, any material which is exposed to enough heat will melt and disintegrate structurally. Fire-resistive materials will burn when exposed to prolonged flame. The national codes define these two properties as follows:

<u>Noncombustible</u> as applied to a building construction material means a material which, in the form in which it is used, is either one of the following:

1. Material of which no part will ignite and burn when subjected to fire. Any material approved as conforming to this definition is considered noncombustible within the meaning of this section.
2. Material having a structural base of noncombustible material as defined in Item 1 above, with a surfacing material not over 1/8 in thick which has a flamespread rating of 50 or less as approved by the Fire Code.

Noncombustible does not apply to surface finish materials. Material requiring to be noncombustible for reduced clearances to flues, heating appliances, or other sources of high temperature shall refer to material conforming to Item 1 above. No material shall be classed as noncombustible which is subject to increase in combustibility or flame-spread rating, beyond the limits herein established, through the effects of age, moisture, or other atmospheric conditions.

It should be noted here that there is a difference between flamespread rating and fire resistance. A flamespread rating compares the time it takes flame to spread on the surface of the tested material against the time it takes for the same amount of flame to spread on untreated, dried red oak of uniform density. On the flamespread rating scale, red oak is given a rating of 100 representing maximum flamespread and, at the other end of the scale, cement asbestos board is given a rating of 0 representing minimum flamespread.

Fire resistance is determined by an Underwriter's Laboratories (UL) test of the material. A constant flame is applied to one side of the material in a controlled laboratory test environment, and the temperature change (Delta factor) is measured on the other side. If, in the test time allotment, the exposed side does not exceed 250^0 F, the material meets specifications for a UL listed flame resistance and receives an according UL flame-resistance rating. Interior and exterior wall

Fire-resistant Materials and Construction

coverings as well as door and window assemblies are all sample-tested and rated for fire resistance in this manner.

701.1 Scope [701.0; 701.1]
Provisions of this chapter shall govern the fire-resistant materials and assemblies used for structural fire resistance and fire-resistant separation of adjacent spaces, including construction, opening protectives, penetrations, fireblocking, and draftstopping.

701.2 Tests

701.2.1. Fire-resistance requirements of code are based on fire-resistance ratings. Materials, thicknesses, and assemblies which have successfully performed under tests made by a recognized laboratory in accordance with the requirements of ASTM E 119 or based on calculations and accepted engineering practice as set forth in Section 709 shall be accepted by the building official for specific ratings.

> **EXCEPTION**: In determining the fire resistance rating of exterior bearing walls, compliance with the ASTM E 119 criteria for unexposed surface temperature rise and ignition of cotton waste due to passage of flame or gases is required only for a period of time corresponding to the required fire-resistance rating of an exterior, nonbearing wall with the same horizontal separation distance and is required in a building of the same type of construction.
>
> When the fire resistance rating determined in accordance with this exception exceeds the fire resistance rating determined in accordance with ASTM E 119, the fire exposure time period, water pressure, and application duration criteria for the hose stream test of ASTM E 119 shall be based upon the fire resistance rating determined in accordance with this exception.

701.2.2. When insulation or other materials which may change the capacity for heat dissipation are added to or subtracted from fire-resistant roof or ceiling assemblies whose fire ratings are listed in the national codes or listed in reference documents, fire test results or other substantiating data shall be submitted to the building official to show that the required fire resistance time period is not reduced.

701.2.3. Thicknesses established by fire tests shall be construed as establishing minimum requirements for fire resistance only and shall not preclude the application of other requirements of code where consideration of strength, durability, or stability require greater thicknesses.

701.3 Column Fire Resistance

Where columns require a fire resistance rating, the entire column, including its connections to beams or girders, shall be protected. Where the column extends through a ceiling, fire resistance of the column shall be continuous from the top of the floor through the ceiling space to the underside of the floor deck above, except as provided in Table 600 (see Appendix), or other such provisions of code.

701.4.1 Fire-Resistance References

To meet the fire resistance requirements of code, it shall be determined that materials, constructions, and assemblies of construction materials have successfully performed under accepted tests as are prescribed in Section 701.2.

701.4.2. Appropriate fire-resistant materials, constructions, and assemblies of constructions as listed in Section 709.7 and the following publications may be accepted as if herein listed:

- FM Specification Tested Products Guide
- GA Fire Resistance Design Manual
- SBCCI PST & ESI Evaluation Report Listing
- UL Fire Resistance Directory

701.4.3 Other fire resistance ratings may be accepted by the building official on evidence of compliance with Section 701.2.

701.5 Exceptions to Fire Resistance

701.5.1 Elevator Frames

Structural members of frames for elevators will not need to have the fire resistance required for structural steel, provided such members are erected within an enclosure of the prescribed fire resistance rating. See Section 705.2.

701.5.2 Lintel Protection

Lintels over openings in walls shall be protected to provide a fire-resistance rating at least equal to that required for beams, except that fire resistance may be omitted from the bottom flange of lintels, shelf angles, and plates that are not a part of the structural frame or that have a span of 6 ft (1829 mm) or less.

701.5.3 Unusable Space
In 1-hr fire-resistant construction, the ceiling may be omitted over unusable crawl space and flooring may be omitted when unusable attic space occurs above it.

702.0 DEFINITIONS [702.0; 702.0]
ANNULAR SPACE - The opening around a penetrating item.
DAMPER, FIRE - A damper arranged to seal off airflow automatically through part of an air duct system, so as to restrict the passage of heat.
DOOR ASSEMBLY, FIRE - A combination of the fire door, frame, hardware, and other accessories which together provide a specific degree of fire protection to the opening.
DOOR, FIRE - A door and its assembly, so constructed and assembled in place as to give protection against the passage of fire.
DRAFTSTOPPING - Building materials installed to prevent the movement of air, smoke, gases, and flame to other areas of the building through large concealed passages, such as attic spaces and floor assemblies with suspended ceilings or open-web trusses.
FIRE AREA - The aggregate floor area enclosed and bounded by fire walls, exterior walls, or fire separation assemblies of a building.
FIRE PARTITION - A vertical assembly of materials having protected openings and designed to restrict the spread of fire.
FIRE PROTECTION RATING - The time in hours, or fractions thereof, that an opening protective assembly will resist fire exposure as determined by the test standard specified in code.
FIRE SEPARATION ASSEMBLY - A horizontal or vertical fire resistance-rated assembly of materials having protected openings and designed to restrict the spread of fire.
FIRE SEPARATION DISTANCE - The distance in feet measured from the building face to the closest interior lot line, to the centerline of a street or public way, or to an imaginary line between two buildings on the same property.
FIRE WINDOW - A window constructed and glazed to give protection against the passage of fire.
FIREBLOCKING - Building materials installed to prevent the movement of flame and gases to other areas of a building through small concealed passages in building components such as floors, walls, and stairs.

FIRE RESISTANCE - That property of materials or their assemblies which prevents or retards the passage of excessive heat, hot gases, or flames under conditions of use.

FIRE-RESISTANCE RATING - The time in hours, or fractions thereof, that materials or their assemblies will resist fire exposure as determined by the fire test specified in code.

FIRE-RESISTIVE JOINT SYSTEM - An assemblage of specific materials or products that are designed, tested, and fire resistance rated in accordance with ASTM E 119 to resist, for a prescribed period of time, the spread of fire through joints made in or between fire resistance-rated assemblies.

JOINT - The linear opening in or between adjacent fire resistance-rated assemblies.

PROTECTED CONSTRUCTION - That in which all structural members are constructed, chemically treated, covered, or protected so that the individual unit or the combined assemblage of all such units has the required fire resistance rating specified for its particular application and includes protected combustible and protected noncombustible construction.

SELF-CLOSING - As applied to a fire door or other opening protective, means normally closed and equipped with an approved device which will ensure closing after having been opened for use.

SHAFT - An enclosed space extending through one or more stories of a building, connected vertical openings in successive floors, or floors and the roof.

SINGLE MEMBRANE PENETRATION - An opening through a single membrane (one side) of a fire resistance-rated wall, roof/ceiling, or floor through ceiling assembly made to accommodate electrical, mechanical, plumbing, environmental, and communication systems.

Smoke barrier: A continuous membrane that will resist the movement of smoke.

Smoke compartment: A space within a building enclosed by smoke barriers or fire separation assemblies on all sides, including top and bottom.

Splice: The result of a factory or field method of joining or connecting two or more lengths of a fireresistive joint system into a continuous assembly.

Fire-resistant Materials and Construction

Through penetration: Specific building materials or assemblies of materials that are designed and installed to prevent the spread of fire through openings that are made in fire resistance rated floors and walls to accommodate through penetrations for electrical, mechanical, plumbing, environmental, and communication systems.

The *F* rating indicates the period of time that the through-penetration fire-stop system is capable of preventing the passage of flame to the unexposed (nonfire) side of the assembly in conjunction with an acceptable hose stream test performance.

The *T* rating indicates the period of time that the through-penetration fire-stop system is capable of preventing the passage of flame and a maximum individual temperature rise of 325^0 F. (163^0 C.) above ambient temperature on the unexposed (nonfire) side of the assembly in conjunction with acceptable hose stream test performance.

Vertical opening: An opening through a floor or roof.

Wall, Fire separation: A fire resistance rated assembly of materials having protected openings which is designed to restrict the spread of fire.

Wall, Fire: A fire resistance rated wall, having protected openings, which restricts the spread of fire and extends continuously from the foundation to or through the roof.

Wall, Party: A fire wall on an interior lot line used or adapted for joint service between two buildings.

703.0 MATERIALS FOR FIRE RESISTANCE [703.0; 703.1]
703.1 Scope
Materials prescribed herein for fire resistance shall conform with the requirements of this chapter.

703.2 Brick
Brick shall be laid in Type M, S, N, or O mortar. Solid clay and shale brick shall conform to ASTM C 216 or ASTM C 62. Hollow clay and shale brick shall conform to ASTM C 652. Concrete brick shall conform to ASTM C 55.

Sand-lime brick shall conform to ASTM C 73. Ceramic glazed structural facing tile and facing brick shall conform to ASTM C 126.

703.3 Clay or Shale Tile
Hollow clay or shale tile shall be laid in Type M, S, N, O, or gypsum mortar. Clay or shale tile used in nonbearing partitions and for fire resistance shall meet the requirements of ASTM C 56.

Clay or shale tile used in exterior walls and in all loadbearing walls shall comply with the requirements of ASTM C 34 and ASTM C 212.

703.4.1 Gypsum
Poured gypsum used for fire resistance and floor and roof construction shall contain not more than 12-1/2% of wood chips, shavings, or fiber, measured in a dry condition, as a percentage by weight of the dry mix.

Gypsum mortar shall be composed of one part gypsum and not more than three parts clean, sharp, well-graded sand, by weight.

703.4.2. Fibered plaster may be used where unsanded or neat gypsum plaster is prescribed.

703.4.3. All plaster mixes for sanded gypsum plasters shall be measured by dry weight.

703.4.4. When gypsum plaster is used with an aggregate, the proportions shall be as required in 2504.

703.5 Gypsum Lath, Wallboard and Sheathing Board
703.5.1. Gypsum lath shall comply with the provisions of ASTM C 37.

703.5.2. Gypsum lath shall be nailed to wood studs or joists in all constructions required to be fire-resistant, with 1-1/8 in (29 mm), 13 ga, 19/64 in (7.5 mm) flat head blued nails at intervals not exceeding 4 in (102 mm) on centers (five nails per lath for support of 16-in (406 mm) lath) or equivalent attachment.

703.5.3. Gypsum wallboard shall comply with the provisions of ASTM C 36.

703.5.4. Gypsum sheathing board shall comply with the provisions of ASTM C 79.

703.5.5. Gypsum veneer base shall comply with the provisions of ASTM C 588.

703.5.6. Gypsum veneer plaster shall comply with the provisions of ASTM C 587.

703.5.7. Exterior gypsum soffit board shall comply with the provisions of ASTM C 931.

703.5.8. Water-resistant gypsum backing board shall comply with the provisions of ASTM C 630.

703.6 Metal or Wire Lath

703.6.1. Metal lath shall comply with the provisions of ASTM C 847. Wherever metal lath or wire lath and plaster are used as required protection against the spread of fire, the weight of lath shall be not less than 2.5 lb per sq yd (1.4 kg/m^2) when used in vertical position, and not less than 2.75 lb per sq yd (1.5 kg/m^2) when used in horizontal position.

Wire lath shall not be lighter than 2-1/2 meshes per inch, or equivalent.

703.6.2. Weight tags shall be left on all metal lath or wire lath until inspected and approved by the building official.

703.6.3. Metal lath for ceilings below wood joists in construction which is required to be fire-resistant shall be attached with 1-1/2 in (38 mm), 11ga, 7/16 in (11.1 mm) head barbed roofing nails spaced at intervals not to exceed 6 in (152 mm) on centers, or equivalent attachment.

703.6.4. Welded wire lath shall comply with the provisions of ASTM C 933.

703.6.5. Woven wire lath shall comply with ASTM C 1032.

703.7 Concrete Block

Concrete masonry units used in exterior walls and in all walls or partitions shall comply with ASTM C 90 and C 129.

703.8 Vermiculite

Vermiculite, when used as an aggregate with plaster, shall conform in particle size to ASTM C 35. The weight of vermiculite shall be not less than 6 pcf (96 kg/m^3) nor more than 10 pcf (160 kg/m^3) as determined by measurement in a cubic-foot box, using the shoveling procedure as outlined in ASTM C 29.

703.9 Perlite

Perlite, when used as an aggregate with plaster, shall conform in particle size to ASTM C 35. The weight of perlite shall be not less than 7-1/2 pcf (120 kg/m^3) nor more than 15 pcf (240 kg/rn^3), as determined by measurement in a cubic-foot box, using the shoveling procedure as outlined in ASTM C 29.

703.10 Glass Block

Glass block shall be labeled to conform to ASTM E 163 or UL 9.

704.0 FIRE-RESISTANT SEPARATIONS AND RATINGS [703.1; 703.5]

All construction materials must bear identification showing the fire performance rating of that material. Such identification is issued and attached or marked on the material by the approved agency having a service for inspection of materials at the factory.

Small openings in fire-rated walls or ceilings are permitted for conduit, piping, duct work, and electrical boxes made of ferrous metal. Maximum openings up to 100 sq in within 100 sq ft of ceiling are permitted. However, fire-resistive floors must be continuous. Floor and roof penetrations as openings for electrical and mechanical equipment must be enclosed. The exception here is that vent and service piping may be installed through fire-resistive floors as long as the assembly does not reduce the required fire resistance rating of the floor. Conduit and pipes cannot be embedded in structural members or required fire-protective coverings.

Protective coverings for penetrations in fire-rated assemblies must meet certain thickness requirements as set forth in chapter 7 in each of the national building codes. These thickness requirements apply to all materials used from fire-resistive paint to sheetrock to concrete. These minimum thicknesses are measured in the material's net thickness and do not include any open or hollow spaces behind the protection.

[In the UBC, section 700.0 covers fire-resistant materials, ratings and construction. The section is formatted as follows:
701.0 Scope
702.0 Definitions
703.1 Fire-resistive Materials and Systems
704.0 Protection of Structural Members
705.0 Projections
706.0 Construction Joints
707.1 Insulation
708.1 Fire Blocks and Draft Stops
710.0 Floor-Ceiling Systems

Fire-resistant Materials and Construction

In BOCA, sections 701.0 through 723.6 cover fire-resistant materials, ratings, and construction. The section is formatted as follows:

701.0 FIRE-RESISTANT MATERIALS and CONSTRUCTION
721.0 FIREBLOCKING and DRAFTSTOPPING
721.2 Fireblocking materials
721.3 Draftstopping materials
721.4 Integrity
721.5 Required inspection
721.6 Fireblocking required
721.6.1 Concealed wall spaces
721.6.2 Connections between horizontal and vertical spaces
721.6.3 Stairways
721.6.4 Ceiling and floor openings
721.6.5 Architectural trim
721.6.6 Combustible finish and trim
721.6.7 Concealed sleeper spaces
721.7.0 Draftstopping required
721.7.1 Floors
721.7.1.1 Use Groups R-l and R-2
721.7.1.2 Use Group R-3
721.7.1.3 Other use groups
721.7.2 Attics and concealed spaces
721.7.2.1 Use Group R
721.7.2.2 Other use groups
721.8 Ventilation
722.0 FIRE-RESISTIVE REQUIREMENTS for PLASTER
722.1 Thickness of plaster
722.2 Plaster equivalents
722.3 Noncombustible furring
722.4 Double reinforcement
722.5 Plaster alternatives for concrete
723.0 THERMAL- and SOUND-INSULATING MATERIALS
723.1 General
723.2 Exposed installations

723.3 Concealed installations
723.3.1 Facings
723.4 Loose-fill insulation testing
723.4.1 Test procedure modifications
723.4.2 Flame spread rating
723.4.3 Smoke-developed rating
723.5 Cellulose insulation
723.6 Protection]

704.1 Occupancy Separation Requirements
704.1.1. The minimum fire resistance of construction separating any two occupancies in a building of mixed occupancy shall be the higher rating required for the occupancies being separated, as specified in Table 704. 1.

Table 704.1
Occupancy Separation Requirements

Large or small assembly	2 hour
Business	1 hour
Educational	2 hour
Factory-industrial	2 hour
Hazardous	See Section 704.1.4
Institutional	2 hour
Mercantile	1 hour
Residential	1 hour
Storage, moderate hazard S1	3 hour
Storage, low hazard S2	2 hour
Automobile parking garages[1]	1 hour
Automobile repair garages	2 hour

Note:
1. See SBC section 411.2.6 for exceptions.

Fire-resistant Materials and Construction 111

704.1.2 Accessory Occupancies
704.1.2.1. Portions of buildings used as accessory offices or for customary nonhazardous uses necessary for transacting the principal business in Group S and Group F Occupancies need not be separated from the principal use. Group F Occupancies producing, using, or storing low-hazard products listed in Section 312.2.2 need not be considered mixed occupancies. Height and area will be governed by the principal intended use.
704.1.2.2. The following occupancies need not be separated from the uses to which they are accessory:
1. A kitchen in a Group A Occupancy does not constitute a mixed occupancy. A fire-resistant separation is not required.
2. Assembly rooms having a floor area of not over 750 sq ft (70 m^2).
3. Administrative and clerical offices and similar rooms which, in area per story, do not exceed 25% of the story area of the major use when not related to Group H Occupancies.
    EXCEPTION: Accessory uses in Group F and S Occupancies conforming to Section 704.1.2.1.
4. Rooms or spaces used for customary storage of nonhazardous materials in Group A, Group B, Group E, Group F, Group M, and Group R, which in aggregate do not exceed one-third of the major occupancy floor area in which they are located.

704.1.2.3 A 1-hr occupancy separation shall be permitted in assembly rooms greater than 750 sq ft (70 m^2) but less than 2,000 sq ft (186 m^2) in area when all of the following are met:
1. The occupant content does not exceed 300 persons calculated in accordance with Table 1003.1 (see Appendix).
2. The assembly room does not constitute the major occupancy classification of the building.
3. The assembly room is not associated with a hazardous, or Group S1 Occupancy.
4. The assembly room is not associated with a kitchen.
5. The assembly room is not a theater or restaurant.

704.1.3 Special Occupancy Separations
704.1.3.1 Assembly and Educational
Fire-resistance separation shall not be required between Sunday school rooms and a church auditorium of Group A - Small Assembly occupancy, or between classrooms in day schools and auditoriums, gymnasiums, cafeterias, and libraries of small assembly occupancy, which are used only as accessory to the education occupancy.
704.1.3.2 Automobile Parking Garages
A separation between an automobile parking garage, used exclusively for the storage of passenger vehicles that will accommodate not more than nine passengers, and any other occupancy having a hr rating of 2 hr or more in Table 704.1 shall be given a fire-resistance of 2 hr fire-resistive rating.
704.1.3.3 Boiler and Machinery Rooms
704.1.3.3.1. Every central heating boiler as defined in the Standard Mechanical Code, installed in any building other than a one or two family dwelling or Group F Occupancy, shall be separated from the rest of the building by not less than 1-hr fire-resistant construction.

704.1.3.3.2. A central heating boiler installed in a Group A or H Occupancy shall be separated from the rest of the building by construction having a fire resistance rating of not less than 2 hr.
704.1.3.3.3 Steam Boilers
Every steam boiler carrying more than 15-psi (103-kPa) pressure with a rating in excess of 10 boiler horsepower (98 kW) installed in a building other than one of Group F Occupancy shall be located in a separate room or compartment, shall not be located under a means of egress, and shall be separated from the rest of the building by construction having at least a 2-hr fire resistance.

This rating may be reduced in accordance with the hazard existing when in the opinion of the building official it is desirable to provide for explosion venting upward.
704.1.3.3.4 Refrigerant System Machinery Rooms
Where required by the Standard Mechanical Code due to refrigerant type, amount, system classification, and occupancy, a Level 2 machinery room shall be of noncombustible construction. A minimum of 1-hr construction shall separate the machinery room from other occupied spaces. A minimum of 3/4-hr C-labeled doors shall be used when separating from other occupancies.

704.1.4 Hazardous Occupancies

704.1.4.1. The separation of a hazardous occupancy from other occupancies shall be in accordance with Table 704.1.4.

Table 704.1.4
Hazardous Occupancy Separation Requirements

Occupancy	H1	H2	H3	H4
A	NP	4	4	4
B	NP	2	2	1
E	NP	4	4	4
F	NP	2	1	1
H1	-	NP	NP	NP
H2	NP	-	1	2
H3	NP	1	-	1
H4	NP	2	1	-
I	NP	4	4	4
M	NP	2	2	2
R1,2,3	NP	4	4	4
S1,2	NP	2	2	2

Note:
NP = H1 Occupancies not permitted to be attached to other occupancies or other H subclassifications.

704.1.4.2. The separation of a hazardous occupancy sub-classification shall only apply to storage areas.

704.1.4.3. Building areas intended for the use, processing, manufacture, or generation of materials having different hazard classifications, all being Group H, need not be separated further from each other within the confines of the Group H Occupancy provided the requirements for each hazard are met.

704.1.4.4. Accessory areas, other than assembly occupancies, that do not exceed 10% of the allowable area for the hazardous occupancy subclassification in Table 500 (see Appendix) and that do not exceed 1500 sq ft (139 m^2) shall not be required to comply with Section 704.1. Where accessory areas are separated from hazardous occupancies by partitions, the partitions shall be not less than 1-hr fire-resistant construction with an opening protection rating not less than 3/4-hr. Opening protection shall be either self-closing or automatic-closing in accordance with Section 705.1.3.2.

704.2 Interior Wall and Partition Fire Separation Requirements

704.2.1 General

704.2.1.1. This section shall apply to the fire separation requirements of interior walls and partitions for the various occupancies and types of construction. Partitions of higher fire-resistance rating required by other sections of code may also serve to meet the requirements of this section.

704.2.1.2. All partitions enclosing vertical openings such as stairways, utility shafts, and elevator shafts which are required to have a fire resistance rating shall extend from floor to floor or from floor to roof. These walls shall be continuous through all concealed spaces such as the space above a suspended ceiling.

The supporting structure shall have a fire resistance rating equal to or greater than the fire resistance rating required for the vertical enclosure. Where the openings are offset at intermediate floors, the offset and floor construction shall have a fire resistance of not less than that required for the enclosing partitions.

704.2.1.3. All other partitions required to have a fire resistance rating shall extend from the top of the floor below to the ceiling above and shall be securely attached thereto. Where said ceiling is not a part of an assembly having a fire resistance rating at least equal to that required for the partition, the partition shall be constructed tight against the floor or roof deck above. The design of the partitions or ceilings and any openings shall be such as to prevent the spread of smoke to the corridor.

704.2.1.4. View panels in 1-hr fire-resistant partitions shall be limited either to 1/4 in (6.4 mm) thick labeled wire glass assemblies installed in steel frames or to labeled glass block panels installed in steel channels. The wired glass shall be limited to 1296 sq in (0.84

m²) with no dimension greater than 54 in (1372 mm). The glass block shall be limited to 120 sq ft (11.1 m²) with no dimension greater than 12 ft (3658 mm). Neither assembly shall exceed 25% of the wall area separating each tenant space from the corridor.

704.2.1.5. Corridor partitions, smoke-stop partitions, horizontal exit partitions, exit enclosures, and fire-rated walls required to have protected openings shall be effectively and permanently identified with signs or stenciling in a manner acceptable to the authority having jurisdiction. Such identification shall be above any decorative ceiling and in concealed spaces. The suggested wording is "Fire and smoke barrier. Protect all openings."

704.2.2 Partition Requirements by Occupancy

704.2.2.1 Group I Restrained

704.2.2.1.1. Any required smoke barrier shall be continuous from outside wall to outside wall, from floor slab to floor slab or roof deck, from smoke barrier to smoke barrier, or a combination thereof, including continuity through all concealed spaces such as those found above suspended ceilings; however, smoke barriers are not required in interstitial spaces designed and constructed with ceilings equivalent to smoke barriers. Barriers shall be of 0.10 in (2.5 mm) thick steel or of 1-hr fire-resistant construction. Fixed wired glass vision panels shall be permitted in such barriers provided they do not individually exceed an area of 1296 sq in (0.84 m²) and are mounted in steel frames. There is no restriction on the total number of such panels in any barrier.

704.2.2.1.2. All interior partitions in Type I and II construction shall be of noncombustible construction.

704.2.2.2 Group I Unrestrained

704.2.2.2.1. Smoke barriers shall have a minimum of 1-hr fire resistance. Such partitions shall form an effective membrane continuous from outside wall to outside wall and from floor slab to floor slab or roof deck thereby including continuity through all concealed spaces, such as those found above suspended ceilings, and including interstitial structural and mechanical spaces.

> **EXCEPTION**: Smoke barriers are not required in interstitial spaces when such spaces are designed and constructed with ceilings that provide resistance to the passage of smoke equivalent to that provided by smoke barriers.

704.2.2.3 Group R Residential

Non-fire-rated partitions may be constructed within one- and two-family dwellings and within individual dwelling units unless required by Table 600 (see Appendix). The tenant separation in a two-family dwelling shall comply with Section 704.3.

704.2.3 Partitions Within Tenant Space

704.2.3.1. Partitions dividing portions of stores, offices, or similar places occupied by one tenant only, which do not establish an exit access corridor serving an occupant load of 30 persons or more, and partial partitions may be temporary or permanent and constructed in accordance with Section 609 without fire resistance, provided that

1. Their location is restricted by their method of construction or by means of permanent tracks, guides, or other approved methods.
2. Flammability shall be limited to materials having an interior finish classification as set forth in Table 803.3 (see Chapter 7) for rooms or areas.

704.2.3.2. Exit access corridors are not required to be rated on any single-tenant floor or in any single-tenant space in a Group B building when the building is protected throughout by an approved automatic sprinkler system and when smoke detectors installed in return air are arranged to shut down mechanical ventilation systems in accordance with the Standard Mechanical Code.

704.3 Tenant Fire Separation

704.3.1. In a building or portion of a building of a single occupancy classification, when enclosed spaces are provided for separate tenants, such spaces shall be separated by partitions with not less than 1-hr fire resistance.

> **EXCEPTION**: In buildings of Group B and S Occupancies non-fire-rated partitions may be used to separate tenants provided no area between partitions rated at 1 hr or more exceeds 3000 sq ft (278.7 m^2).

704.3.2. In buildings with usable crawl spaces, tenant separation walls required to have a fire resistance rating shall extend from the underside of the floor to the ground below. A suitable foundation shall be provided at grade level.

EXCEPTION: The wall need not be extended when the floor above the crawl space has a minimum 1-hr fire-resistance rating.

704.4 Townhouse Fire Separation

704.4.1. Each townhouse shall be considered a separate building and shall be separated from adjoining townhouses by a party wall complying with Section 704.4.2 or by the use of separate exterior walls meeting the requirements of Table 600 (see Appendix) for zero clearance from property lines as required for the type of construction. Separate exterior walls shall include one of the following:

1. A parapet not less than 18 in (457 mm) above roof line.
2. Roof sheathing of noncombustible material or fire-retardant-treated wood, for not less than a 4-ft (1219 mm) width on each side of the exterior dividing wall.
3. One layer of 5/8 in (15.9 mm) Type X gypsum board attached to the underside of roof decking, for not less than a 4-ft (1219 mm) width on each side of the exterior dividing wall.

704.4.2. When not more than three stories in height, townhouses may be separated by a single wall meeting the following requirements:

1. Such wall shall provide not less than a 2-hr fire resistance rating. Plumbing, piping, ducts, electrical systems, or other building services shall not be installed within or through the 2-hr wall, unless such materials and methods of penetration have been tested in accordance with Section 701.2.
2. Such wall shall be continuous from the foundation to the underside of the roof sheathing or shall have a parapet extending not less than 18 in (457 mm) above the roof line. When such wall terminates at the underside of the roof sheathing, the roof sheathing for not less than a 4-ft (1219 mm) width on each side of the wall shall be of noncombustible material, fire-retardant-treated wood, or one layer of 5/8 in (15.9 mm) Type X gypsum wallboard attached to the underside of the roof decking.
3. Each dwelling unit sharing such wall shall be designed and constructed to maintain its structural integrity independent of the unit on the opposite side of the wall.

EXCEPTION: Said wall may be penetrated by roof and floor structural members provided that the fire-resistance rating and the structural integrity of the wall is maintained.

704.5 Fire Wall Extensions and Parapets

704.5.1 Fire Wall Extensions

704.5.1.1. Party walls and fire walls shall extend not less than 3 ft (914 mm) above the roof.

EXCEPTION: Fire walls shall not be required to extend above the roof where the roof is:
1. Noncombustible in Types I, II, and IV construction
2. Noncombustible or fire-retardant-treated wood for an area within 40 ft (12.2 m) of each side of the wall in Types III, V, and VI construction

704.5.1.2. Party walls and fire walls shall extend not less than 18 in (457 mm) past exterior intersecting walls of combustible construction or exterior noncombustible walls with combustible projections or veneers. The party or fire wall shall extend not less than 18 in (457 mm) past any combustible projection or veneer. Party walls or fire walls shall extend to the inside facing of the exterior surface of noncombustible construction.

704.5.1.3. All fire walls shall be in accordance with the requirements of NCMA-TFK Bulletin 95 or equivalency in brick or poured concrete or other nationally recognized tested systems.

705.0 PROTECTION OF OPENINGS [704.0; 706.0]

705.1.1 Protection of Openings in Exterior Walls

705.1.1.1. The provisions of Section 705.1.1 do not apply to Group R3 Occupancies.

705.1.1.2. Every exterior wall within 15 ft (4572 mm) of a property line shall be equipped with approved opening protectives.

EXCEPTIONS:
1. Exterior walls not required by Table 600 (see Appendix) to have a fire resistance rating
2. Show windows fronting on a street or public space
3. Open parking structures meeting the requirements of Section 411.3

705.1.1.3. Where openings in an exterior wall are above and within 5 ft (1524 mm) laterally of an opening of the story below, such

Fire-resistant Materials and Construction 119

openings shall be separated by an approved noncombustible flame barrier extending 30 in (762 mm) beyond the exterior wall in the plane of the floor or by approved vertical flame barriers not less than 3 ft (914 mm) high measured vertically above the top of the lower opening. Such flame barriers are not required when a complete approved automatic sprinkler system is installed.

705.1.1.4. Fresh air intakes shall be protected against exterior fire exposure by means of approved fire doors, dampers, or other suitable protection in accordance with the degree of exposure hazard.

705.1.2 Protection of Openings in Interior Walls

705.1.2.1. Openings in walls and partitions, except in one- and two-family dwellings, shall be protected in accordance with Table 700.

705.1.2.2. Fire dampers shall comply with the requirements of UL 555 and shall bear the label of an approved testing agency. Fire dampers shall be classified and identified for use in either:

1. Static systems that automatically shut down in the event of fire
2. Dynamic systems that operate in the event of fire

705.1.2.3. Fire dampers shall be installed in accordance with the manufacturer's installation instructions in the following locations:

1. Ducts penetrating walls or partitions having a fire resistance rating of 1 hr or more
2. Ducts penetrating shaft walls having a fire resistance rating of 1 hr or more

705.1.2.4. Fire dampers are not required under the following conditions:

1. In duct systems serving only one story, used only for exhaust of air to the outside, and not penetrating a wall or partition having a required fire resistance rating of 2 hr or more or not passing entirely through the enclosure for a vertical shaft.
2. Where branch ducts connect to return risers in which the air flow is upward and subducts at least 22 in (559 mm) long are carried up inside the riser at each inlet.
3. In duct systems of any duct materials or combinations thereof allowed by chapter 6 of the Standard Mechanical Code penetrating 1-hr walls or partitions, where the duct

penetrating the rated wall or partition meets the following minimum requirements:
a. The duct shall not exceed 100 sq in (0.06 m^2)
b. The duct shall be of 0.0217-in (0.55 mm) minimum thick steel
c. The duct shall continue with no duct openings for not less than 5 ft (1.5 m) from the rated wall
d. The duct shall be installed above a ceiling

When wall registers occur at the rated wall, a fire damper shall be provided. See section 610 of the Standard Mechanical Code for additional exceptions. [UMC and NMC.]

705.1.2.5. Unless the air system is designed to provide smoke control or pressurization functions during a fire emergency, smoke dampers with listed operators shall be installed at all duct penetrations of required smoke barriers.

705.1.2.6. Transfer grilles, whether equipped with fusible link-operated dampers or not shall not be used in smoke barriers. See section 610.3 of the Standard Mechanical Code.

705.1.2.7. Where a fire-resistant wall is required due to type of construction only, opening protectives are not required.

705.1.3 Approved Types of Fire Windows, Doors, and Shutters

705.1.3.1. Wall openings required to be protected shall be protected by approved listed and labeled fire doors, windows, and shutters and their accompanying hardware, including all frames, closing devices, anchorage, and sills, in accordance with the requirements of NFiPA 80, except as otherwise specified in code.

705.1.3.2. Openings are classified in accordance with the character and location of the wall in which they are situated. Fire protection ratings for products intended to comply with this section shall be as determined and reported by a nationally recognized testing agency in accordance with ASTM E 152 or ASTM E 163. All such products shall bear an approved label. In each of the following classes, the minimum fire protection ratings are shown.

705.1.3.2.1. Fire doors are classified as 3 hr (A), 1-1/2 hr (B), 1 hr (B), 3/4 hr (C), 1-1/2 hr (D), 3/4 hr (E) or 20 min. The letter designation indicates the classification of opening in a wall or partition assembly for which a door is considered suitable and the relative importance of the door in preventing the spread of fire. These designations are described as follows:

Fire-resistant Materials and Construction 121

1. Class A. Openings in walls that divide a single building into fire areas or fire walls separating buildings.
2. Class B. Openings in enclosures of vertical communications through buildings. They are also suitable for certain other openings in walls or partitions.
3. Class C. Openings in walls or partitions between rooms and corridors or hallways, except as provided in Section 705.1.3.2.2 for 20 min doors.
4. Classes D and E. Openings in exterior walls subject to severe and moderate fire exposure, respectively, from outside of the building.

705.1.3.2.2. Unless otherwise specified, door assemblies in walls required to have a fire resistance rating of 1-hr or less shall have a fire resistance rating of 20 min when tested in accordance with ASTM E 152 without the hose stream.

> **EXCEPTION**: For Group I Unrestrained, corridor doors shall be in accordance with SBC section 409.1.4.

705.1.3.2.3. All doors in smoke barriers, horizontal exits, stairway enclosures, and other doors opening between rooms and fire-rated exit access corridors shall be self-closing and so maintained or shall be provided with approved door holding devices of the fail-safe type which will release the door causing it to close when activated by approved listed smoke detectors. When doors are automatic-closing by smoke detection, there shall be not more than a 10-sec delay before the door starts to close after the smoke detector is actuated.

> **EXCEPTION**: Doors from classrooms in Group E Occupancies, opening directly into a 1-hr fire-rated corridor, may be installed without self-closing devices.

705.1.3.3. The maximum size of fire doors shall not exceed that specified in Appendix C of NFiPA 80.

705.1.3.4. For 1-1/2-hr (B) and 1-hr (B) fire-rated doors used in stairway enclosures the average temperature developed on the unexposed side shall not exceed 450^0 F (232^0 C) at the end of 30 min of standard fire test exposure.

705.1.3.5. Fire doors shall be equipped with an approved closer. See Section 1009.2.2 for doors in horizontal exits.

> **EXCEPTION**: Doors located in common walls separating guest rooms in Group R1 hotels and motels may be installed without automatic or self-closing devices.

705.1.3.6. One-quarter-in (6.4 mm) thick wired glass labeled for fire-protection purposes may be used in approved opening protectives with the maximum sizes shown in Table 705.1.3.6. Other glazing materials which have been tested and labeled to indicate the type of opening to be protected for fire protection purposes may be used in approved opening protectives in accordance with their listing with the maximum sizes tested. For requirements for safety glazing see SBC section 2405.

Table 705.1.3.6
Limiting Size of Wired Glass Panels 1-3

Rating, Opening	Max. Area (sq in)	Max. Height (in)	Max. Width (in)
3 hr, Class A door	0	0	0
1 & 1-1/2 hr, B doors	100	33	12
3/4 hr, Class C door	1296	54	54
1-1/2 hr, Class D door	0	0	0
3/4 hr, Class E door	1296	54	54

1 in = 25.4 mm
1 sq in = 645.16 mm^2

<u>Notes</u>:
1. The glass shall be well-embedded in putty, and all exposed joints between the metal and glass shall be struck and pointed.
2. Devices used to view through fire doors rated at 1-1/2 hrs or less shall be labeled.
3. Wired glass in 20-min doors shall be limited to the amount of glass tested in a door.

705.1.4 Fire Shutters

705.1.4.1. When equipped with fire shutters of the swinging type, at least one in every three openings facing a street in each story shall have such shutters arranged to be readily opened from the outside. Distinguishing marks shall be provided on such shutters.

705.1.4.2. Fire shutters of the rolling type shall be carefully counterbalanced and so arranged that they can be readily opened from the outside.

705.1.5 Opening Protection in Stairway Enclosures

Opening protectives in stairway shafts are limited to self-closing or smoke-actuated, automatic-closing fire door assemblies. If smoke-actuated closures close one door, all doors serving that stairway shall close.

Table 700
Minimum Fire Resistance of Walls, Partitions and Opening Protectives[1] (hr)

Component	Walls and Partitions	Opening Protectives
Shaft enclosures (including stairways, exits, and elevators)		
Four or more stories	2	1-1/2 B
Less than 4 stories	12	1 B[2]
All refuse chutes	2	1-1/2 B
Walls and partitions		
Fire walls[3]	4	3 A
Within tenant space	See Section 704	
Tenant space (see also 704.3)	1	3/4 C

Horizontal exit	2	1-1/2 B
Exit access corridors[4, 5]	1	20 min
Smoke barriers	See SBC section 409.1.2	
Refuse and laundry	1	3/4 C
Chute access rooms	1	1-1/2 B
Incinerator rooms	2	3/4 C
Refuse and laundry chute Termination rooms	1	3/4 C
Hazardous occupancy Control areas	1	3/4 C
High-rise buildings	See SBC section 412	
Covered mall buildings	See SBC section 413	
Assembly buildings	See Note 2	
Bathrooms, restrooms	See Note 6	
Exterior Walls[8]	All	3/4 E

Notes:
1. Table 600 (see Appendix) may require greater fire resistance of walls to ensure structural stability.
2. All exits and stairways in Group A and H Occupancies shall be 2 hr with 1-1/2-hr B door assemblies.
3. See also Section 503.1.2.
4. See Section 704.2.3
5. See SBC section 409 for sprinklered Group I buildings.
6. Fire-rated bathroom and restroom doors are not required when opening onto fire-rated halls, corridors, exit access, provided:
   A. No other rooms open off of the bathroom/restroom.
   B. No gas or electric appliances are located in the bathroom/restroom.
   C. The walls, partitions, floor, and ceiling of the bathroom/restroom have a fire rating at least equal to the rating of the hall, corridor, or exit access.
   D. The bathroom/restroom is not used for any other purpose than it is designed.

7. See Section 704.1.
8. See SBC Tables 503.4.8, 600, and 705.1.

705.2 Protection of Floor Openings
705.2.1 General Requirements
705.2.1.1. Protection of floor openings shall be provided in accordance with these provisions to prevent the spread of fire from story to story.

705.2.1.2. For protection of stairways, see Section 1006. For protection of elevators, see SBC section 3003.

705.2.1.3. For protection of pipe, conduit, cable, wire, tube, duct, and vent penetrations of fire resistance-rated floors, roofs, ceilings, and shaft enclosures, see Section 705.4.

705.2.1.4 Duct Penetrations
705.2.1.4.1. Fire dampers shall be installed in accordance with the manufacturer's installation instructions in ducts penetrating only one floor of a building requiring the protection of vertical openings when the duct is not protected by a shaft enclosure described in section 610.3 of the Standard Mechanical Code.

705.2.1.4.2. Fire dampers shall comply with the requirements of UL 555 and shall bear the label of an approved testing agency. Fire dampers shall be classified and identified for use in either
- Static systems that automatically shut down in the event of fire
- Dynamic systems that operate in the event of fire

705.2.2. Shaft Enclosures
All openings through a floor and penetrations through a floor shall be protected by a shaft enclosure in accordance with Section 705.2.3.

EXCEPTIONS:
1. A shaft enclosure is not required for openings totally within a dwelling unit and connecting four stories or less.
2. A shaft enclosure is not required for a floor opening which
   A. Is not part of the required means of egress
   B. Is not concealed within the building construction
   C. Does not connect more than two stories
   D. Does not connect with a stairway or escalator serving other floors

E. Is not open to any corridors in Group I and R Occupancies or to corridors on nonsprinklered floors in other occupancies
F. Is separated from floor openings serving other floors by construction conforming with required shaft enclosures

3. A shaft enclosure is not required for penetrations by pipe, tube, conduit, wire, cable, duct, and vents protected in accordance with Section 705.3.1.4 or 705.4.4.
4. A shaft enclosure is not required for floor openings complying with the special provisions for covered malls or atriums.
5. A shaft enclosure is not required for approved masonry chimneys where annular space protection is provided at each floor level in accordance with Section 705.4.6.
6. A shaft enclosure is not required for floor openings in an open parking garage.
7. A shaft enclosure is not required for floor openings between a mezzanine and the floor below.
8. A shaft enclosure is not required in fully sprinklered buildings for an escalator opening protected in accordance with one of the following alternatives:

Alternative A. (1) The area of the floor opening between stories shall not exceed twice the horizontal projected area of the escalator, (2) draft curtains and special sprinkler head locations shall meet the requirements of NFiPA 13, and (3) in other than Group B and M Occupancies the escalator openings shall not connect more than four stories.

Alternative B. The opening is protected by approved power-operated automatic shutters at every floor opening. The shutters shall be of noncombustible construction and have a fire resistance rating of not less than 1-1/2 hr. The shutter shall fully close immediately upon the automatic detection of smoke by an approved device and shall completely shut off the floor opening. The escalator shall stop when the shutter begins to close. The shutter shall operate at a speed of not more than 30 ft/min (0.152 mm/min) and shall be equipped with a sensitive leading edge to arrest its progress when in contact with any obstacle and to continue its progress after release.

705.2.3 Shaft Enclosure Construction
705.2.3.1 Fire Resistance
705.2.3.1.1. The fire resistance rating of shaft enclosures shall be in accordance with Table 700 and not less than the floor assembly penetrated, but the rating need not exceed 2 hr.

705.2.3.2 Construction Type
705.2.3.2.1 . Shaft enclosures shall be of noncombustible materials in Types I, II, and IV construction and may be of combustible materials in Types III, V, and VI construction.

705.2.3.3 Enclosure at Bottom
705.2.3.3.1. Shafts which do not extend to the bottom of the building or structure shall be enclosed at the lowest level with construction of the same fire resistance as the lowest floor through which the shaft passes, but not less than the rating required for the shaft enclosure, or shall terminate in a room having a use related to the purpose of the shaft. The room shall be separated from the remainder of the building by construction having a fire resistance rating and opening protectives at least equal to the protection required for the shaft enclosure. For shafts containing refuse or laundry chutes, see Section 705.2.4.

EXCEPTIONS:
1. Separation of the room by fire-resistant construction is not required provided there are no openings in or penetrations through the shaft enclosure to the interior of the building except at the bottom of the shaft. The bottom of the shaft shall be closed off around the penetrating items with materials permitted by the code for draft-stops, or the room shall be provided with an approved automatic sprinkler system.
2. Separation of the room by fire-resistant construction is not required, and the bottom of the shaft may be open provided there are no combustibles in the shaft and there are no other openings or penetrations through the shaft enclosure to the interior of the building.

705.2.3.4 Enclosure at Top

705.2.3.4.1. A shaft that does not extend to or through the underside of the roof deck of the building shall be enclosed at the top with construction of the same fire resistance as the topmost floor penetrated by the shaft, but not less than the rating required for the shaft enclosure.

705.2.3.4.2. When a shaft extends through a roof, the shaft enclosure shall extend at least 36 in (914 mm) above the highest part of the roof that is within 5 ft (1524 mm) of the opening. The enclosure wall or combination of enclosure wall and guardrail shall be at least 42 in (1066 mm) high.

> **EXCEPTION**: When a shaft extends fully through a noncombustible roof, a noncombustible guardrail at least 42 in (1066 mm) high may be used around the opening instead of a wall.

705.2.3.5 Openings in Shaft Enclosures

705.2.3.5.1. Openings in shaft enclosures shall be permitted when limited to those necessary for the purposes of the shaft. Permitted openings in shaft enclosures shall be protected with opening protectives having a fire resistance in accordance with Table 700.

705.2.4 Special Provisions for Refuse and Laundry Chutes

705.2.4.1. Refuse and laundry chutes, access and termination rooms, and incinerator rooms shall be constructed in accordance with Sections 705.2.4.2 through 705.2.4.6.

> **EXCEPTION**: Group R3 Occupancies.

705.2.4.2. Refuse and Laundry Chute Enclosures

705.2.4.2.1. A shaft containing a refuse or laundry chute shall be used for no other purpose and shall be protected by a shaft enclosure in accordance with Section 705.2.3. Refuse chute material shall be noncombustible meeting part 1 of the definition for noncombustible (see Chapter 2).

All openings into the shaft enclosure, including those from access rooms and termination rooms, shall be protected in accordance with Table 700. Such opening protectives shall be self-closing or automatic-closing upon detection of smoke, except that the opening protective between the shaft and the termination room may be closed by a heat-activated device.

705.2.4.3 Refuse and Laundry Chute Access Rooms
705.2.4.3.1. Access openings for refuse and laundry chutes shall be located in rooms or compartments completely enclosed by construction and opening protectives in accordance with Table 700. Access openings to refuse and laundry chutes shall not be located in exit access corridors or exit enclosures.

705.2.4.4 Termination Room
705.2.4.4.1. Refuse and laundry chutes shall discharge into an enclosed termination room completely separated from the remainder of the building by construction and opening protectives in accordance with Table 700. Refuse chutes shall not terminate in an incinerator room.

705.2.4.5 Incinerator Room
705.2.4.5.1. Incinerators shall be enclosed within a room separated from the remainder of the building by construction and opening protectives in accordance with Table 700.

705.2.4.6 Automatic Sprinklers
An approved automatic sprinkler system shall be installed at the top and at alternate floor levels in refuse and laundry chutes and in the termination and incinerator rooms.

705.3 Fireblocking and Draft-stopping
705.3.1 Fireblocking
705.3.1.1. Fireblocking shall be provided in all walls and partitions to cut off all concealed draft openings (both horizontal and vertical) and to form a fire barrier between floors and between the upper floor and the roof space. See also Section 2305.1.

705.3.1.2. Fireblocking shall not be covered or concealed until inspected by the building official.

705.3.1.3. Walls and stud partitions shall be fireblocked at floors, ceilings, and roofs. Fireblocking in noncombustible partitions shall not be required at the ceiling for suspended ceiling systems. Fireblocking shall consist of approved noncombustible materials unless otherwise specified in code. Material shall be securely fastened in place.

705.3.1.4. The annular space around pipes, tubes, conduit, wires, cables, and vents shall be protected in accordance with SBC Tables 705.3.1.4, 705.4.6, and 705.4.4.

705.3.1.5. Chimneys shall be fireblocked in accordance with Section 2305.0.

705.3.1.6. Any openings between the edge of a floor deck and an exterior wall shall be sealed using an approved material or assembly of materials designed and tested for this purpose. The material shall remain in place, sealing the opening, for a time period at least equal to the required fire resistance rating of the floor deck.

705.3.2 Draft-stopping

705.3.2.1. Enclosed attic and floor spaces formed of combustible construction shall be divided in accordance with Section 2305.2.

705.4 Penetrations of Fire-Resistant Assemblies

705.4.1 General

705.4.1.1. Penetrations of pipes, tubes, conduits, wires, cables, ducts, and vents through fire-resistant walls and shaft enclosures and through floor, ceiling, and roof elements of fire-resistant assemblies (including Type III construction) shall be protected in accordance with Sections 705.4.3 through 705.4.8. See Section 104.2.4.

705.4.2 Definitions

For the purpose of this section, through-penetration fire-stop system is defined below:

THROUGH-PENETRATION FIRE-STOP SYSTEM - A system installed to prevent, for a prescribed time period, the passage of flame, heat, and hot gases through openings which penetrate an entire fire-resistant assembly in order to accommodate cables, cable trays, conduits, tubing, pipes, or similar items.

705.4.3 Walls and Shaft Enclosures

Penetrations through fire-resistant walls and shaft enclosures shall be protected in accordance with Sections 705.4.3.1 through 705.4.3.4.

> **EXCEPTION**: Concrete or masonry bearing walls which are not required to have protected openings.

705.4.3.1. Cables and wires without combustible jackets and insulations and noncombustible pipes, tubes, conduits and appliance vents may penetrate fire-resistant walls and shaft enclosures provided they are protected in accordance with Method C, E, or F, as described in Section 705.4.5.

705.4.3.2 Combustible cables and wires; cables and wires with combustible jackets or insulations; and combustible pipes, tubes,

Fire-resistant Materials and Construction 131

conduits and listed appliance vents may penetrate fire-resistant walls and shaft enclosures provided the penetrations are protected in accordance with Method E or F as described in Section 705.4.5.

705.4.3.3. Openings for steel electrical outlet boxes not exceeding 16 sq in (0.010 m²) are permitted provided the area of such openings does not aggregate more than 100 sq in (0.06 m²) for any 100 sq ft (9 m²) of fire-resistant wall area or shaft enclosure wall area. Outlet boxes on opposite sides of the fire-resistant wall or shaft enclosure shall be separated by a horizontal distance of not less than 24 in (610 mm).

> **EXCEPTION**: Openings for electrical outlet boxes of any material are permitted provided such boxes are listed for use in fire-resistant assemblies and are installed in accordance with their listings.

705.4.3.4. Ducts may penetrate fire-resistant walls provided the duct penetrations are protected with approved fire dampers. Fire dampers used for protection of duct penetrations shall be installed in accordance with their listing.

> **EXCEPTION**: Duct penetrations in accordance with Sections 705.1.2 and 705.2.1.4.

705.4.4 Floor, Roof, and Ceiling Penetrations

705.4.4.1. Penetrations through fire resistant floor/ceiling and roof/ceiling assemblies shall be protected in accordance with SBC Table 705.4.4.

705.4.5 Protection Methods

705.4.5.1. The protection methods referenced in this section are defined as follows:

> Method A. Two-hr fire-resistant shaft enclosure, in accordance with Section 705.2.3.
>
> Method B. One-hr fire resistant shaft enclosure, in accordance with Section 705.2.3.
>
> Method C. Protection of the annular space around the penetrating item in accordance with Section 705.4.6.
>
> Method D. Installation of an approved ceiling damper or fire damper in a duct penetration in accordance with its listing.
>
> Method E. Use of an approved through-penetration fire-stop system tested and installed in accordance with Section 705.4.7.

Method F. Penetrations tested in accordance with ASTM E 119 as part of the fire-resistant assembly.

705.4.6 Method C: Annular Space Protection

705.4.6.1. The annular space between the penetrating item and the assembly being penetrated shall be protected in accordance with Sections 705.4.6.1 through 705.4.6.4. The material used to fill the annular space shall prevent the passage of flame and hot gases sufficient to ignite cotton waste when subjected to ASTM E 119 time-temperature fire conditions under a minimum positive-pressure differential of 0.01 in of water (2.5 Pa) at the location of the test specimen for the time period equivalent to the fire resistance rating of the construction penetrated.

EXCEPTIONS:
- In buildings equipped with an approved automatic sprinkler system, annular space protection shall not be required where sprinklers penetrate only the ceiling membrane of fire rated assemblies.
- For nonrated assemblies, any approved materials shall be permitted.

705.4.6.2. In concrete or masonry assemblies, the material shall be permitted to be concrete, grout, or mortar, for the full thickness of the assembly or the equivalent thickness required to provide a fire-resistance rating equal to the required fire resistance rating of the assembly being penetrated, provided the gross cross-sectional area of the penetrating item does not exceed 36 sq in (23,226 mm^2) and the width of the annular space between the penetrating item and the assembly does not exceed 1-1/2 in (38 mm).

705.4.6.3. Where sleeves are used, the sleeves shall be securely fastened to the assembly penetrated. All space between the item contained in the sleeve and the sleeve itself and any space between the sleeve and the assembly penetrated shall be filled with a material complying with Section 705.4.6.1 or 705.4.6.2.

705.4.6.4. Insulation and coverings on the penetrating item shall not pass through the assembly unless these materials comply with Section 705.4.6.1 or 705.4.6.2.

EXCEPTION: For nonrated assemblies any approved materials shall be permitted.

705.4.7 Method E: Through-Penetration Fire-stop Systems

705.4.7.1. Through-penetration fire-stop systems shall be listed and tested in accordance with ASTM E 814 and Sections 705.4.7.1 through 705.4.7.3 and shall be installed in accordance with the listing.

705.4.7.1. The system shall be tested at a positive-pressure differential between the exposed and unexposed surfaces of the test assembly of not less than 0.01 in of water (2.5 Pa).

705.4.7.2. All systems shall have a fire rating of at least 1 hr but not less than the required fire resistance of the assembly being penetrated.

705.4.7.3. Systems for floor penetrations shall have a fire rating of at least 1 hr but not less than the required fire resistance of the floor being penetrated.

705.4.8 Walls, Floors, and Partitions

705.4.8.1. When walls, floors, and partitions are required to have a minimum 1 hr or greater fire resistance rating, cabinets, bathroom components, and lighting and other fixtures shall be so installed such that the required fire resistance will not be reduced.

> **EXCEPTION**: Fixtures which are listed for such installation are permitted.

706.0 COMBUSTIBLES IN FIRE-RATED ASSEMBLIES [705.0; 706.0]

706.1 Plumbing, Electrical, and Air Handling Systems in Fire-Rated Assemblies

706.1.1. In Type I and Type II construction, materials used for piping, conduit raceways, or duct systems which do not qualify as noncombustible in accordance with the requirements of part 1 of the definition of noncombustible material given in Chapter 2 shall neither

1. Penetrate any assembly which is required to have a fire-resistance rating unless such materials and methods of penetration have been tested in accordance with Section 705.4
2. Be concealed within any assembly which is required to have a fire resistance rating unless enclosed by or totally embedded within noncombustible materials or unless such materials and methods have been tested in accordance with Section 701.2

707.0 COMBUSTIBLES IN CONCEALED SPACES [706.0; 707.0]
707.1 Concealed Spaces in Type I, II, and IV Construction
707.1.1. Combustibles shall not be permitted in concealed spaces of Type I, II, or IV construction.
EXCEPTIONS:
1. Materials complying with section 609.1.2 of the Standard Mechanical Code
2. Class A interior finish materials
3. Fire-retardant-treated wood used in accordance with Table 600 (see Appendix) and wood used in accordance with Section 609
4. Floor finish complying with Section 803.6
5. Conduit or raceway systems complying with Section 706
6. Foam plastic insulation complying with SBC section 2603
7. Thermal insulation materials complying with Section 708
8. Combustible piping within partitions or enclosed shafts installed in accordance with the provision of code

Combustible piping may also be used within concealed ceiling space when approved.

707.2 Combustibles in Plenums
The use of combustible materials in plenums shall be restricted in accordance with the Standard Mechanical Code.

708.0 THERMAL INSULATING MATERIALS [707.1; 723.0]
Today's superinsulated homes are so tightly constructed it is as though they were sealed in airtight enclosures. Superinsulated structures can be made to work even more efficiently through the use of products like powered sidewall-vented water heaters that save energy by using outside air, instead of indoor conditioned air, for combustion. New heater-mounted blowers can direct-vent through a side wall up to 40 ft away using standard 3-in polyvinyl chloride (PVC) pipe for both venting and air intake. These new combustion systems precisely control airflow to enhance and regulate the combustion process. Energy factors are 0.62 for 50-gal units, and 0.61 for the 40-gal units. Summer cooling gain and winter heating loss are important factors with code compliant construction. Beware of relative mean average temperature in the national weather zones.

However, superinsulated homes have now revealed problems in being airtight structures. One of the enemies of good indoor air in today's tighter homes is formaldehyde adhesives used in plywood and laminates which release gas over time (off-gassing). But the Engineered Wood Association (EWA) says the builder should take structural engineered wood panels like plywood, oriented strand board (OSB), and composite panels off the list of potential formaldehyde sources. This verdict is based on results of their tests that show a typical 2000 sq ft house contains about 10,000 sq ft of its members' products. According to the EWA, the adhesives used are key. Structural wood panels use waterproof phenol formaldehyde, not urea formaldehyde, which off-gasses. Thermal properties of the new insulating materials and HVAC equipment are striving for economical energy efficiency with no long-term health hazards for the occupants.

708.1 General

708.1.1. Insulating materials, including facings such as vapor retarders and breather papers, similar coverings, and all layers of single and multilayer reflective foil insulations, shall comply with the requirements of this section. Where a flamespread rating or a smoke-developed rating is specified in this section, such rating shall be determined in accordance with ASTM E 84. Any material which is subject to an increase in flamespread rating or smoke-developed rating beyond the limits herein established through the effects of age, moisture, or other atmospheric conditions shall not be permitted.

708.2 Concealed Installation

708.2.1. Insulating materials, when concealed as installed, in buildings of any type construction, shall have a flamespread rating of not more than 75 and a smoke-developed rating of not more than 450.

708.2.2 When such materials are installed in concealed spaces in buildings of Type III, V, or VI construction, the flamespread and smoke-developed limitations do not apply to facings, coverings, and layers of reflective foil insulation that are installed behind and in substantial contact with the unexposed surface of the ceiling, wall, or floor finish.

708.3 Exposed Installation

Insulating materials when exposed as installed in buildings of any type construction shall have a flamespread rating of not more than 25 and a smoke-developed rating of not more than 450.

708.4 Loose-Fill Insulation

Loose-fill insulation materials, which cannot be mounted in the ASTM E 84 apparatus without a screen or artificial supports, shall comply with the flamespread and smoke-developed limits of Sections 708.2 and 708.3.

708.5 Roof Insulation

The use of combustible roof insulation not complying with Section 708.2 or 708.3 shall be permitted in any type construction provided it is covered with approved roof coverings directly applied thereto.

708.6 Duct Insulation

Duct linings and coverings shall conform to the appropriate requirements of the Standard Mechanical Code.

708.7 Foam Plastics

Foam plastics shall comply with Section 2603.

708.8 Cellulose Fiber Thermal Insulation

Cellulose fiber thermal insulation shall be tested in accordance with and shall comply with the requirements of Sections 708.1 through 708.3, 16 CFR parts 1209 and 1404, and ASTM C 739 (see Chapter 2 for definitions). Each package of such insulating material shall be clearly labeled as meeting the requirements of the CPSC and ASTM standards.

709.0 CALCULATED FIRE RESISTANCE [703.0; 716.0]

709.1 General

709.1.1 Scope

These provisions contain procedures by which the fire resistance of specific materials or combinations of materials can be established by calculations. These procedures apply only to the information contained in this section.

709.1.2 Definitions

For the purpose of this section, certain special terms are defined as follows:

CARBONATE AGGREGATE CONCRETE - Concrete made with aggregates consisting mainly of calcium or magnesium carbonate, e.g., limestone or dolomite.

CELLULAR CONCRETE - A lightweight insulating concrete made by mixing a preformulated foam with portland cement slurry and having a dry unit weight of approximately 30 lb/cu ft (480 kg/m^3).

CERAMIC FIBER BLANKET - A mineral wool insulation material made of alumina-silica fibers and weighing 4 to 10 lb/cu ft (64 to 160 kg/m^3).

GLASS FIBER BOARD - Fibrous glass roof insulation consisting of inorganic glass fibers formed into rigid boards using a binder. The board has a top surface faced with asphalt and kraft reinforced with glass fiber.

LIGHTWEIGHT AGGREGATE CONCRETE - Concrete made with aggregates of expanded clay, shale, slag, or slate or sintered fly ash, and weighing 85 to 115 lb/cu ft (1360 to 1840 kg/m^3).

MINERAL BOARD - A rigid, felted thermal insulation board consisting of either felted mineral fiber or cellular beads of expanded aggregate formed into flat rectangular units.

PERLITE CONCRETE - A lightweight insulating concrete having a dry unit weight of approximately 30 lb/cu ft (480 kg/m^3) made with perlite concrete aggregate. Perlite aggregate is produced from a volcanic rock which, when heated, expands to form a glasslike material of cellular structure.

SAND-LIGHTWEIGHT CONCRETE - Concrete made with a combination of expanded clay, shale, slag, or slate or sintered fly ash and natural sand. Its unit weight is generally between 105 and 120 lb/cu ft (1680 and 1920 kg/m^3).

SILICEOUS AGGREGATE CONCRETE - Concrete made with normal-weight aggregates consisting mainly of silica or compounds other than calcium or magnesium carbonate.

VERMICULITE CONCRETE - A lightweight insulating concrete made with vermiculite concrete aggregate which is laminated micaceous material produced by expanding the ore at high temperatures. When added to a portland cement slurry the resulting concrete has a dry unit weight of approximately 30 lb/cu ft (480 kg/m^3).

709.2 Concrete Assemblies
709.2.1 Concrete Walls
709.2.1.1 Cast-in-Place or Precast Walls

709.2.1.1.1. The minimum equivalent thicknesses of cast-in-place or precast concrete walls for fire resistance ratings of 1 hr to 4 hr are shown in SBC Table 709.2.1.1. For solid walls with flat vertical surfaces, the equivalent thickness is the same as the thickness. The values in SBC Table 709.2.1.1 apply to plain, reinforced, or prestressed concrete walls.

709.2.1.1.2. For hollow-core precast concrete wall panels in which the cores are of constant cross section throughout the length, the equivalent thickness may be calculated by dividing the net cross-sectional area (the gross cross section minus the area of the cores) of the panel by its width.

709.2.1.1.3. Where all the core spaces of hollow-core wall panels are filled with loose-fill material, such as expanded shale, clay, or slag, or vermiculite or perlite, the fire resistance rating of the wall is the same as that of a solid wall of the same concrete type and of the same overall thickness.

709.2.1.1.4. The thickness of panels with tapered cross sections shall be that determined at a distance 2T or 6 in (152 mm), whichever is less, from the point of minimum thickness, where T is the minimum thickness.

709.2.1.1.5. The equivalent thickness of panels with ribbed or undulating surfaces shall be determined by one of the following expressions:

for $s \geq 4T$ the thickness to be used shall be T;
for $s \leq 2Tt$: the thickness to be used shall be T_e;
for $4T > s > 2T$, the thickness to be used shall be

$$T + \left(\frac{4T}{s} - 1\right)(T_e - T)$$

where s = spacing of ribs or undulations
T = minimum thickness
T_e = equivalent thickness of panel calculated as net cross sectional area of panel divided by width, in which maximum thickness used shall not exceed 2T

709.2.1.2 Multi-Wythe Walls

709.2.1.2.2. The fire resistance rating for wall panels consisting of two or more wythes may be determined by the formula:

$$R = (R_1^{0.59} + R_2^{0.59} + \ldots + R_n^{0.59})^{1.7} \qquad \text{(Eq. 709.2.1.2)}$$

where R is the fire endurance of assembly in minutes and, R_1, R_2, R_n are the fire endurances of the individual wythes in minutes. Values of $R_n^{0.59}$ for use in Eq. (709.2.1.2) are given in Table 709.2.1.2 (see Appendix).

709.2.1.2.3. The fire resistance ratings of precast concrete wall panels consisting of a layer of foam plastic insulation sandwiched between two wythes of concrete may be determined by use of Eq. (709.2.1.2). Foam plastic insulation with a total thickness of less than 1 in (25 mm) shall be disregarded. The R_n value for thickness of foam plastic insulation of 1 in or greater, for use in the calculation, is 5 min; therefore $R_n^{0.59} = 2.5$.

709.2.1.3 Joints Between Precast Wall Panels

709.2.1.3.1. Joints between precast concrete wall panels which are not insulated as required by this section shall be considered as openings in walls. Uninsulated joints shall be included in determining the percentage of openings permitted by Table 600 (see Appendix).

Where openings are not permitted or are required by code to be protected, the provisions of this section shall be used to determine the amount of joint insulation required. Insulated joints shall not be considered openings for purposes of determining compliance with allowable percentage of openings in Table 600.

709.2.1.4 Walls with Gypsum Wallboard or Plaster Finishes

709.2.1.4.1. The fire resistance rating of cast-in-place or precast concrete walls with finishes of gypsum wallboard or plaster applied to one or both sides may be calculated in accordance with the provisions of this section.

709.2.1.4.2. Where the finish of gypsum wallboard or plaster is applied to the non-fire-exposed side of the wall, the contribution of the finish to the total fire resistance rating shall be determined as follows: The thickness of the finish shall first be corrected by multiplying the actual thickness of the finish by the applicable factor determined from Table 709.2.1.4A (see Appendix) based on the type of aggregate in the concrete. The corrected thickness of finish shall then be added to the actual thickness or equivalent thickness of concrete, and the fire resistance rating of the concrete and finish shall be determined from SBC Table 709.2.1.1 or Table 709.2.1.2 (see Appendix).

709.2.1.4.3. Where gypsum wallboard or plaster is applied to the fire-exposed side of the wall, the contribution of the finish to the total fire resistance rating shall be determined as follows: The time assigned to the finish as established by Table 709.2.1.4B (see Appendix) shall be added to the fire resistance rating determined from Table 709.2.1.1 or Table 709.2.1.2 (see Appendix) for the concrete alone or added to the rating determined in Section 709.2.1.4.2 for the concrete and finish on the non-fire-exposed side.

709.2.1.4.4. For a wall having no finish on one side or having different types or thicknesses of finish on each side, the calculation procedures of Sections 709.2.1.4.2 and 709.2.1.4.3 shall be performed twice; i.e., assume that either side of the wall may be the fire-exposed side. The fire resistance rating of the wall shall not exceed the lower of the two values.

> **EXCEPTION**: For an exterior wall with more than 5 ft (1524 mm) of horizontal separation, the fire shall be assumed to occur on the interior side only.

709.2.1.4.5. When the finish applied to a concrete wall contributes to the fire resistance rating, the concrete alone shall provide not less than one-half the total required fire resistance.

709.2.1.4.6. Finishes on concrete walls which are assumed to contribute to the total fire resistance rating of the wall shall comply with the installation requirements of Section 709.3.1.6.

709.2.2 Concrete Floor and Roof Slabs
709.2.2.1 Reinforced and Prestressed Floors and Roofs
709.2.2.1.1. The minimum thicknesses of reinforced and prestressed concrete floor or roof slabs for fire resistance ratings of 1 hr to 4 hr are shown in SBC Table 709.2.2. 1.

709.2.2.1.2. For hollow-core prestressed concrete slabs in which the cores are of constant cross section throughout the length, the equivalent thickness may be obtained by dividing the net cross-sectional area of the slab, including grout in the joints, by its width.

709.2.2.1.3. The thickness of slabs with sloping soffits shall be determined at a distance 2T or 6 in (152 mm), whichever is less, from the point of minimum thickness, where T is the minimum thickness.

709.2.2.1.4. The thickness of slabs with finned or undulating soffits shall be determined by one of the following expressions, whichever is applicable.

for $s \geq 4T$ the thickness to be used shall be T;
for $s \leq 2Tt$: the thickness to be used shall be T_e;
for $4T > s > 2T$, the thickness to be used shall be

$$T + (\frac{4T}{s} - 1)(T_e - T)$$

where s = spacing of ribs or undulations
T = minimum thickness
T_e = equivalent thickness of panel calculated as net cross sectional area of panel divided by width, in which maximum thickness used shall not exceed 2T

709.2.2.2 Multicourse Floors and Roofs
709.2.2.2.1. This subsection gives information on the fire resistance ratings of floors which consist of a base slab of concrete with a topping (overlay) of a different type of concrete.
709.2.2.2.2. This subsection gives information on the fire resistance ratings of roofs which consist of a base slab of concrete with a topping (overlay) of an insulating concrete or with an insulating board and built-up roofing.
709.2.2.3 Joints in Precast Slabs
Joints between all adjacent precast concrete slabs may be ignored in calculating the slab thickness provided that a concrete topping at least 1 in (25.4 mm) thick is used. Where no concrete topping is used, joints must be grouted to a depth of at least one-third the slab thickness at the joint, but not less than 1 in (25.4 mm), or the joints must be made fire resistant by other approved methods.
709.2.3 Concrete Cover Over Reinforcement
709.2.3.1 Slab Cover
The minimum thickness of concrete cover to the positive reinforcement moment is given in SBC Tables 709.2.3A for

reinforced concrete and 709.2.3B for prestressed concrete. These tables are applicable for solid or hollow-core one-way or two-way slabs with flat undersurfaces. Slabs may be cast-in-place or precast. For precast prestressed concrete not covered elsewhere, the national building codes refer to <u>ASTM Design for Fire Resistance of Precast Prestressed Concrete</u>.

Fire-Resistant Component Systems

The new construction component systems technologies in walls, roofs, and floors have been adopted by all the national building codes as newer ones are being reviewed currently. Panelized construction offers up to 20% savings over conventional wall construction. It also offers high sound rating, easy installation, fire resistance rating, and traditional appearance. The need for specialized component systems is great among school modulars construction and manufactured housing producers. Gypsum wallboard must meet specific fire resistance standards and withstand racking forces from wind and transportation. Joint compounds and spray texture must apply and set quickly. These gypsum panels are classified by Underwriters Laboratories, Inc., for shear resistance and surface-burning characteristics. These panels are made with a cross-fiber construction to provide greater joint strength than conventional wallboard. These will be classified in detail in the upcoming revisions of the national building codes and should appear in this section.

Other new products that are components within these new panelized construction systems include aerosol polyurethane foams that are independently certified as a spray-on drywall adhesive. This is a major cost breakthrough for drywallers since one can of polyurethane foam produces the equivalent of 40 quart tubes of ordinary drywall or panel adhesive. Used with an applicator tool for quick and easy installation, polyurethane foams lower the number of screws used by one-third, decrease mudding labor, and reduce nail-pop callbacks. The product meets ASTM C 557.

Icynene is a polycynene insulation that expands 100:1 as it is sprayed to fill wall, ceiling, and floor spaces. Icynene adheres to everything it touches and minimizes air leakage by sealing gaps and perforations between building materials. The product eliminates the need for plastic vapor barriers and building wraps, as well as a lot of caulking, tapes, sealants, and foams used for other energy-efficient construction. Icynene has no detectable emissions and is recommended for use in homes for environmentally sensitive people.

Chapter 7

Interior Finishes

SECTION 801.0 INTERIOR [801.0; 801.1]

Ease of installation and maintenance, durability, and lower price have prompted builders and homeowners to use the new engineered wood products in such quantities that the national building codes are being revised to include such new advances in building materials. New technology materials like fiber-reinforced cement board have opened new methods of modern construction. Homeowners have also been spurred along by "green" ethics resulting from the scarcity of mature forests, traditional wood products, and the economy of new veneering processes. A new front door made of desirable old-growth redwood may have been made from the stump of a redwood tree logged over a hundred years ago. The natural bird's-eye cabinet in the kitchen may have come from a log which has been rotary peeled, dyed, sliced, and layered to form a block and then run through a planer to be sliced into veneers. These products have the look and feel of expensive, hard-to-find, and endangered wood, but at artificial technology prices. The result has turned out to be a more careful utilization of dwindling precious natural resources.

Even the old standard of the industry, the traditional two-by-four may not be all that it seems today. One of the most profound changes in the home building industry has been the introduction of engineered lumber products that meet structural specifications and acceptably replace solid-wood structural studs, floor joists, beams, and other structural and finish house elements that traditionally have been built with large-dimension lumber. Engineered lumber is extending the use of inferior grades of wood. Inferior-quality small trees or knotty wood are peeled, chopped, or chipped and then reassembled as I-beams, panels, and other products. The up side of this to the environment is that the limited natural timber resource of the planet is extended. Engineered wood products are taking lesser-quality wood and increasing its quality and viability through the engineering process. A majority of the interior materials are engineered wood products.

Only a few products run counter to the general trend toward highly processed materials. Linoleum, which is made from linseed oil and cork, is enjoying a minor comeback. Cotton batts (mixed with polyester to retain fluffiness) and macerated paper have made tiny inroads into the fiberglass-dominated insulation market. And cellulose, a wood by-product, is turning up as a main ingredient in drywall panels.

Even for construction management professionals, it is no longer easy to define what is natural anymore. Everything today is processed to some degree; with the decline in old-growth forests, the rise of engineered wood products was inevitable. But like every other highly processed product, engineered wood products have both pros and cons. Because engineered lumber is stronger and more consistent in quality than traditional lumber, it creates less construction waste on the job site and is produced in nominal sizing before purchase. Engineered wood can also be used to span longer distances, allowing greater design freedom.

But the national building codes have fire-safety concerns about engineered lumber because it burns much more easily and readily than solid lumber. And since petroleum-based adhesives and binders are used to hold the reassembled wood fibers and fragments together, serious questions are raised about these products off-gassing over time and reducing indoor air quality, especially in today's airtight energy-efficient homes. A few "green" builders and architects reject the new synthetic products because they are concerned about the long-term health effects of gases slowly given off by these synthetics, as well as the environmental impact of artificial materials that will be difficult to disassemble and recycle once the building's life is over.

So although the national building codes consist essentially of specific regulations on details of interior design and installation, there is occasional explanatory material in the form of notes to rules. These notes are modified during the code's revision to adapt to these new changes in building materials. Compliance with the code consists in satisfying all requirements and conditions that are stated by the use of the word <u>shall</u>. That word, anywhere in the code, designates a mandatory rule. Failure to comply with any mandatory code interior finish rule constitutes a code violation.

Gypsum board inspections are made after all lathing and gypsum board, interior and exterior, is in place, but before any plastering is applied or before gypsum board joints and fasteners are taped and finished. Final inspections are made after finish grading and the building is completed and ready for occupancy. In finish work, most contracts call for craftsmanship of a standard "workmanlike manner." This statement has been the source of many legal conflicts because opinions differ as to what is an installation according to a "workmanlike manner." The codes place the responsibility for determining what is acceptable and how it is applied by the authority having jurisdiction. This basis in most areas is the result of

- Competent knowledge and experience of installation methods
- What has been the established practice by the qualified journeyman in the particular local area
- What has been taught in the trade schools having certified training courses for apprentices and journeymen

Codes and standards must be carefully interrelated and followed with care and precision. Modern finish work that fulfills these demands should be the objective of all construction builders, designers, and inspectors. The steps in designing construction to code compliance are

1. Determine the occupancy classification for the building's design purposes.
2. Determine the fire-resistive hour rating needed for the occupancy of the building.
3. Select the types of construction that will meet the required fire rating.
4. Select the materials that will meet compliance with the types of construction required.

A good tip in keeping costs down is to remember that in 1-hr fire-rated construction, floors may be omitted under unusable space and ceilings may be omitted over unusable space.

801.1 Scope
Provisions of this chapter shall govern the use of materials as interior finishes by limiting the allowable flamespread and smoke development based on location and occupancy classification.

803.0 RESTRICTIONS ON INTERIOR FINISHES [800.0; 801.1]

803.1.1. Combustible materials may be used as a finish for ceilings, floors, and other interior surfaces of buildings as provided in this section. Show windows in the first story of buildings may be of wood or of unprotected metal framing.

803.1.2. Interior finish shall mean the exposed interior surfaces of buildings including, but not limited to, fixed or movable walls and partitions, columns, and ceilings, interior wainscoting, paneling, or other finish applied structurally or for decoration, acoustical correction, surface insulation, structural fire-resistance, or similar purposes.

Requirements for finishes shall not apply to trim, defined as picture molds, chair rails, baseboards, and handrails; to doors and windows or their frames; nor to materials which are less than 1/28 in (0.9 mm) thick that are cemented to the surface of walls or ceilings, when these materials have flamespread characteristics no greater than paper of this thickness cemented to a noncombustible or fire-retardant-treated wood backing.

803.2 Classification
Interior finish materials other than those applied to floors shall be classified in accordance with ASTM E 84. Such interior finish materials shall be grouped in the following classes in accordance with their flamespread and smoke development:

1. Class A interior finish. Flamespread 0 to 25, smoke development 0 to 450. Any element thereof when so tested shall not continue to propagate fire.
2. Class B interior finish. Flamespread 26 to 75, smoke development 0 to 450.
3. Class C interior finish. Flamespread 76 to 200, smoke development 0 to 450.

Interior Finishes

803.3 Interior Finish Requirements Based on Occupancy

803.3.1. The minimum flamespread classification of interior finish other than floor finish and floor coverings shall be based on the use or occupancy as set forth in Table 803.3.

EXCEPTIONS:
1. Except in Group I Occupancies and in enclosed vertical exits, Class C interior finish material may be used in access to exits and other spaces as wainscoting extending not more than 48 in (1219 mm) above the floor and for tack and bulletin boards covering not more than 5% of the gross wall area of the room. In Group I Occupancies, Class B interior finish material may be used in access to exits as wainscoting extending not more than 48 in (1219 mm) above the floor.
2. The exposed faces of Type III structural members, including decking and planking, where otherwise permitted by code, are excluded from flamespread requirements.

Table 803.3
Minimum Interior Finish Classification

Occupancy	Unsprinklered			Sprinklered		
	Exits[1]	Exits Access	Other Spaces	Exits[1]	Exit Access	Other Spaces
A	A	A	B	B	C	C
B	B	B	C	C	C	C
E	A	B	C	B	C	C
F	C	C	C	C	C	C
H	*Sprinklers required*			B	C	C
I Restrained	A	A	C	A	A	C
I Unrestrained	*Sprinklers required*			B	B	B3
M	B	B	C	C	C	C
R2	B	B	C	C	C	C
S	C	C	C	C	C	C

Notes:
1. In vertical exits of buildings three stories or less in height of other than Group I Restrained, the interior finish may be Class B for unsprinklered buildings and Class C for sprinklered buildings.
2. Class C interior finish materials may be used within a dwelling unit.
3. Rooms with occupancy of four or less persons require Class C interior finish.

803.3.2. For churches or places of worship, nothing in this section shall prevent the use of wood for ornamental purposes, trusses, paneling, or chancel furnishing.

803.3.3. Imitation leather or other material consisting of, or coated with, a pyroxylin or similarly hazardous base shall not be used in Group A Occupancies.

803.4 Foam Plastics

Foam plastics shall not be used as interior finish.

> **EXCEPTION**: Foam plastic trim, defined as picture molds, chair rails, baseboards, handrails, ceiling beams, door trim, and window trim shall be permitted to be used provided:
> 1. The minimum density is 20 lb/cu ft (320 kg/m^3).
> 2. The maximum thickness of the trim is 1/2 in (12.7 mm) and the maximum width is 4 in (102 mm).
> 3. The trim constitutes no more than 10% of the area of any wall or ceiling.
> 4. The flamespread rating does not exceed 75 when tested per ASTM E 84. The smoke-developed rating is not limited.

803.5 Carpet on Walls and Ceilings

803.5.1. Textile materials having a napped, tufted, looped, woven, nonwoven, or similar surface may be used as interior finish on ceilings only when said materials have a flamespread rating of 25 or less in accordance with ASTM E 84.

803.5.2. Textile wall coverings, including materials such as those having a napped, tufted, looped, nonwoven, woven, or similar surface shall comply with one of the following:

1. Textile wallcoverings shall have a flamespread rating of 25 or less in accordance with ASTM E 84 and shall be protected by automatic sprinklers.
2. Textile wallcoverings shall meet the acceptance criteria of SBC Standard Test Method for Evaluating Room Fire Growth Contribution of Textile Wallcovering when tested using the product mounting system, including adhesive, of actual use.

[Both UBC and BOCA have similar flamespread tests.]

803.6 Floor Finish

803.6.1. In buildings of Type I or Type II construction, floor finish, if of combustible material, shall be applied directly upon the floor construction, except that a floor finish of wood, linoleum, rubber, tile, or cork may be secured to a subfloor of wood. Where wood sleepers are used for laying wood floors or subfloors in such buildings, they shall be fireblocked so that there will not be an open space extending under any permanent partition.

Where wood sleepers are used and the space between the floor slab and the underside of the floor or subfloor is more than 2-1/2 in (64 mm), such space shall be filled with noncombustible material so that such space is not more than 2-1/2 in (64 mm).

803.6.2. Combustible insulating boards may be used for sound deadening or insulating of floors, except that in buildings required to be of Type I or Type II construction, such insulating board shall not be more than 1/2 inch (12.7 mm) thick and shall be cemented directly to the floor slab or secured to wood sleepers fireblocked as called for above and covered with approved finish flooring.

803.7 Floor Covering

803.7.1. Finished floors or floor covering materials of a traditional type, such as wood, vinyl, linoleum, terrazzo, and other resilient floor covering materials, are exempt from the requirements of this section. Carpet-type floor coverings shall be tested as proposed for use including underlayment.

803.7.2. Carpet materials used on floors of exit access corridors and enclosed exits in other than Group I Occupancies shall satisfactorily withstand a minimum critical radiant flux of 0.22 W/cm^2 when tested in accordance with the NFiPA 253.

EXCEPTION: Buildings equipped with an approved automatic sprinkler system.

803.7.3. Interior floor finish materials used on the floor of exit access corridors and enclosed exits in Group I Occupancies shall satisfactorily withstand a minimum critical radiant flux of 0.45 W/cm^2 when tested in accordance with NFiPA 253.

803.7.4. All carpet required by code to meet critical radiant flux limits established by NFiPA 253 shall have been tested by an approved laboratory. A copy of the test report representing the style shall be provided to the building official upon request. The test report shall identify the carpet by manufacturer or supplier and style name and shall be representative of the current construction of the carpet.

803.7.5. The carpet shall be identified by a handtag or other suitable method to indicate the manufacturer or supplier, the style, and the critical radiant flux level.

803.8 Application of Interior Finish

803.8.1. When walls and ceilings are required by any provision in code to be of fire-resistant, noncombustible, or fire-retardant-treated wood construction and the finish material is applied to furring strips not exceeding 1-3/4 in (44 mm) thick applied directly against such surfaces, the intervening spaces between such furring strips shall be filled with inorganic or Class A materials or shall be fireblocked not to exceed 8 ft (2438 mm) in any direction.

803.8.2. Where walls and ceilings are required to be of fire-resistant, noncombustible, or fire-retardant-treated wood construction and walls are set out or ceilings are dropped distances greater than specified in Section 803.8.1, Class A finish materials shall be used except where the finish materials are protected on both sides by automatic fire extinguishing systems or are attached to a noncombustible or fire-retardant-treated wood backing or to furring strips installed as specified in Section 803.8.1.

The hangers and assembly members of such dropped ceilings that are below the main ceiling line shall be of noncombustible or fire-retardant-treated wood materials.

803.8.3. Wall and ceiling finish materials of all Class A, B, or C materials, as permitted, may be installed directly against the wood decking or planking of heavy timber construction or to wood furring strips applied directly to the wood decking or planking installed and fire-stopped as specified in Section 803.8.1.

803.8.4 Interior finish materials shall be cemented or otherwise fastened in place so that they will not readily become detached when subjected to room temperatures of 300°F (149°C) for 25 min.

803.9 Interior Plastic Signs
Applications using approved plastic interior signs shall comply with SBC section 2604.15. Applications using approved plastic interior signs in covered mall buildings shall comply with Section 413.13.

804.0 ACOUSTICAL CEILING SYSTEMS [803.1; 804.1]
804.1 General
The quality, design, fabrication, and erection of metal suspension systems for acoustical tile and lay-in panel ceilings in buildings or structures shall conform to good engineering practices, the provisions of this chapter, and other applicable requirements of code.

804.2 Materials and Installations
804.2.1. Acoustical materials complying with the interior finish requirements of Section 803 shall be installed in accordance with the manufacturer's recommendations and applicable provisions for applying interior finish.

804.2.2. Suspended acoustical ceiling systems shall be installed in accordance with the provisions of ASTM C 635 and ASTM C 636.

804.2.3. Acoustical ceiling systems which are part of a fire-resistant construction shall be installed in the same manner used in the assembly tested and shall comply with the provisions of Section 701.0 of code. If the weight of lay-in ceiling panels, used as a part of fire-resistant floor-ceiling or roof-ceiling assemblies, is not adequate to resist an upward force of 1 psf (48 Pa), wire or other approved devices shall be installed above the panels to prevent upward displacement under such upward force.

Interior Design Matrix

The most sophisticated (and currently emerging) interior construction design systems incorporate matrix networking, in which specialized materials, components and production phases are utilized in a variety of project-oriented configurations in computer assisted design (CAD). The builder works with small teams from the various areas of specialization in the project, toward common ends of interlinking completion events. This system is the computer evolution of the old handwritten finish schedules of the past.

This is a highly responsive and event-oriented operations system, producing a matrix network that will suffice as divisional factors in the project's interior design, thereby reducing the time required to manage multiple production schedules. This is also one of the basic elements in fast-tracking construction of franchise anchor stores. However, matrix networking requires sophisticated manufacturers and builders with skills in cross-referencing the national building codes for different locales. Design matrix networking is a process of determining:

- The interior tasks and activities to be performed
- The time, in events, to be devoted to each
- The job logic sequence of each interior activity
- An optimistic and pessimistic completion time frame

This type of project interior design and production controlling is an area of opportunity for most small- and medium-sized projects running budgets of 10 to 50 million dollars [that is a medium-sized construction project in California, according to the Small Business Administration (SBA) statistics]. The development of cost-efficient processes for assessing national building codes compliance, materials performance expectations, production measured by milestone indications of success, coupled with remote (field) data collection systems linked to a host processing computer back at the office, is fast becoming the difference between success and failure in the builder's business.

Chapter 8

Fire Protection Systems

SECTION 901.0 FIRE PROTECTION [901.0; 901.0]

Fire protection standpipe systems are wet or dry systems of piping, valves, outlets, and related equipment designed to provide water at specified pressures and installed exclusively for the fighting of fires. The national building codes' definitions of the different classifications of the types of systems are as follows:

- **Class I.** A standpipe system equipped with 2-1/2 in outlets.
- **Class II.** A wet standpipe system directly connected to a pressurized water supply and equipped with 1-1/2 in outlets and hose.
- **Class III.** A standpipe system directly connected to a pressurized water supply and equipped with 2-1/2 in outlets or 2-1/2 in and 1-1/2 in outlets when a 1-1/2 in hose is required. Hose connections for Class III systems may be made through 2-1/2 in hose valves with easily removable 2-1/2 in by 2-1/2 in reducers.

Fire doors must be installed with automatic or self-closure hardware which is equipped with heat-actuated devices on each side of the wall at the top of the doorway opening. If the ceiling is more than 3 ft above the opening, a fusible link must be installed and located at the ceiling on each side of the wall. Signs required by the Fire Code must be attached to all fire doors stating: "Fire door - Do not block." Glass or glazed openings of 100 sq in maximum are permitted if they are heat tempered glass or wireglass.

901.1 Scope

Provisions of this chapter shall govern the application, design, installation, testing, and maintenance of automatic sprinklers, standpipes, and fire alarms.

902.0 DEFINITIONS [902.0; 902.1]

The following words and terms shall, for the purposes of this chapter and as stated elsewhere in code, have the meanings shown herein. Refer to Chapter 2 for general definitions.

ALARM-INDICATING APPLIANCE - An electromechanical appliance that converts energy into audible or visible form for perception as an alarm signal.

FIRE ALARM BOX, MANUAL - A manually operated, alarm-initiating device that activates a fire-protective signaling system.

FIRE-DETECTION SYSTEM, AUTOMATIC - A fire-protective signaling system containing automatic detecting devices that activate a fire alarm signal.

FIRE-PROTECTIVE SIGNALING SYSTEM - Electrically operated circuits, instruments, and devices, together with the necessary electrical energy, designed to transmit alarms, supervisory, and trouble signals necessary for the protection of life and property.

SMOKE DETECTOR - An approved listed detector sensing either visible or invisible particles of combustion.

SMOKE DETECTOR, MULTIPLE-STATION - Single-station smoke detectors which are capable of being interconnected such that actuation of one causes all integral or separate audible alarms to operate.

SMOKE DETECTOR, SINGLE-STATION - An assembly incorporating the detector, control equipment, and alarm-sounding device in one unit, which is operated from a power supply either in the unit or obtained at the point of installation.

STANDPIPE - An arrangement of piping, valves, hose outlets, and allied equipment installed in a building or structure with outlets located in such a manner that water can be discharged through the hose and nozzles for the purpose of extinguishing a fire. Standpipes are classified as one of four types as follows:

> **Class 1**. For use by fire departments and those trained in handling heavy fire streams [2-1/2 in (64 mm) hose].
>
> **Class 2**. For use primarily by the building occupants until the arrival of the fire department [1-1/2 in (38 mm) hose].

Class 3. For use by either fire departments and those trained in handling heavy hose streams [2-1/2 in (64 mm) hose] or by the building occupants [1-1/2 in (38 mm) hose].
Combined systems. One where the water piping serves both 2-1/2 in (64 mm) outlets for fire department use and outlets for automatic sprinklers.

STANDPIPE, DRY - A system having no permanent water supply. A filled standpipe having a small water supply connection to keep the piping full but requiring water to be pumped into the system shall be considered a dry standpipe.

STANDPIPE, WET - A system having the supply valve open and water pressure maintained at all times.

903.0 SPRINKLERS [904.0; 904.1]
903.1 Approved Equipment and Layout
Only approved sprinklers and devices shall be used in automatic sprinkler systems, and the complete layout of the system shall be submitted to the building official for approval before installation.

903.2 Requirements
Every automatic sprinkler system required by code shall conform to NFiPA 13, as modified by NFiPA 231 and NFiPA 23 IC, except that a single water supply of adequate pressure, capacity, and reliability, equal to the primary supply required by those standards, may be permitted by the building official. Automatic sprinkler systems installed in lieu of or as an alternate to other requirements, as permitted by code, shall be considered required systems and shall comply with NFiPA 13.

903.3 Material
Piping shall be as specified in NFiPA 13.

903.4 Hose Threads
All hose threads in connections shall be uniform with that used by the fire department of the applicable governing body.

903.5 General
Approved automatic sprinkler equipment meeting the requirements of Section 903 shall be installed in buildings as follows:

1. Basements having floor areas exceeding 2500 sq ft (232 m^2) when used as workshops or for manufacture, repair, sale, or storage of combustible materials or when used as

lounges or nightclubs regardless of the size. See Section 503.4.1, Exception 2.
2. In buildings which do not have suitable access, as set forth in Section 1405, to each story above grade on at least one accessible side of the building. Openings which are glazed with security glazing designed to withstand breakage shall not be considered as access openings.
3. See chapter 36 of the Standard Fire Prevention Code.
4. See SBC sections 407.1.3, 411.7.6, and 411.7.7.
5. The spray finishing booth, area, or room shall comply with chapter 10 of the Standard Fire Prevention Code.

903.6 Garages

Approved automatic sprinkler systems shall be provided in the following garages:
1. Enclosed parking garages over 65 ft (19.8 m) high and exceeding 10,000 sq ft (929 m²) per floor.
2. Repair garages two stories or more high and exceeding 10,000 sq ft (929 m²) in a single floor area.
3. One-story repair garages exceeding 15,000 sq ft (1394 m²).
4. Basement garages or repair garages in a basement.
    EXCEPTION: Group R3 occupancies.
5. Garages used for the storage of commercial trucks and having an area exceeding 5000 sq ft (465 m²).
6. Bus garages when used as passenger terminals for four or more buses or when used for bus storage or loading of four or more buses.

903.7 Other Occupancy Sprinkler Requirements

903.7.1 Group M

An approved automatic sprinkler system shall be provided in stores and similar occupancies where stocks of combustible materials are on display for public sale and where the story floor area exceeds 15,000 sq ft (1394 m²).

903.7.2 Group A

903.7.2.1. An approved automatic sprinkler system shall be provided in Group A-1 Occupancies over areas which could be used for the display, sale, or storage of combustible materials when such display, sale, or storage floor area exceeds 15,000 sq ft (1394 m²).

903.7.2.2 Stages shall be provided with an approved automatic sprinkler system. Such sprinklers shall be provided throughout the

stage and in dressing rooms, workshops, storerooms, and other accessory spaces contiguous to such stages.
EXCEPTIONS:
1. Sprinklers are not required where stages are 1000 sq ft (93 m²) or less in area and 50 ft (15.2 m) or less in height and curtains, scenery, or other combustible hangings are not retractable vertically. Combustible hangings shall be limited to a single main curtain, borders, legs, and a single backdrop.
2. Sprinklers are not required under stage areas less than 4 ft (1219 mm) in clear height used exclusively for chair or table storage and lined on the inside with 5/8 in (15.9 mm) Type X gypsum wallboard or approved equal.

903.7.2.3. Buildings or portions thereof used for the specific purpose of sound stages for motion picture or television productions and greater than 1000 sq ft (93 m²) shall be protected with an approved automatic sprinkler system.

903.7.3 High-Piled Combustible Stock
An approved automatic sprinkler system shall be provided throughout buildings required to have sprinkler protection by chapter 36 of the Standard Fire Prevention Code.
> **EXCEPTION**: Automatic sprinkler systems may be provided only in the storage area of the building when the storage is separated from the remainder of the building by a minimum 2-hr fire-resistant separation.

903.7.4 Hazardous Production Material (HPM) Facility
An approved automatic sprinkler system shall be provided throughout buildings containing Group H (HPM) facilities as defined in SBC section 408, shall be designed in accordance with NFiPA 13, and shall not be less than that required for the special fire hazard areas shown in Table 903.7.4.

Table 903.7.4
Hazardous Production Material Facilities

Special Fire Hazard Area Requirements

Location	NFIPA Hazard Group
Fabrication areas	Ordinary Hazard Group 3
HPM service corridors	Ordinary Hazard Group 3
HPM separate inside storage rooms without dispensing	Ordinary Hazard Group 3
HPM separate inside storage rooms with dispensing	Extra Hazard Group 2
Exit access corridors	Ordinary Hazard Group 3

903.7.5 Group R1 - Residential Occupancy
An approved automatic sprinkler system shall be provided throughout Group R1 Occupancies that are three or more stories in height.
> **EXCEPTION**: An automatic sprinkler system shall not be required when exterior exit stairs complying with Section 1006.2 are provided for the guest rooms.

903.7.6 Group R2 - Residential Occupancy
An approved automatic sprinkler system shall be provided throughout Group R2 Occupancies that are three or more stories in height.
> **EXCEPTION**: Three-story buildings which are not required to have an automatic sprinkler system by other provisions of the code and provided with exterior exit stairs complying with Section 1006.2.2. An automatic

sprinkler system complying with NFiPA 13R shall be permitted for buildings not exceeding four stories in height provided the automatic sprinkler system shall not be considered as an alternate to other requirements of the code. See Section 903.2.

903.8 Supervision

903.8.1. Where an automatic sprinkler system is provided either as a requirement or as an alternate to another requirement of code, the system shall be supervised by one of the following methods:
- Approved central station system in accordance with NFiPA 71
- Approved proprietary or remote station system, or an approved supervisory service in accordance with NFiPA 72, which will cause the actuation of an audible appliance at a constantly attended location

> **EXCEPTION**: Supervisory facilities in accordance with Section 903.8 shall not be required for extinguishing systems in one-and two-family dwellings.

903.8.2. In HPM facilities, as defined in SBC section 408, all valves shall be provided with supervisory tamper switches. In addition to the requirements of Section 903.8.1, the closing of a valve shall activate an audible and visual signal at the emergency control station.

903.8.3. When a building's fire-protective signaling system is provided, actuation of the sprinkler system shall cause the building alarm to sound.

904.0 STANDPIPES [903.0; 903.1]

904.1 Approval

904.1.1. Unless otherwise provided herein, standpipe system design, installation, and testing requirements shall comply with NFiPA 14.

904.1.2 The complete layout of the standpipe and hose system shall be submitted to the building official before installation.

904.2 Type of System

Standpipe systems shall be one of the following types:
1. Wet standpipe system having supply valve open and water pressure maintained at all times

2. Dry standpipe system so arranged through the use of approved devices as to admit water to the system automatically by opening a hose valve
3. Dry standpipe system arranged to admit water to the system through manual operation of approved remote-control devices located at each hose station

904.3 Where Required
904.3.1. Standpipes shall be provided in all buildings in which the highest floor is greater than 30 ft (9144 mm) above the lowest level of fire department vehicle access.

EXCEPTION: Standpipes are not required in Group R3 buildings.

904.3.2. Stages greater than 1000 sq ft (93 m^2) in area shall be provided with a standpipe on each side of the stage.
904.3.3. Covered malls shall be provided with standpipe connections in accordance with SBC section 413.9.
904.3.4. Standpipes shall be provided in public assembly halls more than 5000 sq ft (165 m^2) in area used for exhibition or display purposes.
904.3.5. Standpipes shall be provided in nonsprinklered Group A buildings having an occupant load exceeding 1000 persons.
904.3.6. Standpipes shall be provided in buildings in which the highest floor is 30 ft (9144 mm) or less above the lowest level of fire department vehicle access and exceeding 10,000 sq ft (929 m^2) in area per story when any portion of the building's interior area is more than 200 ft (61 m) of travel from the nearest point of fire department vehicle access.

EXCEPTIONS:
1. Standpipes are not required in Group R2 Occupancies with 8 units or less, or in R3 and S2 Occupancies.
2. Standpipes are not required in buildings protected throughout with automatic sprinklers installed in accordance with NFiPA 13.

904.4 Class and Type System
904.4.1. Standpipes required by Sections 904.3.1 and 904.3.5 shall be Class 1 wet standpipes.

EXCEPTIONS:
- Buildings without approved automatic sprinkler protection shall be permitted to use any Class 1 standpipe system listed

in Section 904.2 if the highest floor surface used for human occupancy is 75 ft (22.9 m) or less above the lowest level of fire department vehicle access
- Buildings protected with an approved automatic sprinkler system shall be permitted to use any Class I standpipe system listed in Section 904.2 if the highest floor surface used for human occupancy is 75 ft (22.9 m) or less above the lowest level of fire department vehicle access

904.4.2. Standpipes required by Section 904.3.2 shall be Class 3 wet standpipes.

904.4.3. Standpipes required by Section 904.3.4 shall be Class 2 wet standpipes.

904.4.4. Standpipes required by Section 904.3.6 shall be Class 1 dry filled standpipes having a small water supply connection to keep the piping full but requiring water to be pumped into the system for fire-fighting purposes or any Class 1 standpipe permitted by Section 904.2.

904.4.5. In buildings requiring standpipes in accordance with Section 904.3, dry standpipes having no permanent water supply may be installed when, in the opinion of the building official and the fire official, a constant and automatic water supply is not necessary.

904.5 Hose Connection Location

904.5.1. A standpipe hose connection shall be located at each floor level at every exit stairway and on each side of the wall adjacent to the exit opening of a horizontal exit. In nonsprinklered buildings, additional standpipe hose connections shall be provided to limit the travel distance to the nearest hose connection to 150 ft (45.7 m). For large open area buildings, refer to Section 904.5.2.

904.5.2. Hose connections for standpipes required by Section 904.3.6 shall be located such that all portions of the building are within 150 ft (45.7 m) of a hose connection or 200 ft (61 m) from the nearest point of fire department vehicle access.

904.6 Standpipes During Construction

See Section 3311.3 for requirements.

904.7 Supervisory Facilities

Where a building's fire-protective signaling system is provided, the closing of any standpipe water supply valve, including any valves associated with a fire pump installation, shall cause an audible supervisory signal to sound at the fire alarm annunciator or at a constantly attended location. If the building does not have a fire-protective signaling system, locks shall be provided on all valves and shall be of a type acceptable to the building official.

904.8 Water Supply

904.8.1. Standpipe piping may be used to supply water for automatic sprinkler systems.

904.8.2. For nonsprinklered buildings, and sprinklered buildings having floor surfaces used for human occupancy located more than 75 ft (22.9 m) above the lowest level of fire department vehicle access, the water supply shall meet the requirements of NFiPA 14.

904.8.3. For sprinklered buildings having floor surfaces used for human occupancy located 75 ft (22.9 m) or less above the lowest level of fire department vehicle access, the required water supply shall meet minimum water pressure requirements of NFiPA 13 and shall be:

1. Five hundred gallons per minute (31.6 L/s) for light hazard occupancy as defined in NFiPA 13.
2. One thousand gallons per minute (63.1 L/s) for ordinary hazard occupancy as defined in NFiPA 13.
3. In no case shall the water supply be less than the automatic sprinkler demand including hose stream allowance.

904.9 Signs

If control valves are located in a separate room, a sign shall be provided on the entrance door. The lettering shall be a contrasting color at least 4 in (102 mm) high and shall read: <u>Standpipe Control Valve</u>.

905.0 FIRE ALARMS [403.5; 904.0]

905.1 Manual Fire Alarm Systems

905.1.1. A fire-protective signaling system in accordance with NFiPA 72 shall be installed in all the following occupancies:
- Group A having an occupant load of 1000 persons or more
- Group B having an occupant load of 500 or more persons or more than 100 persons above or below the street floor
- Group E
- Group F being two stories or more in height and having an occupant load of 500 or more persons above or below the street floor level
- Group H

- Group I
- Group M having an occupant load of 500 or more persons or more than 100 persons above or below the street floor level
- Group R
- R1   Occupancies having accommodations for more than 15 guests
- R2 apartment houses that are four or more stories in height, and dormitories or rooming houses having more than 15 sleeping accommodations

    EXCEPTION: Where each guest room has a direct exit to the outside of the building and the building is three stories or less in height.

905.1.2. Manual fire alarm boxes shall be located not more than 5 ft (1524 mm) from the entrance to each exit. Except in Group I Occupancies, the manual fire alarm boxes required in Section 905.1 may be omitted in buildings equipped with an automatic fire-detection or automatic sprinkler system covering all areas. Actuation of the automatic fire-detection or automatic sprinkler system shall activate the fire-protective signaling system. Provisions shall be made to manually activate the fire-protective signaling system at a minimum of one centrally located station.

905.1.3. Each floor shall be zoned separately. No one zone may exceed 15,000 sq ft (1394 m²). A zone indicator panel shall be located at grade level at the normal point of fire department access or at a constantly attended building security control center.

   EXCEPTION: Automatic sprinkler system zones shall not exceed the area permitted by NFiPA 13.

905.1.4. Alarm-indicating appliances listed for the purpose shall be provided. Visible and audible alarm-indicating appliances shall be provided in occupancies housing the hearing impaired. Audible alarm-indicating appliances shall provide a distinctive sound which shall not be used for any purpose other than that of a fire alarm. Such devices shall provide a sound pressure level of 15 decibels actual dBA above the average ambient sound level in every occupied space within the building.

The minimum sound pressure levels shall be: 70 dBA in buildings of Group R Occupancy, 90 dBA in mechanical equipment rooms, and 60 dBA in all other occupancy classifications. The maximum sound pressure level for audible alarm-indicating appliances shall not exceed 130 dBA at the minimum hearing distance from the audible appliance. Visible alarm-indicating appliances, where required, shall be so located as to notify all occupants in every occupied space within the building.

905.1.5. Upon completion of the fire-protective signaling system, all alarm-initiating devices and circuits, alarm-indicating appliances and circuits, supervisory signal-initiating devices and circuits, signaling line circuits, and primary and secondary power supplies shall be subjected to a 100% acceptance test in accordance with NFiPA 72 and 72E.

905.1.6. The alarm-indicating appliances shall be automatically activated by all the following where provided:

1. Smoke detectors, other than single-station smoke detectors, as required by Section 905.2. Activation of the fire-protective signaling system by smoke detectors shall be by either two cross-zoned smoke detectors within a single protected area or a single smoke detector monitored by an alarm verification zone or an approved equivalent method.
2. Sprinkler water-flow devices.
3. Manual fire alarm boxes.
4. Other approved types of automatic fire-detection device suppression systems.

905.1.7. Required fire-protective signaling systems shall include visible alarm-indicating appliances in public and common areas.

905.2 Automatic Fire Detection

905.2.1. Approved single-station or multiple-station smoke detectors shall be installed in accordance with NFiPA 74 within every dwelling; every dwelling unit within an apartment house, condominium, or townhouse; and every guest or sleeping room in a motel, hotel, or dormitory. Where more than one detector is required to be installed within an individual dwelling unit, the detectors shall be wired in such a manner that the actuation of one alarm will actuate all the alarms in the individual unit.

905.2.2. In dwellings and dwelling units, a smoke detector shall be mounted on the ceiling or wall at a point centrally located in the corridor or area giving access to each group of rooms used for sleeping purposes. Where the dwelling or dwelling unit contains more than one story, detectors are required on each story including basements, but not including uninhabitable attics, and shall be located in close proximity to the stairway leading to the floor above.

905.2.3. In dwelling units with split levels and without an intervening door between the adjacent levels, a smoke detector installed on the upper level shall suffice for the adjacent lower level provided that the lower level is less than one full story below the upper level.

905.2.4. Smoke detectors connected to a fire protective signaling alarm system shall be installed in accordance with NFiPA 72 and 72E.

905.2.5. In dwelling units, smoke detectors shall be hardwired into an alternating current (ac) electrical power source and shall be equipped with a monitored battery backup in all new construction. A monitored battery power source shall be permitted in existing construction.

Chapter 9

Means of Egress

SECTION 1001.0 EGRESS [1001.0; 1001.0]

Exiting, or egress, from a building is of prime concern to the national building codes since an emergency situation in a building with a congestion of people without rapid means of escape may lead to injury or death. Therefore the reader will find that the means-of-egress chapters in all the national building codes are extensive and quite comprehensive. All exit doors are required to be a minimum of 20 sq ft which measures out to 3 ft 0 in by 6 ft 8 in for a standard exit doorway. All escape or rescue windows must have a minimum net clear openable area of 5.7 sq ft. The minimum net clear openable height dimension must be 24 in.

The number of exits required from any story of a building is determined by using the occupant load of that story plus the percentages of the occupant loads of floors which exit through the level under consideration. Every building or usable portion of a building must have at least one exit. The size and number of required exits of a building or structure is dependent upon the number of people that will occupy the specific spaces within the building or structure.

Occupant load is factored from minimum egress and access requirements for the building. To determine the occupant load permitted for the building or structure, divide the floor area by the square feet per occupant required in occupant load tables. Each occupancy has a different rated usage, but the national building codes factor the usage of a building or structure as if the total occupant load will be full at all times. Even though this may not always be the case, the codes must assume full usage of a building to ensure public safety in the event of an emergency with the building full of people.

When windows are provided as a means of escape or rescue, they must have a finished sill height not more than 44 in above finished floor. Bars, grilles, grates, or other similar security devices may be installed on emergency escape or rescue windows or doors provided the devices are equipped with approved release mechanisms which are openable from the inside without the use of a key or special knowledge or effort, and the building is equipped with smoke detectors that meet code compliance.

Egress Terms and Definitions

EXIT - A continuous and unobstructed means of egress to a public way and includes intervening aisles, doors, doorways, gates, corridors, exterior exit balconies, ramps, stairways, smokeproof enclosures, horizontal exits, exit passageways, and exit courts and yards. Elevators or escalators cannot be used as a required exit. Exit doors must be clearly marked as such.
EXIT COURT - A yard or court providing access to a public way for one or more required exits.
EXIT ENCLOSURE - A fire-safe exit passageway, usually an emergency exit. If a floor is more than 75 ft (262.5 m) above the finished grade, one of the required exits must be a smokeproof exit enclosure.
EXIT PASSAGEWAY - An enclosed exit connecting a required or exit court with a public way.
EXIT POSITIONING - If only two exits are required, they must be placed a distance apart equal to not less than one-half of the length of the maximum overall diagonal dimension of the building or area to be served measured in a straight line between exits. The exception to this is that exit separation may be measured along a direct line of travel within the exit corridor when exit enclosures are provided as a portion of the required exit and are interconnected by a 1-hr fire-resistive corridor conforming to the fire code. Enclosure walls cannot be less than 30 ft (105 m) apart at any point in a direct line of measurement.

Where three or more exits are required, at least two exits must be placed a distance apart equal to not less than one-half of the length of the maximum overall diagonal dimension of the building or area to be served, measured in a straight line between exits, and the additional exits must be arranged a reasonable distance apart so that if one becomes blocked, the others will be available for egress.

EXIT WIDTH - The total width of exits in inches which cannot be less than the total occupant load served by an exit multiplied by 0.3 for stairways and 0.2 for other exits nor less than specified elsewhere by the adopted code. The widths of exits must be divided approximately equally among the separate exits.

HORIZONTAL EXIT - An exit from one building into another building on approximately the same level or through or around a wall constructed as required for a 2-hr rated occupancy separation and which completely divides a floor into two or more separate areas so as to establish an area of refuge affording safety from fire or smoke coming from the area from which escape is made.

PUBLIC WAY - Any street, alley, or similar parcel of land essentially unobstructed from the ground to the sky which is deeded, dedicated, or otherwise permanently appropriated for public use and having a clear width of not less than 10 ft.

SPECIAL DOORS - Revolving, sliding, and overhead doors cannot be used as required exits unless they meet special requirements set forth in the code such as emergency release push panels. Power-actuated doors which comply with emergency egress requirements of the adopted code can be used for exit purposes.

Such doors when swinging must have two guide rails installed on the swing side projecting out from the face of the door jambs for a distance not less than the widest door leaf. Guide rails cannot be less than 30 in (1 m) with solid or mesh panels to prevent penetration into the door swing and must be capable of resisting a horizontal load at the top of the rail of not less than 50 lb per lineal foot. These guide rails are required to prevent people from walking into the swinging doors.

1001.1 Scope
Provisions of this chapter shall govern the design, construction, and arrangement of elements to provide a safe means of egress from buildings and structures.
1001.1.2. In every building hereafter elected, the means of egress shall comply with the minimum requirements of this chapter.
1001.1.3. Means of egress shall consist of continuous and unobstructed paths of travel to the exterior of a building. Means of

egress shall not be permitted through kitchens, closets, restrooms, and similar areas nor through adjacent tenant spaces.

> **EXCEPTION**: Means of egress shall be permitted through a kitchen area serving adjoining rooms constituting part of the same dwelling unit or guest room.

1001.1.4. When unusually hazardous conditions exist, the building official may require additional means of egress to assure the safety of the occupants.

1001.2 Alterations
A building shall not hereafter be altered so as to reduce the capacity of the means of egress to less than required by this chapter nor shall any change of occupancy be made in any building unless such building conforms with the requirements of this chapter.

1001.3 Exit Construction
Stairways, ramps, and passageways used for required exits shall be of noncombustible construction except where otherwise specifically permitted by Sections 1007, 1010, and 1013.

1002.0 DEFINITIONS [1002.0; 1002.0]
The following words and terms shall, for the purposes of this chapter and as stated elsewhere in code, have the meanings shown herein. Refer to Chapter 2 for general definitions.

ALTERNATING TREAD STAIRWAY - A stairway having a series of steps between 50^0 0.87 radian (rad) and 70^0 (1.22 rad) from horizontal, usually attached to a center support rail in an alternating manner so that the user never has both feet on the same level at the same time. The initial tread of the stairway begins at the same elevation as the platform, landing, or floor surface.

DOOR BALANCED - A door equipped with double-pivoted hardware so designed as to cause a semi-counterbalanced swing action when opening.

ILLUMINATION UNIFORMITY RATIO - The illumination uniformity ratio as determined by the following formula: Maximum illumination at any point divided by minimum illumination at any point.

PANIC HARDWARE - A door latching assembly incorporating a device which releases the latch upon the application of a force in the direction of exit travel.

1003.0 OCCUPANT LOAD AND MEANS OF EGRESS CAPACITY [1003.0; 1003.0]

1003.1 Occupant Load

For determining the means of egress required, the minimum number of persons for any floor area shall in no case be taken to be less than specified in Table 1003.1.

1003.1.2. The occupant load of any occupancy may be determined in accordance with Table 1003.1 when the necessary aisles and means of egress are provided as approved by the building official. An aisle, egress and seating diagram shall be provided to the building official to substantiate the occupant load.

Table 1003.1
Minimum Occupant Load

Use	Area per Occupant, sq ft
Assembly without fixed seats	
Concentrated (includes among others, auditoriums, churches, dance floors, lodge rooms, reviewing stands, stadiums)	7 net
Waiting space	3 net
Unconcentrated (includes among others conference rooms, exhibit rooms, gymnasiums, lounges, skating rinks, stages, platforms)	15 net
Assembly with fixed seats	See note 1
Bowling alleys (allow 5 persons for each alley, including 15 ft of runway, and other spaces in accordance with the appropriate listing herein)	7 net
Business areas	100 gross
Courtrooms other than fixed seating areas	40 net
Educational (including educational uses above the twelfth grade)	
Classroom areas	20 net

Shops and other vocational areas	50 net
Industrial areas	100 gross
Institutional	
Sleeping areas	120 gross
Inpatient treatment and ancillary areas	240 gross
Outpatient areas	130 gross
Resident housing areas	120 gross
Library	
Reading rooms	50 net
Stack areas	100 gross
Malls	See Section 413
Mercantile	
Basement and grade floor areas open to public	30 gross
Areas on other floors open to public	60 gross
Storage, stock, shipping areas not open to public	300 gross
Parking garage	200 gross
Residential	200 gross
Restaurants (without fixed seats)	15 net
Restaurants (with fixed seats)	See Note 1
Storage area, mechanical	300 gross

1 sq ft = 0.0929 m²

Notes:
1. The occupant load for an assembly area having fixed seats installed shall be determined by the number of fixed seats. The capacity of seats without dividing arms shall equal one person per 18 in (457 mm); for booths, one person per 24 in (610 mm).
2. See Section 202 for definitions of gross and net floor areas.
3. The occupant load of floor areas of the building shall be computed on the basis of the specific occupancy classification of the building. Where mixed occupancies occur, the occupant load of each occupancy area shall be computed on the basis of that specific occupancy.

1003.2 Measurement of Means of Egress

1003.2.1. The width of the means of egress shall be determined from occupants served in accordance with Table 1004 (see Appendix).

1003.2.2. The width shall be measured in the clear at its narrowest point. Handrails may project 1-1/2 in (89 mm) and door jambs 1 in (25.4 mm) on each side of the measured width; however, the clear width of doorways shall not be reduced.

1003.2.3. Objects projecting from walls with their leading edges between 27 and 80 in (686 and 2032 mm) above the finished floor shall protrude no more than 4 in (102 mm) into walks, corridors, passageways, or aisles. Free standing objects mounted on posts or pylons may overhang 12 in (305 mm) maximum from 27 to 80 in (686 and 2032 mm) above the ground or finished floor.

1003.2.4. There shall be a minimum headroom of 6 ft 8 in (2032 mm), excluding stops, from the walking surface to the lowest part of any structural member, fixture, or furnishing.

>**EXCEPTION**: Sloping ceilings permitted by Section 1203.2.

1003.3 Capacity of Means of Egress

1003.3.1. The width of the means of egress shall be not less than the required capacity based on occupant load from Table 1003.1 (see Appendix).

1003.3.2. The capacity of exit stairways constructed in accordance with Section 1007 shall be not less than the minimum required herein. Exit stairways shall be permitted to be used as a required exit from all floors which they serve.

>[If, for example, three stairways are required to serve the third floor of a building and a like number are required for the second floor, the total number of stairways required will be three, not six, and the capacity of the stairway will be determined by the floor having the highest occupant load, and not the total occupant load of the building.]

1003.3.3. The required capacity of an exit access corridor shall be defined as the occupant load using the corridor for exit access divided by the required number of exits to which the corridor connects, but not less than the required capacity of the exit element to which the corridor leads.

1003.3.4. The aggregate width of passageways, aisles, or corridors serving as access to exits shall be at least equal to the required width of the exit. Where all travel to any exit is along the same access to the exit, the width of the access shall be at least equal to the exit. Where there are several accesses to an exit, each shall have a width suitable for the travel which it may be called on to accommodate.

1003.3.5. When exits serve more than one floor, only the occupant load of each floor, considered individually, need be used in computing the required capacity of the exits at that floor. At no point along the exit path may the exit width be decreased. When an exit from an upper floor and a lower floor converge at an intermediate floor, the capacity of the exit from the intermediate floor shall be not less than the sum of the required capacities of such upper and lower floors.

1003.3.6. The minimum width of exit access corridors shall be in accordance with Table 1004.

1003.3.7. Exit access corridors shall have fire-resistance ratings as specified in Table 700 (see Chapter 6).

1003.4 Elevators, Escalators, and Moving Walks

1003.4.1. Elevators, escalators, and moving walks shall not be used as a component of a required means of egress.

　　　EXCEPTION: Elevators shall be permitted for the purpose of providing egress for people with physical disabilities only when permitted by Section 1004.3.

**1004.0 ARRANGEMENT AND NUMBER OF
　　　　EXITS [1004.0; 1004.0]**

1004.1 Arrangement of Exits

Exits shall be so located that the distance from the most remote point in the floor area, room, or space served by them to the nearest exit, measured along the line of travel, shall be not more than the travel distance specified in Table 1004 (see Appendix).

1004.1.2. Where two or more exits or exit access doors are required, at least two of the exits or exit access doors shall be placed a distance apart equal to not less than one-half of the length of the maximum overall diagonal dimension of the building or area to be served measured in a straight line between such exits or exit access doors. The two exits or exit access doors shall be so located and constructed to minimize the possibility that both may be blocked by any one fire or other emergency condition.

> **EXCEPTION**: When exit enclosures are provided as a portion of the required exit and are interconnected by a corridor conforming to the requirements for 1-hr rated construction, the exit separation may be measured along a direct line of travel within the corridor.

1004.1.3. Where open stairways or ramps are permitted as part of the path of travel to required exits, such as between mezzanines, balconies, and the floor below, the travel distance shall include

1. The distance to reach the stair or ramp
2. The line of travel on a stair measured in the plane of the stair nosing
3. The distance from the end of the stair or ramp to the exit

1004.1.4. In one-story Group F and Group S buildings equipped with automatic heat and smoke vents complying with this section and sprinklered, the travel distance may be increased to 400 ft (122 m). Smoke and heat vents shall be constructed and installed in a manner approved by the building official.

1004.1.4.1. Smoke and heat vents shall open automatically by activation of a heat responsive device rated at $100°$ to $220°F$ ($38°$ to $104°C$) above ambient. The releasing mechanism shall be capable of operation such that the vent will be fully open when the vent is exposed to a time-temperature gradient that reaches an air temperature of $500°F$ ($260°C$) within 5 min. Vents shall be capable of being opened by an approved manual operation.

1004.1.4.2. Curtain boards shall be provided to subdivide a vented building. Curtain boards shall be constructed of material that will resist the passage of smoke and be consistent with the building type of construction. Curtain boards location and depth shall comply with Table 1004.1.4 (see Appendix).

> **EXCEPTION**: When a smoke and heat venting system complies with the guidelines of NFiPA 204M.

1004.1.4.3. The maximum spacing of roof vents and vent area shall comply with Table 1004.1.4.

> **EXCEPTION**: When a smoke and heat venting system complies with the guidelines of NFiPA 204M.

1004.2 Minimum Number of Exits

1004.2.1. There shall be not less than two approved independent exits, accessible to each tenant area, serving every story, except in Group R3 Occupancies and as modified in Section 1018.

1004.2.2. The minimum number of exits for all occupancies, except as modified by Section 1018, based on occupant load, shall be as follows:

Minimum Number of Exits	Occupancy Load per Story
2	1 - 500
3	501 - 1000
4	more than 1000

1004.2.3. Sufficient exit facilities shall be provided so that the aggregate capacity of all such exits, determined in accordance with this chapter, shall be not less than the occupant load as determined from Section 1003.1.

1004.2.4. It shall be unlawful to occupy any part of a building by a greater number of persons than that for which means of egress capacity, as prescribed in this chapter, has been provided.

1004.3 Accessible Means of Egress

1004.3.1. Spaces required by chapter 11 of the SBC to be accessible shall be provided with not less than one accessible means of egress. Where more than one means of egress is needed from any required accessible space, each accessible portion of the space shall be served by not less than two accessible means of egress.

1004.3.2. Each accessible means of egress shall be continuous from each required accessible occupied area to a public way and shall include accessible routes, ramps, exit stairs, elevators, horizontal exits, or smoke barriers.

1004.3.2.1. An exit stair to be considered part of an accessible means of egress shall have a clear width of at least 48 in (1219 mm) between handrails and shall either incorporate an area of refuge within an enlarged story-level landing or shall be accessed from either an area of refuge complying with Section 1004.3.5 or a horizontal exit.

EXCEPTIONS:
- Exit stairs serving a single dwelling unit or guest room

- Exit stairs serving occupancies protected throughout by an approved automatic sprinkler system
- The clear width of 48 in (1219 mm) between handrails is not required for exit stairs accessed from a horizontal exit

1004.3.2.2. An elevator to be considered part of an accessible means of egress shall comply with the requirements of Section 211 of ANSI/ASME A17.1 and standby power shall be provided. The elevator shall be accessed from either an area of refuge complying with Section 1004.3.5 or a horizontal exit.

> **EXCEPTION:** Elevators are not required to be accessed by an area of refuge or a horizontal exit in occupancies equipped throughout with an automatic sprinkler system.

1004.3.3. In buildings where a required accessible floor is four or more stories above or below a level of exit discharge serving that floor, at least one elevator shall be provided to comply with Section 1004.3.2.2 and shall serve as one required accessible means of egress.

> **EXCEPTION:** In fully sprinklered buildings, the elevator shall not be required on floors provided with a horizontal exit and located at or above the level of exit discharge.

1004.3.4. Platform (wheelchair) lifts shall not serve as part of an accessible means of egress.

> **EXCEPTION:** Within a dwelling unit.

1004.3.5 Areas of Refuge

1004.3.5.1. Every required area of refuge shall be accessible from the space it serves by an accessible means of egress. The maximum travel distance from any accessible space to an area of refuge shall not exceed the travel distance permitted for the occupancy. Every required area of refuge shall have direct access to an exit complying with Section 1004.3.2.1 or an elevator complying with Section 1004.3.2.2.

> **EXCEPTION:** Areas of refuge are not required in open parking garages.

1004.3.5.2. Each area of refuge shall be sized to accommodate one wheelchair space of 30 in X 48 in (762 mm X 1219 mm) for each 200 occupants, or portion thereof, based on the occupant load served by the area of refuge. Such wheelchair spaces shall not reduce the required means of egress width.

1004.3.5.3. Access to any of the required wheelchair spaces in an area of refuge shall not be obstructed by more than one adjoining wheelchair space.

1004.3.5.4. Each area of refuge shall be separated from the remainder of the story by a smoke barrier having at least a 1-hr fire-resistance rating. Smoke barriers shall extend to the roof or floor deck above. Doors in the smoke barrier shall have a 20-min fire-resistance rating, except those in horizontal exits and exit enclosures. Doors shall be self-closing or automatic-closing by smoke detection. HVAC openings in smoke barriers where 20-min fire-rated doors are permitted shall be ducted and provided with a smoke-actuated damper designed to resist the passage of smoke.

1004.3.5.5. Every area of refuge shall be provided with a two-way communication system between the area of refuge and a central control point.

EXCEPTION: Buildings four stories or less in height.

1004.3.5.6. In each area of refuge provided with a two-way emergency communication system, instructions on the use of the area under emergency conditions shall be posted adjoining the communication system. The instructions shall include

1. Directions to other means of egress
2. Advice that persons able to use the exit stairs do so as soon as possible unless they are assisting others
3. Information on planned availability of assistance in the use of stairs or supervised operation of elevators and how to summon such assistance
4. Directions for use of the emergency communication system

1004.3.5.7. Each area of refuge shall be identified by a sign stating "Area of refuge" and the international symbol of accessibility. The sign shall be located at each door providing access to the area of refuge. The sign shall be illuminated as required for exit signs where exit sign illumination is required. Tactile signs complying with Council of American Building Officials/American National Standards Institute (CABO/ANSI) A117.1 shall be located at each door to an area of refuge.

1004.3.5.8. At all exits and elevators serving a required accessible space, but not providing an approved accessible means of egress, signs shall be installed indicating the location of accessible means of egress.

1004.3.5.9. Every area of refuge that is not protected by an automatic sprinkler system shall be designed to prevent the intrusion of smoke.

EXCEPTION: Areas of refuge located within a stair enclosure.

1005.0 SPECIAL EXIT REQUIREMENTS [1005.0; 1005.0]

1005.1 Boiler, Incinerator, Furnace Rooms
Except in one- and two-family dwellings, two egress doors shall be provided from all boiler, incinerator, and furnace rooms that exceed 500 sq ft (46 m^2) in area and where the largest installed piece of fuel-fired equipment exceeds 400,000 British thermal units (Btu) (117 kW) input capacity. Egress doors shall be separated by a horizontal distance not less than one-half the maximum horizontal dimensions of the room. A 6-in (152 mm) sill (dike) shall be provided where oil fire equipment is used.

Interior openings between a Group H Occupancy and a boiler, incinerator, furnace, or similar room shall not be permitted. The maximum distance of travel to an egress door shall not exceed 50 ft (1.5 m).

1005.2 Dead-End Pockets or Hallways
Exits and exit access shall be so arranged that dead-end pockets or hallways in excess of 20 ft (6096 mm) long shall not occur.

1005.3.1 Exit Access Corridors
It shall be prohibited to use exit access corridors, separated from building use areas by fire-rated partitions and providing access to exit, for return or exhaust air from adjoining air conditioned spaces through louvers or other devices mounted in corridor doors, partitions, or ceilings.

1005.3.2. Except in Group I or Group R Occupancies, Section 1005.3.1 may be waived by the building official, providing corridors are equipped with approved smoke detectors arranged to automatically stop supply, return, and exhaust air and close louvers or other devices mounted within the corridor doors, partitions, or ceilings.

1005.4 Emergency Egress Openings
1005.4.1. Every sleeping room on the first and second story of Group R Occupancies shall have at least one operable exterior

window or exterior door approved for emergency egress or rescue. The units must be operable from the inside to a full clear opening without the use of separate tools or keys. Where windows are provided as a means of egress or rescue, they shall have a sill height of not more than 44 in (1118 mm) above the floor.

1005.4.2. The minimum net clear opening height dimension shall be 22 in (559 mm). The minimum net clear opening width dimension shall be 20 in (508 mm). The net clear opening area shall in no case be less than 4 sq ft (0.37 m²).

1005.4.3. Each egress window from sleeping rooms must have a minimum total glass area of not less than 5 sq ft (0.47 m²) in the case of a ground-floor window and not less than 5.7 sq ft (0.53 m²) in the case of a second-story window.

1005.5 Smokeproof Enclosures

1005.5.1. Where the floor surface of any story is located more than 75 ft (23 m) above the lowest level of fire department vehicle access, each of the required exits for the building shall be a smokeproof enclosure.

1005.5.2. A minimum 2-hr fire-resistant construction shall be used for smokeproof enclosures. In each case, openings into the required 2-hr construction shall be limited to those needed for maintenance and operation and shall be protected by self-closing 1-1/2-hr fire-resistance-rated devices. The supporting frame shall be protected as set forth in Chapter 6. [UBC chapter 6; BOCA chapter 6.]

1005.5.3. Group B buildings exceeding 15,000 sq ft (1395 m²) per floor and complying with the area of refuge (compartmentation) option described in SBC section 412.9 are exempt from the smokeproof enclosure requirements.

1005.5.4. Stairs in smokeproof enclosures shall be of non-combustible construction.

1005.5.5. A smokeproof enclosure shall exit into a public way or into an exit passageway, yard, open court, or open space having direct access to a public way. The exit passageway shall be without other openings and shall have walls, floors, and ceiling of 2-hr fire-resistance construction.

1005.5.6. A stairway in a smokeproof enclosure shall not continue below the grade level unless an approved barrier is provided at the ground level to prevent persons from accidentally continuing into the basement.

1005.5.7. Access to the stairway shall be by way of a vestibule or by way of an open exterior balcony of noncombustible materials.

>**EXCEPTION**: Access by way of a vestibule or an open exterior balcony is not required when the stairway meets the requirements of Sections 1005.5.9.2, 1005.5.9.9, and 1005.5.9.10 and is located in a fully sprinklered building.

1005.5.8 Smokeproof Enclosures by Natural Ventilation

1005.5.8.1. Where a vestibule is provided, the door assembly into the vestibule shall have a 1-1/2-hr fire-resistance rating and the door assembly from the vestibule to the stairs shall have not less than a 20-min fire-resistance rating. The doors shall have closing devices as specified in Section 1005.5.9.10. Wired glass 1/4 in (6.4 mm) thick may be installed not to exceed 100 sq in (0.065 m^2) with neither dimension exceeding 12 in (305 mm).

1005.5.8.2. The vestibule shall have a minimum of 16 sq ft (1.49 m^2) of opening in a wall facing an exterior court, yard, or public way at least 20 ft (6096 mm) wide. The vestibule shall be a minimum of 44 in (1118 mm) wide and 72 in (1829 mm) in the direction of travel.

1005.5.8.3. Where access to the stairway is by means of an open exterior balcony, the door assembly to the stairway shall have a 1-1/2-hr fire-resistance rating. Doors shall have closing devices as specified in Section 1005.5.9.10.

1005.5.9 Smokeproof Enclosures by Mechanical Ventilation

1005.5.9.1. Stair pressurization systems shall be independent of other building ventilation systems.

1005.5.9.2. Equipment and ductwork for stair pressurization shall comply with one of the following:
- Be located exterior to the building and be directly connected to the stairway or connected to the stairway by ductwork enclosed in 2-hr fire-resistance-rated construction
- Be located within the stair enclosure with intake or exhaust air directed to the outside or through ductwork in 2-hr construction
- Be located within the building if separated from the remainder of the building, including other mechanical equipment, with 2-hr construction

1005.5.9.3. The door from the building into the vestibule shall have a 1-1/2-hr fire-resistance rating and have closing devices as specified in Section 705.1.3.2.3. The door from the vestibule to the stairway shall have a minimum 20-min fire-resistance rating and have closing devices as specified in Section 705.1.3.2.3. Wired glass, if provided, shall not exceed 100 sq in (0.065 m²) and shall be set in a steel frame. The door shall be provided with a drop sill or other provision to minimize air leakage.

1005.5.9.4. Where access to the stairway is by means of an open exterior balcony, the door assembly to the stairway shall have a 1-1/2-hr fire-resistance rating. Doors shall have closing devices as specified in Section 1005.5.9.10.

1005.5.9.5. The vestibule shall have a minimum dimension of 44 in (1118 mm) wide and 72 in (1829 mm) in the direction of exit travel.

1005.5.9.6. The vestibule shall be provided with not less than one air change per minute, and the exhaust shall be 150% of the supply. Supply air shall enter and exhaust air shall discharge from the vestibule through separate, tightly constructed ducts used only for that purpose. Supply air shall enter the vestibule within 6 in (152 mm) of the floor level.

The top of the exhaust register shall be located at the top of the smoke trap but no more than 6 in (152 mm) down from the top of the trap and shall be entirely within the smoke trap area. Doors, when in the open position, shall not obstruct duct openings. Duct openings may be provided with controlling dampers if needed to meet the design requirements but are not otherwise required.

1005.5.9.7. For buildings where such air changes would result in excessively large duct and blower requirements, a specially engineered system may be used. Such an engineered system shall provide 2500 cu ft/min (1.2 m³/s) exhaust from a vestibule when in emergency operation and shall be sized to handle three vestibules simultaneously. The smoke detector located outside each vestibule shall release to open the supply and exhaust duct dampers in that affected vestibule.

1005.5.9.8. The vestibule ceiling shall be at least 20 in (508 mm) higher than the door opening into the vestibule to serve as a smoke and heat trap and to further provide an upward moving air column. The 20-in (508 mm) height requirement may be reduced proportionally if the minimum vestibule size described in Section 1005.5.9.5 is enlarged so as to maintain the same volume in the smoke trap area above the door when justified by design and test.

In any case, minimum ceiling height shall not be less than 7 ft 6 in (2286 mm).

1005.5.9.9. The stair shaft shall be provided with mechanical supply and exhaust air. There shall be a minimum of 2500 cu ft/min (1.2 m³/s) discharge through a dampered relief opening or an exhaust fan at the top of the stair shaft.

The supply shall be sufficient to provide a minimum positive pressure of 0.05 inches water column (in w.c.) (12.5 Pa) in addition to the maximum anticipated stack pressure, relative to other parts of the building measured with all doors closed.

The combined positive pressure shall not exceed 0.35 in w.c. (87 Pa). The air supply shall be taken directly from outside of the building. The stair pressure shall be static pressures measured at the level of discharge from the stair.

> **EXCEPTION**: The minimum positive pressure shall be increased to 0.15 in w.c. (37 Pa) in an unsprinklered building.

1005.5.9.10. The activation of the ventilating equipment shall be initiated by a smoke detector installed outside the vestibule door in an approved location.

When the closing device for the stair shaft and vestibule doors is activated by smoke detection or power failure, the closing devices on all doors in the smokeproof enclosure at all levels shall be activated and the mechanical equipment shall operate at the levels specified in Sections 1005.5.9.6 and 1005.5.9.9.

1005.6 Mezzanines

1005.6.1. For all public mezzanines, egress must be provided as follows:

Two means of egress shall be provided from any mezzanine with an occupant load or travel distance to an exit or to a point where there is a choice of more than one means of egress which exceeds that shown in Table 1005.6.

Table 1005.6
Single Exit Criteria for Mezzanines

Use	Occupant Load	Maximum Travel Distance, ft
Assembly	50	75
Business	30	75
Courtrooms	50	75
Educational		
Classroom	50	75
Shops and vocational	50	75
Industrial	50	75
Institutional		
Sleeping area	6	75
Inpatient treatment areas	10	75
Outpatient treatment areas	10	75
Library		
Reading rooms	50	75
Stack areas	30	75
Mercantile		
Basement/grade levels	50	75
Other floors	50	75
Storage and shipping	10	75
Parking garages	30	75
Residential		
Dwellings	10	75
Hotels and apartments	10	75
Dormitories	10	75
Hazardous	3	25 [1]
Storage	30	100

1 ft = 0.305 m

Note:
1. Maximum area with one exit or exit access shall be 200 sq ft.

1005.6.2. If any required means of egress is through the room below, the occupant load of the mezzanine shall be added to the occupant load of the room in which it is located.

1005.6.3. Egress stairways from mezzanines shall conform with the requirements of Section 1007. They may be open and may descend to the floor of the room in which they are located when all the following conditions are met:

1. The space beneath the mezzanine is totally open and unencumbered by partitioned rooms or spaces.

    EXCEPTION: The space beneath the mezzanine may be enclosed provided the enclosed space is protected throughout with a smoke detection system in accordance with NFiPA 72E which sounds an alarm in the mezzanine.

2. The travel distance from the most remote point on the floor of the mezzanine to the building exit or to a protected egress corridor, exit court, horizontal passageway, enclosed stairway, or exterior exit balcony, inclusive of travel on the stairway, does not exceed 75 ft (22.9 m) where a single means of egress is permitted, or the limits of Table 1004 (see Appendix) where multiple means of egress are required.

3. The occupant load of the mezzanine is added to the occupant load of the story or room in which it is located for purposes of determining the egress requirements of such story or room.

4. The mezzanine is not occupied for sleeping purposes, unless there are exterior windows accessible to the mezzanine and located not more than two stories above grade.

1005.7 Burglar Bars

Each sleeping room or room with a required exit door in a residential occupancy that has burglar bars installed shall have at least one emergency egress window or door that is operable from the inside without the use of a key, tool, special knowledge, or effort.

1006.0 STAIRWAY PROTECTION [1006.0; 1006.0]
1006.1 Enclosed Stairways

1006.1.1. Exit stairways between floors shall be enclosed in or separated by fire-resistant construction in accordance with Section 705.2 and Table 700.

EXCEPTIONS:
- Stairways serving and contained within a dwelling, dwelling unit, or hotel suite
- Exterior stairways conforming to Section 1006.2
- In open automobile parking garages when the stair is on an open side, as defined in SBC section 411.3.2
- In open parking garages, having all sides open

1006.1.2. Except in one- and two-family dwellings, basement stairways located under stairways from upper stories shall be completely enclosed by construction providing fire-resistance not less than required for the stair enclosure above the basement but in no case less than 1-hr fire-resistance.

1006.1.3. A stairway enclosure shall not be used for any purpose other than means of egress. Openings in exit enclosures other than unexposed exterior openings shall be limited to those necessary for exit access to the enclosure from normally occupiable or habitable rooms and for egress from the enclosure.

1006.1.3.1. Penetrations into and openings through a stairway enclosure assembly are prohibited except for required exit doors, ductwork, and equipment necessary for independent stair pressurization, sprinkler piping, standpipes, and electrical conduit serving the stairway and terminating at a steel box not exceeding 16 sq in (0.010 m^2).

Such penetrations shall be protected in accordance with Section 705.4. There shall be no penetrations or communicating openings, whether protected or not, between adjacent stair enclosures.

1006.1.3.2. Exterior walls of an enclosed stairway shall comply with the requirements of Table 600 (see Appendix) for exterior walls. Where nonrated walls or unprotected openings are used to enclose the exterior of the stairway, the building enclosure walls within 10 ft (3048 mm) horizontally of the nonrated wall or unprotected opening shall be constructed as required for stairway enclosures, including opening protectives, but need not exceed 1-hr fire-resistance with 3/4-hr opening protective.

[This construction extends vertically from the ground to a point 10 ft (3048 mm) above the topmost landing of the stairway or to the roof line, whichever is lower.]

1006.1.4. The space under a stairway may be used if it is separated from the stairway by fire-resistant construction as required by Section 1006.1.

EXCEPTION: Separation is not required from those stairways exempted from enclosure in Section 1006.1.1.

1006.2 Exterior Exit Stairs

1006.2.1. Exterior stairways conforming to the requirements for interior stairways in all respects, except as to enclosures and except as herein specifically modified, may be accepted as an element of a required means of egress in buildings not exceeding six stories or 75 ft (22.9 m) in height for other than Group E and I Unrestrained buildings.

1006.2.2. Exterior stairways shall be permitted where at least one door from each tenant opens onto a roofed-over open porch or balcony served by at least two stairways so located as to provide a choice of independent, unobstructed means of egress directly to the ground, except a single stairway shall be allowed when a single exit is permitted by Sections 1020, 1025, 1026, and 1027. Such porches and stairways shall comply with the requirements for interior exit stairways as specified in Sections 1007 and 1014. Porches and balconies shall be not less than 4-1/2 ft (1372 mm) wide.

The stairways shall be located so that the entrances and all portions of the stairways on each level are a distance apart equal to not less than one half of the length of the maximum overall diagonal dimension of the building or area to be served measured in a straight line between such stairways. The maximum travel distance from any tenant space to the nearest stairway shall be as specified in Table 1004 (see Appendix). Porches and stairways shall be located at least 10 ft (3048 mm) from adjacent property lines and from other buildings on the same lot, unless openings in such buildings are protected by 3/4-hr fire-resistant doors or windows.

1006.2.3. Handrails and guardrails shall be as specified in Sections 1007.5 and 1015, respectively.

1006.2.4. Exterior stairs shall be separated from the interior of the building by walls with a fire-resistance rating of not less than 1 hr with fixed or self-closing opening protectives as required for enclosed stairs. This protection shall extend vertically from the ground to a point 10 ft (3048 mm) above the topmost landing or the roof line, whichever is lower, and horizontally 10 ft (3048 mm) from each side of the stairway. Openings within the 10 ft (3048 mm) horizontal extension of the protected walls beyond the stairway shall be equipped with fixed 3/4-hr assemblies.
EXCEPTIONS:
- Exterior stairways may be unprotected when serving an exterior exit access balcony which has two exterior stairways, remotely located as required in Section 1006.2.2.
- Such protection is not required in two-story buildings where there is a second exit remotely located as required in Section 1006.2.2.

1006.2.5. All required exterior stairways shall be located so as to lead directly to a street or open space with direct access to a street. When located on the rear of the building such stairways may lead through a passageway at grade complying with Section 1010.

1006.2.6. Exterior stairways shall not project beyond the street lot line.

1007.0 STAIRWAY CONSTRUCTION [1007.0; 1007.0]
1007.1 General
1007.1.1. Exterior and interior exit stairways shall be constructed of noncombustible materials throughout in the following buildings:
- All buildings of Type I and Type II construction
- All Group A-1 and Group I buildings
- All Group A-2 and Group E buildings three stories or more in height
- All other buildings three stories or more in height or occupied by more than 40 persons above or below the first story at street or grade level, except one- and two-family dwellings and buildings of Type VI construction

1007.1.2. Stairways located in a required fire-resistant enclosure shall have closed risers. All other stairways shall be permitted to have open risers.

1007.1.3. Interior stairs constructed of wood, except those with open risers, shall be fireblocked as specified in Section 705.3.

1007.1.4. Closets shall not be located beneath stairs unless such stairs are protected as required by Section 1006.1.

> **EXCEPTION**: Protection is not required for those stairways exempted from enclosure in Section 1006.1.1.

1007.1.5. The underside of interior stairways, if of combustible construction, shall be protected to provide not less than 1-hr fire resistance.

> **EXCEPTION**: When located within a dwelling unit.

1007.1.6. Enclosed exit stairways that continue beyond the floor of discharge shall be interrupted at the floor of discharge by partitions, doors, or other effective means.

> **EXCEPTION**: Stairs that continue one-half story beyond the level of exit discharge need not be interrupted by physical barriers where the exit discharge is clearly obvious.

1007.2 Not Used

1007.3 Treads and Risers

1007.3.1. Treads and risers of stairs shall be so proportioned that the sum of two risers and a tread, exclusive of projection of nosing, is not less than 24 in (610 mm) nor more than 25 in (635 mm). The height of riser shall not exceed 7-3/4 in (197 mm), and treads, exclusive of nosing, shall be not less than 9 in (229 mm) wide.

> **EXCEPTION**: Special stairs in Section 1007.8.

1007.3.2. Every tread less than 10 in (254 mm) wide shall have a nosing, or effective projection, of approximately 1 in (25.4 mm) over the level immediately below that tread.

1007.3.3. Tread depth shall be measured horizontally between the vertical planes of the foremost projection of adjacent treads and at a right angle to the tread's leading edge.

> **EXCEPTION**: Tread depth of special stairs in Section 1007.8 shall be measured on a line perpendicular to the centerline of tread.

1007.3.4. Treads shall be of uniform depth and risers of uniform height in any stairway between two floors. There shall be no variation exceeding 3/16 in (4.8 mm) in the depth of adjacent treads or in the height of adjacent risers, and the tolerance between the largest and smallest riser or between the largest and smallest tread shall not exceed 3/8 in (9.5 mm) in any flight. The uniformity of winders and other tapered treads, complying with Sections 1007.8.1, 1007.8.2, and 1007.8.3, shall be measured at consistent distances from the narrower end of the treads.

> EXCEPTION: Where the bottom or top riser adjoins a sloping public way, walk, or driveway having an established grade and serving as a landing, a variation in height of the riser of not more than 3 in (76 mm) for every 3 ft (914 mm) of stairway width is permitted.

1007.4 Landings
1007.4.1. A flight of stairs shall not have a vertical rise of more than 12 ft (3658 mm) between floors or landings.

1007.4.2. The width of landings shall be not less than the width of stairways they serve. Every landing shall have a minimum dimension measured in the direction of travel equal to the width of the stairway. Such dimension need not exceed 4 ft (1219 mm) when the stair has a straight run.

1007.4.3. Stairway landings shall have guardrails as specified in Section 1015 on any open and unenclosed edges.

> EXCEPTION: The top element of a guardrail at the inside open or unenclosed edge of any intermediate landing where the stairs reverse direction may be at the same height as the stairway handrails when the horizontal distance between the stair flights is 1 ft (305 mm) or less and when a fully continuous handrail as specified in Section 1007.5 is provided.

1007.5 Handrails
1007.5.1. Stairways having four or more risers above a floor or finished ground level shall be equipped with handrails located not less than 30 in (762 mm) nor more than 38 in (965 mm) above the leading edge of a tread.

> EXCEPTIONS:
> - Handrails that form part of a guardrail may be 42 in (1067 mm) high.
> - As required for Group I Unrestrained in 1024.1.4.

1007.5.2. Stairways shall have handrails on each side.

> EXCEPTIONS:
> - Aisle stairs provided with a center handrail need not have additional handrails.

- Stairs within dwelling units, spiral stairs and aisle stairs serving seating only on one side may have a handrail on one side only.

1007.5.3. Handrails shall have either a circular cross section with a diameter of 1-1/4 to 2 in (32 to 51 mm), or a noncircular cross section with a perimeter dimension of at least 4 in (102 mm) but not more than 6-1/4 in (159 mm) and a largest cross-section dimension not exceeding 2-1/4 in (57 mm). Edges shall have a minimum radius of 1/8 in (3 mm).

1007.5.4. Gripping surfaces shall be continuous, without interruption by newel posts or other obstructions.

1007.5.5. Where a wall or guardrail exists, handrails shall extend at least 12 in (305 mm) beyond the top riser and at least 12 in (305 mm) plus the depth of one tread beyond the bottom riser. At the bottom of the wall, the handrail shall continue to slope for a distance of the depth of one tread from the bottom riser; the remainder of the extension shall be horizontal.

EXCEPTION: Handrails within a dwelling unit.

1007.5.6. The clear space between handrail and wall shall be a minimum of 1-1/2 in (38 mm).

1007.5.7. When the required width of a flight of stairs exceeds 88 in (2235 mm), one or more intermediate handrails, continuous between landings, substantially supported and terminating at the upper end in newels or standards, shall be provided and there shall be not more than 88 inches (2,235 mm) between such adjacent handrails.

1007.5.8. Handrails, where required along open-sided flights of stairs, shall be of construction adequate in strength, durability, and attachment for their purpose as prescribed in Section 1608.2. They shall include intermediate rails or ornamental patterns such that a 6-in (152 mm) diameter sphere cannot pass through any openings.

EXCEPTIONS:
- Openings between intermediate rails shall be permitted for specific occupancy groups as described in Section 1018.
- Stairways which are part of or connected to the facilities described in Section 1026.4.2, and not more than 6 ft (1829 mm) above the grade below, shall be required to have only one intermediate railing located between 14

and 18 in (356 and 457 mm) above the leading edge of the tread.

1007.5.9. On monumental stairs, handrails shall be located along the most direct path of egress travel.

1007.6 Width

1007.6.1. Stairs shall be clear of all obstructions except projections not exceeding 3-1/2 in (89 mm) at or below handrail height on each side.

1007.6.2. The stair width shall not decrease in the direction of exit travel.

1007.6.3. The minimum width of any stair serving as a means of egress shall be in accordance with Table 1004 (see Appendix).

1007.7 Headroom

Stairs shall have a minimum headroom clearance of 6 ft 8 in (2032 mm) measured vertically from a line connecting the edge of the risings. Such headroom shall be continuous above the stair to the point where the line intersects the landing below, one tread depth beyond the bottom riser. This minimum shall be maintained for the full width of the stair and landing.

1007.8 Special Stairs

1007.8.1. Winders shall have a minimum tread depth of 6 in (152 mm) at the narrow edge and shall have a minimum tread depth of 11 in (279 mm) at a point 12 in (305 mm) from the narrow edge. Winders shall be permitted to be used as a component in the means of egress within a dwelling unit.

1007.8.2. Spiral stairways shall be permitted to be used as a component in the means of egress within dwelling units and from a mezzanine not more than 250 sq ft (23 m^2) and serving not more than five occupants. A spiral stairway shall have a 7-1/2 in (190 mm) minimum clear tread depth at a point 12 in (305 mm) from the narrow edge. The risers shall be sufficient to provide a headroom of 6 ft 6 in (1981 mm) minimum, but riser height shall not be more than 9-1/2 in (114 mm). The minimum stairway width shall be 26 in (660 mm).

1007.8.3. Circular stairways shall be permitted to be used as a component in the means of egress providing the minimum depth of tread is not less than 11 in (279 mm) measured 12 in (305 mm) from the smaller radius, and the smaller radius is not less than twice the width of the stairway.

Means of Egress

In Group R3 Occupancies, circular stairs may have a minimum tread depth of 9 in (229 mm) with 1 in (25.4 mm) of nosing, and the smaller radius may be less than twice the width of the stairway.

1007.8.4. Alternating tread stairways shall have a minimum projected tread exclusive of nosing of 8-1/2 in (216 mm) within a minimum total tread depth of 10-1/2 in (267 mm). The rise to the next alternating tread surface shall be a maximum of 8 in (203 mm). Distance between handrails shall be a minimum of 17 in (432 mm) and a maximum of 24 in (610 mm). A minimum distance of 6 in (152 mm) shall be provided between the stair handrail and any other object. A minimum of 12 in (305 mm) shall be provided between the stair handrails of adjacent alternating tread stairways.

1007.8.5. Alternating tread stairways meeting the requirements of Section 1007.8.4 shall be permitted to be used as a component in a means of egress from a mezzanine of not more than 250 sq ft (23 m^2) in area serving not more than five occupants in F, H, I Occupancies, within dwelling units of R2 and R3, and S Occupancies.

1008.0 ACCESS TO ROOF [1008.0; 1008.0]

Buildings four stories or more in height, except those with a roof slope greater than 4:12, shall be provided with a stairway to the roof. Such a stairway shall be marked at street and floor levels with a sign indicating that it continues to the roof. Where roofs are used for roof gardens or for other purposes, stairways shall be provided as required for such use or occupancy.

1009.0 HORIZONTAL EXITS [1009.0; 1009.1]
1009.1 General
1009.1.1. Horizontal exits shall not comprise more than one-half of the required exits from any building or floor area and shall not serve as the only exit. The walls of horizontal exits shall have a fire-resistance rating of 2 hr using materials dependent on the type of construction.

> **EXCEPTION**: Horizontal exits comprising more than one-half of the required exits shall be permitted in accordance with Sections 1024.1.1 and 1024.2.7.

1009.1.2. Ramps meeting the requirements of Section 1013 shall be used where there is a difference of level between connected areas.

1009.1.3. The area into which a horizontal exit leads shall be provided with exits adequate to meet the requirements of this chapter, but not including the added capacity imposed by persons entering it through horizontal exits from another area. At least one of its exits shall lead directly to the exterior.

1009.2 Doors

1009.2.1. The width of horizontal exits shall be not less than required for exit doorways. The exit capacity of horizontal exits shall be as specified in Section 1003.3.

1009.2.2. All fire doors in horizontal exits shall be self-closing or automatically closing when activated by a smoke detector. All opening protectives in horizontal exits shall be consistent with the fire-resistance rating of the wall with a minimum 1-1/2-hr rating.

1009.2.3. Doors in horizontal exits shall be kept unlocked and unobstructed.

1009.3 Capacity of Refuge Area

The refuge area of a horizontal exit shall be either public areas or spaces occupied by the same tenant, and each such area of refuge shall be adequate to house the total occupant load of both connected areas. The capacity of areas of refuge shall be computed on a net floor area allowance of 3 sq ft (0.28 m^2) for each occupant to be accommodated therein, not including areas of stairs, elevators, and other shafts or courts.

> **EXCEPTION**: Area for Group I shall be computed in accordance with Sections 1024.1.1.2 and 1024.2.7.2.

1010.0 EXIT DISCHARGE [1010.0; 1010.1]

1010.1 General

1010.1.1. Unless directly connected to a public way or to a space leading to a public way, required exits shall be connected to an exit court, exit passageway, lobby, or vestibule leading to a public way.

1010.1.2. The minimum width of such courts, passageways, lobbies, and vestibules shall be 44 in (1118 mm) but not less than the required width of the exits to which they are connected. There shall be no reduction of width in the direction of exit travel.

1010.1.3. The minimum clear ceiling height shall be 8 ft (2438 mm).

Means of Egress   197

1010.1.4. The slope of the floor of exit discharge elements shall not exceed 1:12.

1010.2 Exit Courts
Exit courts 10 ft (3048 mm) or less in width shall have a minimum fire-resistance rating of 1 hr with 3/4 hr opening protectives.

1010.3 Exit Passageways
Exit passageways shall be constructed in accordance with Sections 704.2.1.2, 1006.1.3, and 1006.1.3.1, with a fire-resistance rating equivalent to shaft enclosures in Table 700.

> **EXCEPTIONS**: A maximum 50% of the required number of exits and 50% of the required exit capacity shall be permitted to discharge through areas on the level of exit discharge provided all the following are met:
> - Such exits discharge to a free and unobstructed way to the exterior of the building, such way being readily visible and identifiable from the point of discharge at the exit.
> - The entire area is separated from areas below by construction having a fire-resistance rating not less than that required for the exit enclosure.
> - The area is protected throughout by an approved automatic sprinkler system.
> - Any other portion of the level of discharge with access to the area of discharge is protected throughout by an approved automatic sprinkler system or separated from the area of discharge in accordance with the requirements for the enclosure of exits.

1010.4 Vestibules
An exit may discharge into an interior vestibule which meets the following criteria:
1. The depth from the exterior of the building is not greater than 10 ft (3048 mm) and the length is not greater than 20 ft (6096 mm).
2. The vestibule is separated from the remainder of the level of exit discharge by construction providing protection equivalent to that provided by 1/4-in (6 mm) thick labeled wired glass in steel frames.

1011.0 FIRE ESCAPES [1011.0; 1011.0]
1011.1 General
1011.1.1. Fire escapes shall not be permitted except as approved by the building official for existing buildings when more adequate exit facilities cannot be provided. Fire escapes shall not provide more than 50% of the required exit capacity.

1011.1.2. When located on the front of the building and projecting beyond the building line, the lowest landing shall be not less than 7 ft (2134 mm) nor more than 12 ft (3658 mm) above grade, equipped with a counterbalanced stairway to the street. In alleyways and thoroughfares less than 30 ft (9144 mm) wide, the clearance under the lowest landing shall be not less than 12 ft (3658 mm).

1011.2 Design
1011.2.1. The fire escape shall be designed to support a live load of 100 lb/sq ft (4.8 kPa) and shall be constructed of steel or other approved noncombustible materials. Fire escapes may be constructed of wood not less than 2 in (51 mm) thick on buildings of Type VI construction.

1011.2.2. Stairs shall be at least 22 in (559 mm) wide with risers not more and treads not less than 8 in (203 mm) and with landings at the foot of stairs not less than 40 in (1016 mm) wide by 36 in (914 mm) long, located not more than 8 in (203 mm) below the access window or door.

1011.2.3. All openings located within 10 ft (3048 mm) of fire escapes shall be protected with approved opening protectives of at least 3/4-hr fire resistance.

> **EXCEPTION**: Fire escape ladders as set forth in Section 1011.3.

1011.3 Fire Escape Ladder Devices
A self-contained fire escape ladder device may be used when authorized by the building official in Group R Occupancies not exceeding five stories, when said device conforms to the following:
- The exit ladder serves an occupant load of 10 or less or a single dwelling unit or guest room.
- The access is adjacent to an opening as specified for emergency egress or rescue from a balcony. The exit ladder shall not pass in front of any building opening at or below the unit being served.

Means of Egress

- The exit ladder shall be so installed that the descending face is adjacent to the building wall, and each ladder device shall be offset or staggered not less than 24 in (610 mm) from the ladder above.
- The availability of the activation device for the exit ladder is accessible only from the opening on the balcony served.
- An alarm sounds when the exit ladder is activated.

1012.0 DOORS [1012.0; 1012.0]
1012.1 General
1012.1.1. An egress door used as an exit door shall provide a clear opening of not less than the widths shown in Table 1004 (see Appendix). The maximum leaf width of the door shall not exceed 48 in (1219 mm). Egress doors used in the exit access shall provide a clear opening of not less than 32 in (813 mm) wide.

>**EXCEPTIONS**:
>- Resident sleeping room doors within Group I Restrained Occupancies
>- Storage closets less than 10 sq ft (0.93 m^2) in area
>- Revolving doors
>- Interior egress door within a dwelling unit which is not required to be adaptable or accessible

1012.1.2. Egress doors shall be of the side swinging type. Doors shall swing in the direction of egress for the following:

>1. When serving an occupant load of 50 or more
>2. When serving a high-hazard occupancy

The following prescribed maximum forces applied to the latch side shall perform their respective functions:

>- A 15-lb (67 N) force shall release a latch.
>- A 30-lb (133 N) force shall set door in motion.
>- A 15-lb (67 N) force shall swing door fully open.

EXCEPTIONS TO SECTION 1012.1.2:

1. As permitted for specific occupancies in Section 1018.
2. Revolving doors conforming with Section 1012.3.
3. Horizontal sliding doors conforming with Section 1012.4 when used in elevator lobbies; areas of refuge set forth in Section 1004.3.5; smoke barriers; or any room or space, other than Group H, with an occupant load of less than 50.

1012.1.3. Every room or tenant space shall be provided with a minimum of one means of egress. Every room or tenant space which has an occupant load of 50 or more persons or in which the travel distance from the most remote point to the entrance to the exit access from the room or tenant space exceeds 75 ft (23 m) shall have not less than two egress doors.

EXCEPTION: Doors in an HPM facility shall comply with SBC section 408.3.9.

1012.1.4. The floor surface on both sides of a door shall be at the same elevation. The floor surface over which the door swings shall extend from the door in the closed position a distance equal to the door width. Thresholds at doorways shall not exceed 3/4 in (19.1 mm) in height for exterior sliding doors or 1/2 in (127 mm) for other doors. Raised thresholds and floor level changes greater than 1/4 in (6.4 mm) at doorways shall be beveled with a slope no greater than 1:2.

EXCEPTIONS:
- Exterior doors not on an accessible route as defined in Section 1102
- Variations in elevation due to differences in finish materials, but not more than 1/2 in (12.7 mm)
- Doorways complying with Section 1012.1.6

1012.1.5. Doors opening onto exit stairs or other approved exits shall not obstruct the travel along any required exit. Doors opening onto exit access corridors or onto a landing shall not reduce the corridor width or the landing width to less than one-half the required width during the opening process. When fully open, the door shall not project more than 7 in (178 mm) into the required width of a corridor or a landing.

1012.1.6. Exit doorways shall not open immediately upon a flight of stairs. A landing of at least the width of the door shall be provided, which is the same elevation as the finished floor from which it is exiting.

1012.1.7. In Group R3 Occupancies, a landing shall be provided on the exterior side of all egress door openings. The landing width shall be no less than the width of the door it serves, and the depth shall be not less than 36 in (914 mm). The landing may be one step lower than the inside floor level, but not more than 7 in (178 mm) lower.

Means of Egress

1012.1.8. Required exit doors shall be openable from the inside without the use of a key, tool, special knowledge, or effort. Manually operated flush bolts or surface bolts are prohibited. All hardware must be direct-acting requiring no more than one operation.

Double-cylinder dead bolts, requiring a key for operation on both sides, are prohibited on required means of egress doors unless the locking device is provided with a key which cannot be removed when the door is locked from the inside.

1012.1.9. For required width of doorways serving exit stairways and the exit capacity of doorways, see Sections 1003.2 and 1003.3.

1012.1.10. Special locking arrangements shall be permitted in accordance with Section 1012.6 for the applicable occupancy and Section 1018.

1012.2 Power-Operated Doors

1012.2.1. Where required doors are operated by power which is activated by a photo electric device, floor mat, wall switches, or other approved device as well as doors with power-assisted manual operation, the design, installation, and maintenance shall be such that, in the event of power failure, the door may be manually opened to permit exit travel. These doors shall be openable as is required for other non-power operable doors.

1012.2.2. Power-operated sliding doors may be used provided the sliding leaf is equipped with an emergency swing (panic release) feature.

> **EXCEPTION**: Horizontal sliding doors conforming with Section 1012.4.

1012.2.3. Power-operated doors shall comply with ANSI/BHMA A156.10.

1012.3 Revolving Doors

1012.3.1. Each revolving door shall be capable of collapsing into a book-fold position with parallel egress paths providing an aggregate width of 36 in (914 mm).

1012.3.2. A revolving door shall not be located within 10 ft (3048 mm) of the foot or top of stairs or escalators or the entrance or exit of a moving walk. A dispersal area shall be provided between the stairs or escalators or either end of the moving walk and the revolving doors.

1012.3.3. The turning speed of a revolving door shall not exceed the maximum permitted by Table 1012.3.3.

Table 1012.3.3
Maximum Speed for Revolving Doors

Inside Diameter ft - in	Power-Driven-Type Speed Control, rpm	Manual-Type Speed Control, rpm
6-6	11	12
7-0	10	11
7-6	9	11
8-0	9	10
8-6	8	9
9-0	8	9
9-6	7	8
10-0	7	8

1012.3.4. Each revolving door shall have a conforming side-hinged swinging door in the same wall as the revolving door and within 10 ft (3048 mm).

> **EXCEPTION**: A revolving door may be used without an adjacent swinging door for street floor elevator lobbies if a stairway, escalator, or door from other parts of the building does not discharge through the lobby and the lobby does not have any occupancy or use other than as a means of travel between elevators and street.

1012.3.5. A revolving door to be credited as a component of a means of egress shall comply with Sections 1012.3.1 through 1012.3.4 and the following conditions:

1. Revolving doors shall not be given credit for more than 50% of the required exit capacity.
2. Each revolving door shall be credited with no more than a 50-person capacity.
3. Each revolving door shall be capable of being collapsed when a force of not more than 130 lb [578 newtons (N)] is applied within 3 in (76 mm) of the outer edge of a wing.

1012.3.6. A revolving door not used as a component of a means of egress shall have a collapsing force of not more than 180 lb (801 N).

> **EXCEPTION**: A revolving door may have a collapsing force set in excess of 180 lb (801 N) if the collapsing force is reduced to not more than 130 lb (578 N) when at least one of the following is satisfied:
> - There is a power failure, or power is removed to the device holding the wings in position.
> - There is an actuation of the automatic sprinkler system when such system is provided.
> - There is an actuation of a smoke-detection system which is installed to provide coverage in all areas within the building which are within 75 ft (23 m) of the revolving doors.
> - There is the actuation of a manual control switch which reduces the holding force to below the 130 lb (578 N) level. Such a switch shall be in an approved location and shall be clearly identified.

1012.4 Horizontal Sliding Doors.

Approved and listed horizontal sliding doors complying with the following conditions may be used in a means of egress when specifically permitted by code.

1. The doors shall be power-operated and shall be capable of being operated manually in the event of power failure.
2. The doors shall be openable by a simple method from both sides without special knowledge or effort.
3. The force required to operate the door shall not exceed 30 lb (133 N) to set the door in motion and 15 lb (67 N) to close the door or open it to the minimum required width.
4. The door shall be openable with a force not to exceed 15 lb

(67 N) when a force of 250 lb (1112 N) is applied perpendicular to the door adjacent to the operating device.
5. The door assembly shall comply with the applicable fire protection rating and, when rated, shall be self-closing or automatic-closing by smoke detection, be installed in accordance with NFiPA 80, and comply with Section 705.1.3.
6. The door assembly shall have an integrated standby power supply.
7. The door assembly power supply shall be electrically supervised.

1012.5 Special Doorway Requirements
A door, when opening or when fully open, shall not project beyond the building line.

1012.6 Special Locking Arrangements
1012.6.1. Except in Group A Occupancies, doors in buildings protected throughout by an approved supervised automatic smoke-detection system or automatic sprinkler system may be equipped with approved, listed locking devices which shall
- Unlock upon actuation of the approved supervised automatic smoke-detection system or automatic sprinkler system.
- Unlock upon loss of power controlling the locking device.
- Initiate an irreversible process which will free the latch within 15 sec whenever a force of not more than 15 lb (67 N) is applied to a release device and not relock until the door has been opened. Operation of the release device shall activate a signal in the vicinity of the door for assuring those attempting to exit that the system is functional.

    EXCEPTION: The building official may approve a delay not to exceed 30 sec provided that reasonable life safety is assured.

1012.6.2. Signs shall be provided on the door adjacent to the release device which read: "Push. This door will open in 15 seconds. Alarm will sound." Sign letters shall be at least 1 in (25.4 mm) high.

1012.6.3. Emergency lighting shall be provided at the door.

1013.0 RAMPS [1013.0; 1013.1]
1013.1 General

Where changes in elevations exist in exit access corridors, exits and exit outlets, ramps shall be used when the difference in elevation is 12 in (305 mm) or less. Ramps in the means of egress shall conform to Sections 1013.2 through 1013.8.

1013.2 Slope
The maximum slope in the direction of travel shall be 1:12. The maximum cross slope shall be 1:48.
>   EXCEPTIONS:
>   - Maximum slope in direction of travel shall be 1:8 for a 3 in (76 mm) rise maximum and 1:10 for a 6 in (152 mm) rise maximum.
>   - Aisles in Group A Occupancies. See Section 1019.

1013.3 Rise
Maximum rise for a single ramp run shall be 30 in (762 mm).
>   EXCEPTION: Aisles in Group A Occupancies.

1013.4 Landings
Ramps shall have landings at the top and bottom and at doors opening onto the ramp. Angle of ramp slopes and landings shall not be steeper than 1:48.

1013.5 Handrails
Ramps steeper than 1:20 shall be provided with handrails along both sides of a ramp segment and shall conform with the requirements in Sections 1007.5.3, 1007.5.4, and 1007.5.6. If handrails are not continuous, they shall extend at least 12 in (305 mm) beyond the top and bottom of the ramp segment and shall be parallel with the floor or ground surface. Handrails shall be not less than 34 in (864 mm) nor more than 38 in (965 mm) above the ramp surface.
>   EXCEPTIONS:
>   Handrails are not required when the total rise is 6 in (152 mm) or less.
>   - Aisles in Group A Occupancies. See Section 1019.

1013.6 Drop-Offs
Ramps and landings with drop-offs at the sides shall have a curb with a minimum 4-in (102 mm) height, wall, railing, or a guardrail.

1013.7 Slip Resistance
Ramps shall have a slip-resistant surface.

1013.8 Water Accumulation

Exterior ramps and landings shall be designed so water will not accumulate on their surfaces.

1014.0 BALCONIES, PORCHES, AND GALLERIES [1014.0; 1014.0]

1014.1 General

1014.1.1. Any exterior balcony, porch, or gallery may serve as a means of egress if it complies with all the requirements as to width, arrangement, headroom, travel distance, and materials of the construction that are specified in this chapter for means of egress and provided it complies with the requirements of the following paragraphs of this section.

> **EXCEPTION**: Protection of openings onto exterior balconies in the building walls is not required, unless regulated by Section 1006.2.4.

1014.1.2. All porches, balconies, raised floor surfaces, or landings located more than 30 in (762 mm) above the floor or grade below shall have guardrails as specified in Section 1015.

1014.1.3. Balconies or other open spaces serving as a means of egress shall be maintained as a required path of travel without obstruction so as to maintain the required minimum width of exit travel.

1014.1.4. Exterior balconies used as an exit access from buildings four or more stories in height shall be of noncombustible construction. See Sections 1404.2 and 1404.3 for fire-protection requirements of balconies not used as a means of egress.

1014.1.5. See Section 3206 for projections over public property.

1015.0 GUARDRAILS [1015.0; 1015.0]

1015.1 General

All unenclosed floor and roof openings, open and glazed sides of landings and ramps, balconies or porches which are more than 30 in (762 mm) above finished ground level or a floor below shall be protected by a guardrail. Guardrails shall form a vertical protective barrier not less than 42 in (1067 mm) high.

Open guardrails shall have intermediate rails or ornamental pattern such that a 6-in (152 mm) diameter sphere cannot pass through any opening. A bottom rail or curb shall be provided that will reject the passage of a 2-in (51 mm) diameter sphere.

Construction of guardrails shall be adequate in strength, durability, and attachment for their purpose as described in Section 1608.2.
> **EXCEPTIONS**:
> - Guardrails are not required on the loading side of loading docks
> - Guardrails shall be permitted in conformance with requirements for specific occupancies in Section 1018.

1015.2 Glass
Glass guardrail components shall comply with SBC section 2405.5.

1016.0 EXIT ILLUMINATION AND SIGNS [1016.0; 1016.1]
1016.1 Means of Egress Illumination
1016.1.1. Means of egress shall be illuminated at all times when the building is occupied, with light of intensity of not less than 1 footcandle (fc) (11-1x) at the floor level, except theaters which shall have not less than 1/5 fc (2-1x) in aisles. For purposes of illumination, means of egress shall consist only of the exits and aisles, corridors, passageways, ramps, escalators, and lobbies leading to the exits.

1016.1.2. An independent and separate source of emergency power shall be provided for means of egress illumination in occupancies with the occupant load listed in Table 1016. Such emergency power shall be automatically actuated and emergency illumination provided for a period of 1-1/2 hr in the event of failure of normal lighting. Emergency lighting facilities shall be arranged to provide initial illumination that is not less than an average of 1 fc (11-1x) and a minimum at any point of 0.2 fc (2- 1x) measured along the path of egress at the floor level.

Measurements shall be taken at intervals of 2 ft 0 in (305 mm) between light sources. Illumination levels may decline to 0.6 fc (6-1x) average and a minimum at any point of 0.1 fc (1-1x) at the end of the emergency lighting duration. A maximum illumination uniformity ratio of 40:1 shall not be exceeded.
> **EXCEPTION**: The decline in the illumination level does not apply to occupancy Groups H and I or to Group B and R high rises.

1016.2 Exit Signs
1016.2.1. Exits shall be marked by an approved sign readily visible from any direction of exit access. Access to exits shall be marked by

readily visible signs in all cases where the exit or way to reach it is not immediately visible to the occupants. Sign placement shall be such that no point in the exit access is more than 100 ft (30 m) from the nearest visible sign. Every exit sign shall be suitably illuminated by a reliable light source. Externally and internally illuminated signs shall be visible in both the normal and emergency lighting mode.

1016.2.2. All exit and directional signs shall have letters at least 6 in (152 mm) high with a minimum stroke of 3/4 in (19 mm). The word EXIT shall have letters having a width not less than 2 in (51 mm) except the letter I and the minimum spacing between letters shall be not less than 3/8 in (10 mm).

Signs larger than the minimum established in this paragraph shall have letter widths, strokes, and spacing in proportion to their height. Each door to an exit stairway shall have a tactile sign posted that states EXIT and complies with CABO/ANSI Al17.1.

EXCEPTION: Group R3 Occupancies.

1016.2.3. Externally illuminated signs shall be illuminated by not less than 5 fc (54-1x) and shall employ a contrast ratio of not less than 0.5.

1016.2.3.1. The visibility of an internally illuminated sign shall be the equivalent of an externally illuminated sign. The 0.5 contrast ratio shall be derived from luminance measurements obtained in units of foot-lamberts (ftl). Approved self-luminous or electroluminescent signs which operate in the 5000 to 6000 angstrom (5×10^{-10} to 6×10^{-10}) range and which provide evenly illuminated letters shall have a luminance of not less than 0.06 ftl (0.21 cd/m).

1016.2.3.2. All exit signs for egress elements shall be provided with an emergency source of power supply in accordance with Section 1016.1.2.

EXCEPTION: Approved self-luminous signs.

1016.2.3.3. Where a main entrance serves as an exit and is visible to the occupants, an exit sign is not required over the main entrance door.

1016.2.4. Where exit lights or signs or the exits themselves are not visible from the exit approach, directional signs indicating the way of egress shall be provided. The level at which there is direct exit to the exterior shall also be clearly indicated.

1016.2.5. An independent and separate source of power shall be provided for exit signs in occupancies at the occupant load listed in Table 1016.

1016.2.6. Signs installed as projections from a wall or ceiling within the means of egress shall provide vertical clearance no less than 7 ft (2134 mm) from the walking surface.

Table 1016
Special Power for Exit Signs and Illumination

Occupancy	Minimum Occupant Load
Group A	All
Group I	All
Group H	All[1]
Group R	Greater than 100
Groups B and M	Greater than 150
Group E	Greater than 300

Note:
1. Individual rooms 500 sq ft (46.5 m²) or less in mixed occupancy and buildings 500 sq ft (46.5 m²) or less are exempted.

1016.3 Stair Identification

An approved sign shall be located at each floor level landing in all enclosed stairways of buildings four or more stories in height. The sign shall indicate the floor level and the availability of roof access from that stairway and an identification of the stairway. The sign shall also state the floor level of and direction to exit discharge. The sign shall be located approximately 5 ft (1524 mm) above the floor landing in a position which is readily visible when the door is in the open or closed position.

1017.0 EXIT OBSTRUCTIONS [1017.0; 1017.0]

1017.1. Where floor space is occupied by tables, chairs, or other movable furniture, aisles not less than 36 in (914 mm) clear width shall be maintained to provide ready access to egress doors.

1018.0 SPECIAL EGRESS REQUIREMENTS BY OCCUPANCY [1018.0; 1018.0]

1018.1. The general requirements of Chapter 9 apply to all occupancies except as modified for specific occupancies in accordance with Sections 1019 through 1027.

1019.0 ASSEMBLY [1019.0; 1019.0]

1019.1 Means of Egress Capacity

1019.1.1. The minimum aggregate width of main entrance doorways for Group A Occupancies shall be sufficient to accommodate 50% of the occupant load. Each level of a Group A Occupancy shall have access to a main exit and shall be provided with additional exits of sufficient width to accommodate one-half of the total occupant load served by that level.

1019.2 Foyers and Lobbies

1019.2.1. In every Group A - Large Assembly Occupancy, a foyer consisting of a space at a main entrance of the auditorium or place of assembly shall be provided. Such foyer, if not directly connected to a public street by all the main entrances or exits, shall have a straight and unobstructed corridor or passage to every such main entrance and exit.

1019.2.2. The width of a foyer at any point shall be not less than the combined width of aisles, stairways, and passageways tributary thereto.

1019.2.3. In theaters and similar Group A Occupancies, where persons are admitted to the building at times when seats are not available and are allowed to wait in a lobby or similar space, such use of lobby or similar space shall not encroach upon the required clear width of exits. Such waiting areas shall be separated from the required exits by substantial permanent partitions or by fixed rigid railings not less than 42 in (1067 mm) high.

1019.3 Interior Balcony and Gallery
1019.3.1 Means of Egress
For balconies or galleries of Group A Occupancies having a seating capacity of over 50, at least two means of egress shall be provided, one from each side of every balcony or gallery, leading directly to a street or exit court.

1019.3.2. Two means of egress shall be required from theater balconies when the occupancy exceeds 50.

1019.3.3 Enclosure and Capacity
All interior stairways and other vertical openings shall be enclosed and protected as provided in this chapter, except that stairs may be open between balcony and main assembly floor in occupancies such as theaters, churches, and auditoriums. The means of egress capacity required for balconies or galleries shall be determined on the same basis as those required for the occupancy use.

1019.3.4 Travel Distance
The maximum travel distance for balconies or galleries from any seat to an exit shall be determined on the same basis as the building occupancy.

1019.4 Stages
1019.4.1. Where two means of egress are required, they shall be separate with at least one means of egress on each side of the stage.

1019.4.2. The means of egress from lighting and access catwalks, galleries, and gridirons shall meet the requirements for Group F Occupancies.

EXCEPTIONS:
- A minimum width of 22 in (559 mm) shall be permitted for lighting and access catwalks.
- A second means of egress is not required from these areas where a means of escape to a floor or to a roof is provided.
- Ladders, alternating tread stairs, or spiral stairs shall be permitted in the means of escape.

1019.4.3. Each tier of dressing rooms shall be provided with two exits.

1019.4.4. Stairways from stage and dressing rooms need not be enclosed.

1019.5 Tents
Tent exits, aisles, seating, etc., shall conform with the requirements for places of assembly. All exits shall be kept free and clear of obstructions while the tent is occupied by the public.

1019.6 Projection Rooms
The projection room shall be provided with not less than one exit having a minimum opening of not less than 30 in (762 mm) wide and 80 in (2032 mm) high.

1019.7 Doors
1019.7.1. A key locking device may be used from the egress side on the main exterior exit doors on Group A-2 having an occupancy of 300 or less, subject to the following:
1. There is a readily visible durable sign on or adjacent to the door stating: This exit to remain unlocked when this building is occupied. The sign shall be in letters no less than 1 in (25.4 m) high on a contrasting background.
2. The locking device must be of a type that will be readily distinguishable as locked.
3. The main exit door is a single door or one pair of doors.
4. When unlocked, the door or both leaves of the pair must be free.

The use of the key locking device may be revoked by the building official for due cause.

1019.7.2. Each door in a means of egress from an area of Group A Occupancy may be provided with a latch or lock only if it is panic hardware or fire exit hardware, which releases when pressure of no more than 15 lb (67 N) is applied to the releasing devices in the direction of the exit travel.

Such releasing devices may be bars or panels extending not less than one-half the width of the door and placed at heights suitable for the service required, but not less than 30 in (762 mm) nor more than 44 in (1118 mm) above the floor. Whenever panic hardware is used on a labeled fire door, the panic hardware shall be labeled as fire exit hardware.

1019.7.3. If balanced doors are used and panic hardware is required, the panic hardware shall be of the push-pad type and the pad shall not extend more than one-half the width of the door measured from the latch side.

Means of Egress   213

1019.8 Stairway Construction

1019.8.1. In buildings of Group A Occupancy, flights of less than three risers shall not be used in interior or exterior stairways, passageways, aisles, at the entrance, or elsewhere in connection with required exits. To overcome lesser differences in level, gradients not exceeding 1:8 may be used. See Section 1019.10 for additional aisle and stair information in assembly occupancies.

1019.8.2. Aisles in Group A Occupancies with a gradient exceeding 1:8 shall consist of a series of risers and treads extending across the full width of the aisles and shall be illuminated. Such aisles shall comply with Sections 1019.8.2.1 through 1019.8.2.3.

1019.8.2.1. Tread depths shall be a minimum of 11 in (279 mm) and be uniform within each aisle.

> **EXCEPTIONS:**
> - Nonuniformities shall not exceed 3/16 in (4.8 mm) between adjacent treads.
> - Where seating is on stepped platforms, one tread in each seat platform may have a greater width to accommodate access to seats.

1019.8.2.2. On aisle stairs where the gradient must be the same as the gradient of adjoining seating areas, the riser height shall be not less than 4 in (102 mm) nor more than 8 in (203 mm) and it shall be uniform within each flight.

> **EXCEPTION:** Riser height may be nonuniform but only to the extent necessitated by changes in the gradient of the adjoining seating area to maintain adequate sight lines. Where nonuniformities exceed 3/16 in (4.8 mm) between adjacent risers, the exact location of such nonuniformities shall be indicated with a distinctive marking stripe on each tread at the nosing or leading edge adjacent to the nonuniform risers.

1019.8.2.3. A contrasting marking stripe shall be provided on each tread at the nosing or leading edge such that the location of each tread is readily apparent when viewed in descent. Such stripe shall be a minimum of 1 in (25.4 mm) wide and a maximum of 2 in (51 mm) wide.

> **EXCEPTION:** The marking stripe may be omitted where tread surfaces are such that the location of each tread is readily apparent when viewed in descent.

1019.9 Guardrails

1019.9.1. Assembly aisles located more than 30 in (762 mm) above the floor or grade below shall have guardrails in accordance with Section 1015.

1019.9.2. Where an elevation change of 30 in (762 mm) or less occurs between an aisle parallel to the seats (cross aisle) and the adjacent floor or grade below, guardrails not less than 26 in (660 mm) above the aisle floor shall be provided.

> **EXCEPTION**: Where the backs of seats on the front of the cross aisle project 24 in (610 mm) or more above the finish adjacent floor of the aisle, a guardrail need not be provided.

1019.9.3. Guardrails on a balcony, loge, or gallery immediately in front of the first row of fixed seats and which are not at the end of an aisle shall be not less than 26 in (660 mm) high. Guardrails 42 in (1067 mm) high and the width of the aisle shall be located at the front edge of a balcony, loge, or gallery where the aisle terminates. When the slope of the aisle is less than 1:8, the guardrail may be 36 in (914 mm) high where the aisle terminates.

1019.9.4. Guardrails are not required on the audience side of stages, raised platforms, and other raised floor areas such as runways, ramps, and side stages used for entertainment or presentations.

1019.9.5. Permanent guardrails are not required at vertical openings in the performance area of stages.

1019.9.6. Guardrails are not required where the side of an elevated walking surface is to be open for the normal functioning of special lighting or for access and use of other special equipment.

1019.10 Assembly Aisles and Seating

1019.10.1 General

Provisions in this section shall apply to all assembly aisles and seating except for special provisions relating to seating for reviewing stands, grandstands, and bleachers.

1019.10.1.2. Every portion of any building which contains seats, tables, displays, equipment, or other material shall be provided with aisles leading to exits.

1019.10.2 Aisle Width

1019.10.2.1. Aisle width shall provide sufficient egress capacity for the number of persons accommodated by the catchment area served by the aisle. See Section 1019.10.4. The catchment area served by an aisle is that portion of the total space that is naturally served by that section of the aisle. In establishing catchment areas the assumption shall be made that there is a balanced use of all means of egress, with the number of persons in proportion to egress capacity.

1019.10.2.2. Where aisles converge to form a single path of egress travel, the required egress capacity of that path shall be not less than the combined required capacity of the converging aisles.

1019.10.2.3. Those portions of aisles, where egress is possible in either of two directions, shall be uniform in required width.

1019.10.2.4. In all balconies and galleries having more than 20 rows of seats, there shall be provided a cross aisle not less than 4 ft (1219 mm) wide leading directly to an exit.

1019.10.2.5. The minimum clear width of aisles shall be

1. Forty-eight inches (1219 mm) for stairs having seating on each side
2. Thirty-six inches (914 mm) for stairs having seating on only one side
3. Twenty-three inches (584 mm) between a stair handrail or guardrail and seating when the aisle is subdivided by a handrail (see Section 1007.5)
4. Forty-two inches (1067 mm) for level or ramped aisles having seating on both sides
5. Thirty-six inches (152 mm) for level or ramped aisles having seating only on one side
6. Twenty-three inches (584 mm) between a stair handrail and seating when an aisle does not serve more than five rows on one side

1019.10.2.6. The minimum clear width of aisles in existing buildings shall be in accordance with Section 1402.

1019.10.3 Clear Width of Rows

1019.10.3.1. Where seating rows have 14 or fewer seats, the row minimum clear width shall be not less than 12 in (305 mm) measured as the clear horizontal distance from the back of the row ahead and the nearest projection of the row behind. Where chairs have automatic or self-rising seats, the measurement shall be made with seats in the raised position.

Where any chair in the row does not have an automatic or self-rising seat, the measurements shall be made with the seat in the down position. For seats with folding tablet arms, row spacing shall be determined with the tablet arm down.

1019.10.3.2. For rows of seating served by aisles or doorways at both ends there shall be no more than 100 seats per row and the row minimum clear width of 12 in (305 mm) shall be increased by 0.3 in (7.6 mm) for every additional seat beyond 14, but the minimum clear width need not exceed 22 in (559 mm).

1019.10.3.3. For rows of seating served by an aisle or doorway at one end only, the minimum clear width of 12 in (305 mm) between rows shall be increased by 0.6 in (15.2 mm) for every additional seat beyond seven, but the minimum clear width need not exceed 22 in (559 mm).

1019.10.3.4. For rows of seating served by an aisle or doorway on one end only, the path of travel shall not exceed 30 ft (9144 mm) from any seat to a point where a person has a choice of two paths of travel to two exits.

1019.10.4 Means of Egress Capacity
The width of aisles and other means of egress shall provide sufficient capacity in accordance with the following formulas where clear width is measured to walls, edges of seating, and tread edges except for permitted projections:

1. At least 0.3 in (7.6 mm) of width for each person served shall be provided on stairs having riser heights of 7 in (178 mm) or less and tread depths 11 in (279 mm) or greater, measured horizontally between tread nosings.
2. At least 0.005 in (0.127 mm) of additional stair width for each person shall be provided for each 0.10 in (2.5 mm) of riser height above 7 in (178 mm).
3. Where egress requires stair descent, at least 0.075 in (1.9 mm) of additional width for each person shall be provided on those portions of stair width having no handrail within a horizontal distance of 30 in (762 mm).
4. Level or ramped means of egress, with slopes less than 1:8, shall have at least 0.22 in (5.6 mm) of clear width for each person served.
5. Doorways shall have at least 0.2 in (5.1 mm) of clear width per person served.

1019.10.5 Travel Distance
Exits and aisles shall be so located that the travel distance to an exit door shall not be greater than 200 ft (61 m) measured along the line of travel. Travel distance may be increased to 250 ft (76 m) in sprinklered buildings.

1019.10.6 Aisle Slope
Aisles shall not have a slope of more than 1:8.

1019.10.7 Aisle Termination
1019.10.7.1. Dead-end aisles which terminate only at one end with a cross aisle, foyer, doorway, or vomitory giving access to an exit shall be not greater than 20 ft (6096 mm) long.

> **EXCEPTION:** A longer dead-end aisle is permitted where seats served by the dead-end aisle are not more than 24 seats from another aisle, measured along a row of seats having a minimum clear width of 12 in (305 mm) plus 0.6 in (15.2 mm) for each additional seat above seven in the row.

1019.10.7.2. Each end of a cross aisle shall terminate at an aisle, foyer, doorway, or vomitory giving access to an exit.

1019.10.8 Aisle Obstructions
There shall be no obstructions in the required width of aisles except for handrails as provided in Sections 1007.5 and 1019.10.2.5.

1019.10.9 Seat Stability
In places of assembly used regularly for theatrical or similar performances, or for the display of motion pictures, the seats shall be securely fastened to the floor. In restaurants, cafeterias, gymnasiums, and similar multipurpose places of assembly, the seats shall not be required to be fastened to the floor.

All other Group A Occupancies seating more than 200 persons shall have seats fastened to the floor. All seats in balconies or galleries shall be secured to the floor except that in railed-in enclosures, boxes, or loges, with level floors and having no more than 14 seats, the seat need not be fastened to the floor or have separating arms.

1019.10.10 Other Provisions
Other stair and ramp provisions are found in Sections 1007 and 1013.

1019.11 Grandstands, Bleachers, and Reviewing Stands
1019.11.1 General

These provisions shall apply to buildings or structures of an assembly occupancy which provides permanent, temporary, or portable seating facilities.

1019.11.1.2 Definitions
For the purpose of this section, certain special terms are defined as follows:

AISLE ACCESSWAY - That portion of an exit access that leads to an aisle.
SMOKE-PROTECTED ASSEMBLY SEATING - Seating served by a means of egress that is not subject to smoke accumulation within or under a structure.

1019.11.2 Smoke-Protected Assembly Seating
1019.11.2.1. The lowest portion of the roof shall not be less than 15 ft (4572 mm) above the highest aisle or aisle accessway.
1019.11.2.2. All enclosed areas shall be equipped with an approved automatic sprinkler system.
   EXCEPTIONS:
   - The floor area used for performances or entertainment is restricted to low fire hazard use and the roof is more than 50 ft (15 m) above the floor level.
   - Press boxes and storage facilities with area less than 1000 sq ft (93 m²) in outdoor seating facilities when all seating and means of egress are essentially open to the outside.

1019.11.2.3. All means of egress shall be provided with smoke-actuated ventilation or natural ventilation designed to maintain the smoke level at least 6 ft (1829 mm) above the floor of the means of egress.

1019.11.3 Travel Distance
The travel distance shall comply with Table 1004 (see Appendix). The distance shall be measured along the line of travel to an exit. Where aisles are required, the distance shall be measured along the aisles and aisle accessway without travel over or on the seats.
   EXCEPTIONS:
   - <u>Smoke-protected assembly seating</u>. The travel distance from each seat to the nearest entrance to a vomitory or concourse shall not exceed 200 ft (61 m). The travel

distance from the entrance to the vomitory or concourse to a stair, ramp or walk on the exterior of the building shall not exceed 200 ft (61 m).
- Outdoor assembly seating. The travel distance from each seat to the building exterior shall not exceed 400 ft (122 m). The travel distance shall not be limited in facilities of Type I or II construction.

1019.11.4 Aisles

Aisles shall be provided in all seating facilities except that an aisle may be omitted when all the following conditions exist:
1. Seats are without backrests.
2. The rise from row to row does not exceed 6 in (152 mm) per row.
3. The row spacing does not exceed 28 in (711 mm) unless the seat boards and footboards are at the same elevation.
4. The number of rows does not exceed 16 rows.
5. The first seating board is not more than 12 in (305 mm) above the ground or floor below or a cross aisle.
6. Seat boards have a continuous flat surface.
7. Seat boards provide a walking surface with a minimum width of 11 in (279 mm).
8. Egress from seating is not restricted by rails, guards, or other obstructions.

1019.11.5 Aisle Width

1019.11.5.1. The aisle width shall provide sufficient egress capacity for the number of persons accommodated by the catchment area served by the aisle. The catchment area served by an aisle is that portion of the total space that is naturally served by that section of the aisle. In establishing catchment areas, the assumption shall be made that there is a balanced use of all means of egress, with the number of persons in proportion to egress capacity.

1019.11.5.2. When bench-type seating is used, the number of persons shall be based on one person for each 18 in (457 mm) of length of the bench.

1019.11.5.3. Where aisles converge to form a single path of egress travel, the required egress capacity of that path shall be not less than the combined required capacity of the converging aisles.

1019.11.5.4. Where egress is possible in either of two directions, aisles shall be uniform in required width.

1019.11.5.5. The minimum clear width of aisles shall be
1. Forty-eight inches (1219 mm) for stairs having seating on each side
2. Thirty-six inches (914 mm) for stairs having seating on only one side
3. Twenty-three inches (584 mm) between a stair handrail or guardrail and seating when the aisle is subdivided by a handrail
4. Forty-two inches (1067 mm) for level or ramped aisles having seating on both sides
5. Thirty-six inches (914 mm) for level or ramped aisles having seating only on one side
6. Twenty-three inches (584 mm) between a stair handrail and seating when an aisle does not serve more than five rows on one side

1019.11.5.6. The minimum clear width of aisles in existing buildings shall be in accordance with Section 3402.

1019.11.6 Aisle Termination

1019.11.6.1. Aisles shall terminate at an aisle, foyer, doorway, or vomitory giving access to an exit.

> **EXCEPTION**: Dead-end aisles terminating at a cross aisle, foyer, doorway, or vomitory giving access to an exit at only one end and meeting any of the following conditions shall be permitted:
> 1. Where dead-end aisles do not exceed 20 ft (6096 mm) in length.
> 2. Where there are not more than 24 seats between aisles. The aisle accessway serving those seats shall have a minimum clear width of 12 in (305 mm) plus 0.6 in (15.2 mm) for each additional seat above seven in the row.
> 3. For smoke-protected assembly seating where there are not more than 40 seats between aisles. The aisle accessway serving those seats shall have a clear minimum width of 12 in (305 mm) plus 0.3 in (7.6 mm) for each additional seat above seven in the row.
> 4. For smoke-protected assembly seating, dead ends in vertical aisles do not exceed a distance of 21 rows.

5. When seats are without backrests, dead ends in vertical aisles do not exceed a distance of 16 rows.

1019.11.6.2. Each end of a cross aisle shall terminate at an aisle, foyer, doorway, or vomitory giving access to an exit.

1019.11.7 Aisle Walking Surfaces

Aisles with a slope not exceeding 1:8 shall consist of a ramp having a slip-resistant walking surface. Aisles with a slope exceeding 1:8 shall consist of a series of risers and treads extending across the full width of aisles and complying with the following requirements.

1. Tread depths shall be a minimum of 11 in (279) mm) and be uniform within each aisle.

    EXCEPTION: Nonuniformities shall not exceed 3/16 in (4.8 mm) between adjacent treads.

2. On aisle stairs where the slope must be the same as the slope of adjoining seating areas, the riser height shall be not less than 4 in (102 mm) nor more than 8 in (203 mm) and it shall be uniform within each flight. Riser heights not exceeding 9 in (229 mm) shall be permitted where they are necessitated by the slope of adjacent seating areas to maintain sight lines.

    EXCEPTION: Riser height may be nonuniform but only to the extent necessitated by changes in the slope of the adjoining seating area to maintain adequate sight lines. Where nonuniformities exceed 3/16 in (4.8 mm) between adjacent risers, the exact location of such nonuniformities shall be indicated with a distinctive marking stripe on each tread at the nosing or leading edge adjacent to the nonuniform risers.

3. A contrasting marking stripe shall be provided on each tread at the nosing or leading edge such that the location of each tread is readily apparent when viewed in descent. Such stripe shall be a minimum of 1 in (25.4 mm) wide and a maximum of 2 in (51 mm) wide.

    EXCEPTION: The marking stripe may be omitted where tread surfaces are such that the location of each tread is readily apparent when viewed in descent.

1019.11.8 Aisle Handrails

1019.11.8.1. Ramped aisles having a slope exceeding 1:15 and aisle stairs shall be provided with handrails located either at the side or within the aisle width.

> **EXCEPTIONS:**
> - Handrails are not required for ramped aisles having a slope not exceeding 1:8 and having seating on both sides.
> - Handrails are not required if, at the side of the aisle, there is a guardrail that complies with graspability requirements for handrails.

1019.11.8.2. Where there is seating on both sides of the aisle, handrails located within the aisle shall be discontinuous with gaps or breaks at intervals not exceeding five rows to facilitate access to seating and to permit crossing from one side of the aisle to the other. These gaps or breaks shall have a clear width of at least 22 in (559 mm) and not greater than 36 in (914 mm), measured horizontally; the handrail shall have rounded terminations or bends.

1019.11.8.3. Where handrails are provided in the middle of aisle stairs, there shall be an additional, intermediate handrail located approximately 12 in (305 mm) below the main handrail.

1019.11.9 Rows

1019.11.9.1. Seating rows shall have aisle accessways with minimum clear width measured in accordance with Section 1019.11.9.2 and increased, for row length, in accordance with Sections 1019.11.9.3 and 1019.11.9.4.

1019.11.9.2. The minimum clear width of aisle accessways shall be not less than 12 in (305 mm) measured as the clear horizontal distance from the back of the row or guardrail ahead and the nearest projection of the row behind.

Where chairs have automatic or self-rising seats, the measurement shall be made with seats in the raised position. Where any chair in the row does not have an automatic or self-rising seat, the measurement shall be made with the seat in the down position.

1019.11.9.3. For rows of seats served by aisles or doorways at both ends, there shall be no more than 100 seats per row and the minimum clear width of 12 in (305 mm) for aisle accessways shall be increased by 0.3 in (7.6 mm) for every additional seat beyond 14, but the minimum clear width need not exceed 22 in (559 mm).

EXCEPTION: For smoke-protected assembly seating the row length limits, beyond which the aisle accessway minimum clear width of 12 in (305 mm) must be increased, shall be in accordance with Table 1019.11.9.3.

Table 1019.11.9.3
Smoke-Protected Assembly Seating
12-in Aisle Accessway Row Length Limits

Total No. of Seats in the Space	No. of Seats per Row Permitted to Have a Minimum 12-in (305 mm) Clear Width Aisle Accessway	
	Aisle or Doorway at Both Ends of Row	Aisle or Doorway at One End of Row
<4,000	14	7
4,000	15	7
7,000	16	8
10,000	17	8
13,000	18	9
16,000	19	9
19,000	20	10
≥20,000	21	11

1019.11.9.4. For rows of seats served by an aisle or doorway at one end only, the aisle accessway minimum clear width of 12 in (305 mm) shall be increased by 0.6 in (15.2 mm) for every additional seat beyond seven, but the minimum clear width need not exceed 22 in (559 mm).

EXCEPTION: See exception to Section 1019.11.9.3.

1019.11.9.5. For rows of seats served by an aisle or doorway on one end only, the path of travel shall not exceed 30 ft (9144 mm) from any seat to a point where a person has a choice of two directions of egress travel.

> **EXCEPTION**: For smoke-protected assembly seating, the path of travel shall not exceed 50 ft (15 m) from any seat to a point where a person has a choice of two directions of egress travel.

1019.11.10 Capacity of Means of Egress

1019.11.10.1. The minimum clear width of aisles and other means of egress shall comply with Section 1019.11.10.2 in the case without smoke-protected assembly seating and with Section 1019.11.10.3 in the case of smoke-protected assembly seating. The clear width shall be measured to intermediate handrails, edges of seating, tread edges, and walls.

> **EXCEPTION**: Outdoor assembly seating otherwise complying with the requirements for smoke-protected seating shall have means of egress capacities determined by either the provisions of Section 1019.11.10.3 or 1019.11.10.4.

1019.11.10.2. Without smoke-protected assembly seating. The minimum clear width of aisles and other means of egress shall provide sufficient capacity in accordance with the following:

1. At least 0.3 in (7.6 mm) of width for each person served shall be provided on stairs having riser heights 7 in (178 mm) or less and tread depths 11 in (279 mm) or greater, measured horizontally between tread nosings.
2. At least 0.005 in (0.127 mm) of additional stair width for each person shall be provided for each 0.10 in (2.5 mm) of riser height above 7 in (178 mm).
3. Where egress requires stair descent, at least 0.075 in (1.9 mm) of additional width for each person shall be provided on those portions of stair width having no handrail within a horizontal distance of 30 in (762 mm).
4. Level or ramped means of egress with slopes not exceeding 1:10 shall have at least 0.2 in (5.1 mm) of clear width for each person served. Ramps with slopes exceeding 1:10 shall have at least 0.22 in (5.6 mm) of clear width per person.
5. Doorways shall have at least 0.2 in (5.1 mm) of clear width per person served.

1019.11.10.3. Smoke-Protected Assembly Seating

The minimum clear width of aisles and other means of egress for smoke-protected assembly seating shall provide sufficient capacity in accordance with the following table. The number of seats specified shall be within a single assembly space and interpolation shall be permitted between the specific values shown.

1019.11.10.4. Outdoor Smoke-Protected Assembly Seating

The minimum clear width of aisles and other means of egress, in inches, shall be not less than the total occupant load served by the egress element multiplied by 0.08 when the egress is by stairs and multiplied by 0.06 when the egress is by ramps, corridors, tunnels, or vomitories.

1019.11.11 Guardrails

Guardrails shall be located along open-sided walking surfaces and elevated seating facilities which are located more than 30 in (762 mm) above the floor or ground below. Guardrails shall be not less than 42 in (1067 mm) in height measured vertically above the leading edge of the tread, adjacent walking surface, or adjacent seat boards.

> **EXCEPTION**: Guardrails at the front-row seats, which are not located at the end of an aisle and where there is no cross aisle, may have a height of not less than 26 in (660 mm).

1019.11.12 Bleacher Footboards

Bleacher footboards shall be provided for all rows of seats above the third row or beginning at such a point where the seating plank is more than 2 ft (610 mm) above the ground or floor below. When the same platform is used for both seating and footrests, footrests are not required, provided each level or platform is not less than 24 in (610 mm) wide. When projected on a horizontal plane, there shall be no horizontal gaps exceeding 1/4 in (6.4 mm) between footboards and seat boards. At aisles, there shall be no horizontal gaps exceeding 1/4 in (6.4 mm) between footboards.

1020.0 BUSINESS [1020.0; 10120.1]
1020.1 Single Exit

A single exit is permitted in Group B Occupancies when meeting the following conditions:
- Maximum two stories in height.

- Each floor area served by that exit does not exceed 3500 sq ft (325 m²).
- There are no more than 40 persons above the street floor.
- The maximum distance of travel to the exit does not exceed 75 ft (23 m).

1020.2 Locking

A key locking device may be used from the egress side on the main exterior exit doors on Group B Occupancies subject to the following:

- There is a readily visible durable sign on or adjacent to the door stating: "This exit to remain unlocked when this building is occupied." The sign shall be in letters no less than 1 in (25.4 mm) high on a contrasting background.
- The locking device must be of a type that will be readily distinguishable as locked.
- The main exit door is a single door or one pair of doors.
- When unlocked, the door or both leaves of the pair must be free.

The use of the key locking device may be revoked by the building official for due cause.

1021.0 EDUCATIONAL [1021.0; 1021.1]

1021.1 Special Exit Requirement

Rooms used for first-grade children and younger shall be located on the floor of exit discharge. Rooms used for second-grade children shall not be located more than one story above the floor of exit discharge.

1021.2 Panic and Fire Exit Hardware

Each door in a means of egress from an area of Group E Occupancy having an occupant load of 100 or more may be provided with a latch or lock only if it is panic hardware or fire exit hardware, which releases when a force of no more than 15 lb (67 N) is applied to the releasing devices in the direction of the exit travel.

Such releasing devices may be bars or panels extending not less than one-half the width of the door and placed at heights suitable for the service required, but not less than 30 in (762 mm) or more than 44 in (1118 mm) above the floor. Whenever panic hardware is used on a labeled fire door, the panic hardware shall be labeled as fire exit hardware.

1021.2.2. If balanced doors are used and panic hardware is required, the panic hardware shall be of the pushpad type and the pad shall not extend more than one-half the width of the door measured from the latch side.

1022.0 FACTORY-INDUSTRIAL [1022.0; 1022.0]
1022.1 Travel Distance
For allowable increase in travel distance, see Section 1004.1.4.
1022.2 Doors
Egress doors shall conform to the requirements of Section 1012.1.2 except in factory areas with an occupant load of 10 or less.
1022.3 Locks
A key locking device may be used from the egress side on the main exterior exit doors on Group F Occupancies subject to the following:
- There is a readily visible durable sign on or adjacent to the door stating: "This exit to remain unlocked when this building is occupied." The sign shall be in letters no less than 1 in (25.4 mm) high on a contrasting background.
- The locking device must be of a type that will be readily distinguishable as locked.
- The main door is a single door or one pair of doors.
- When unlocked, the door or both leaves of the pair must be free.

The use of the key locking device may be revoked by the building official for due cause.

1022.4 Handrails and Guardrails
Handrails and guardrails shall be installed in accordance with Sections 1007.5 and 1015, respectively.

> **EXCEPTION**: In areas not accessible to the public in Group F, the clear distance between rails measured at right angles to the rails shall not exceed 21 in (533 mm).

1023.0 HAZARDOUS [1023.0; 1023.0]
1023.1 Doors
All egress doors in Group H Occupancies shall swing in the direction of exit travel.

1023.2 Handrails and Guardrails

Handrails and guardrails shall be installed in accordance with Sections 1007.5 and 1015.

> **EXCEPTION**: In areas not accessible to the public in Group H, the clear distance between rails measured at right angles to the rails shall not exceed 21 in (533 mm).

1024.0 INSTITUTIONAL [1024.0; 1024.0]
1024.1 Group I Unrestrained Occupancy
1024.1.1 Horizontal Exits

Horizontal exits meeting the requirements of Section 1009 may comprise two-thirds the required exits from any building or floor area in Group I Unrestrained Occupancies.

1024.1.1.2. The capacity of areas of refuge shall be computed in accordance with Section 1009 and the area for each occupant as follows:

- Thirty square feet (2.8 m²) per patient for hospitals and nursing homes.
- Fifteen square feet (1.4 m²) per resident for ambulatory Group I. Unrestrained uses.
- Six square feet (0.6 m²) per occupant on stories not housing bed or litter patients in Group I Unrestrained uses.
- Three square feet (0.3 m²) per occupant in all other cases.

1024.1.2 Doors and Corridors

1024.1.2.1. Doors shall be not less than 44 in (1118 mm) clear width in the following:

1. Doorways to areas housing bedridden patients.
2. Doorways between patient rooms and exits.
3. Exterior exit doorways.

> **EXCEPTION**: Exit doors not subject to use for patient care shall be not less than 36 in (914 mm) clear width.

1024.1.2.2. Corridors, ramps, or passageways shall be a minimum of 8 ft (2438 mm) clear width in the following:

1. All areas occupied by patients
2. All means of egress from patient areas

1024.1.3 Locks

Patient rooms or tenant space egress doors in Group I Occupancies shall not be lockable except in places of restraint or detention.

1024.1.4 Handrails

All stairs or changes in grade in hospitals, nursing homes, convalescent homes, and similar occupancies shall be equipped with handrails located not less than 30 in (762 mm) nor more than 38 in (965 mm) above the leading edge of a tread.

>**EXCEPTION**: Handrails that form part of a guardrail may be 42 in (1067 mm) high.

1024.1.5 Institutional Illumination

Each building housing a Group I Unrestrained occupancy equipped with or requiring the use of life-support systems shall have illumination for the means of egress and emergency lighting equipment supplied by the life safety branch of the electrical system described in chapter 3, NFiPA 99.

1024.1.6 Smokeproof Enclosure

The smokeproof enclosure required by Section 1005.5 may be omitted when all required exit stairways are pressurized in accordance with SBC section 412.11(5).

1024.2 Group I Restrained Occupancy

1024.2.1 Mixed Use

Refer to SBC section 409.2.1 for means of egress requirements for areas classified as a different occupancy and traversing other use areas.

1024.2.2 Subclassification of Occupancy

1024.2.2.1. Group I Restrained shall be categorized as one of the following Use Conditions:

1. **Use Condition 1 : Free Egress**. Free movement is allowed from sleeping areas and other spaces where access or occupancy is permitted to the exterior by means of egress meeting the requirements of code. Group I Restrained qualifying for Use Condition 1 may be classified as a Group R Occupancy.
2. **Use Condition 2 : Zoned Egress**. Free movement is allowed from sleeping areas and any other occupied smoke compartment to one or more other smoke compartments.
3. **Use Condition 3 : Zoned Impeded Egress**. Free movement is allowed within individual smoke compartments, such as within a residential unit comprised of individual sleeping rooms and group activity space, with

egress impeded by remote-control release of means of egress from such smoke compartment to another smoke compartment.
4. **Use Condition 4 : Impeded Egress**. Free movement is restricted from an occupied space. Remote-controlled release is provided to permit movement from all sleeping rooms, activity spaces, and other occupied areas within the smoke compartment and to other smoke compartments.
5. **Use Condition 5 : Contained**. Free movement is restricted from an occupied space. Staff-controlled manual release at each door is provided to permit movement from all sleeping rooms, activity spaces, and other occupied areas within the smoke compartment and to other smoke compartments.

1024.2.2.2. To be classified as Use Condition 3 or 4, the arrangement, accessibility, and security of the release mechanism used for emergency egress shall be such that with the minimum available staff, at any time, lock mechanisms can be released within 2 min.

1024.2.3 Capacity of Means of Egress

The capacity of any required means of egress shall be based on the provisions of Sections 1003.2 and 1003.3.

1024.2.4 Number of Exits

1024.2.4.1. A minimum of two exits located remote from each other shall be accessible from each floor, fire compartment, or smoke compartment of the building.

1024.2.4.2. At least one of the required exits shall be accessible from each fire compartment and each required smoke compartment into which residents may be moved in a fire emergency with the exits so arranged that egress shall not require the occupants to return through the compartment from which egress originates.

1024.2.5 Arrangement of Means of Egress

1024.2.5.1. Every sleeping room shall have a door leading directly to an exit access corridor.

EXCEPTIONS:
- If there is an exit door opening directly to the outside from the room at the ground level.
- Where individual occupant sleeping rooms adjoin a dayroom or group activity space which is utilized for access to an exit, such

sleeping rooms may open directly to the day space and may be separated in elevation by up to a full story height.

1024.2.5.2. All exits may discharge through the level of exit discharge. The requirements of Section 1010 may be waived provided that not more than 50% of the exits discharge into a single fire compartment.

1024.2.5.3. Exits may discharge into a fenced or walled courtyard. Enclosed yards or courts shall be of sufficient size to accommodate all occupants, a minimum of 50 ft (15 m) from the building with a net area of 15 sq ft (1.4 m²) per person.

1024.2.5.4. No exit or exit access shall contain a corridor, hallway, or aisle having a pocket or dead end exceeding 50 ft (15 m) for Use Conditions 2, 3, and 4 and 20 ft (6096 mm) for Use Condition 5.

1024.2.5.5. The distance which must be traversed before two separate and distinct paths of travel to two exits are available shall not exceed 50 ft (15 m).

> **EXCEPTION:** 100 ft (30 m) shall be permitted in buildings completely protected by an approved automatic sprinkler system.

1024.2.5.6. A sally port may be permitted as a means of egress where there are provisions for continuous and unobstructed passage through the sally port during an emergency exit condition.

1024.2.5.7. Aisles, corridors, and ramps required for access or exit shall be at least 4 ft (1219 mm) wide.

1024.2.6 Measurement of Travel Distance to Exits

Travel distance shall be determined in accordance with Section 1004, but shall not exceed

1. One hundred feet (30 m) between any room door required as exit access and an exit
2. One hundred fifty feet (46 m) between any point in any room and an exit
3. Fifty feet (15 m) between any point in a sleeping room and the door of that room

   > **EXCEPTION:** The travel distance above may be increased by 50 ft (15 m) in any room other than a sleeping room when the building is protected throughout by an approved automatic sprinkler system or smoke control system.

1024.2.7 Horizontal Exits
1024.2.7.1. Horizontal exits may comprise 100% of the exits required. At least 6 sq ft (0.6 m²) of accessible space per occupant shall be provided on each side of the horizontal exit for the total number of people in adjoining compartments. Every fire compartment for which credit is allowed in connection with a horizontal exit shall not be required to have a stairway or door leading directly outside, provided the adjoining fire compartments have stairways or doors leading directly outside.
1024.2.7.2. The capacity of areas of refuge shall be computed in accordance with Section 1009 and the area for each occupant shall be as follows:

1. Six square feet (0.6 m²) per occupant for Group I Restrained uses
2. Three square feet (0.3 m²) per occupant in all other cases

1023.2.8 Doors
1024.2.8.1. Egress doors shall conform to the requirement of Section 1012.1.2 except in Group I Restrained when used as a place of detention.
1024.2.8.2. Doors to resident sleeping rooms shall be at least 28 in (711 mm) clear width.
1024.2.8.3. Doors in a means of egress may be of the horizontal sliding type provided the force needed to slide the door to its fully open position does not exceed 50 lb (222 N) with a perpendicular force against the door of 50 lb (222 N).
1024.2.9 Locks
1024.2.9.1. Locking devices may be used in Group I Restrained Occupancies.
1024.2.9.2. Doors may be locked in accordance with the applicable use condition.
1024.2.9.3. Doors from areas of refuge to the exterior may be locked with a key lock in lieu of locking methods described in Section 1024.2.9.5. The keys to unlock such doors shall be available at all times, and the locks shall be operable from both sides of the door.
1024.2.9.4. Any remote release in a means of egress shall be provided with reliable means of operation, remote from the resident living areas, to release locks on all required doors.

> EXCEPTION: Provisions for remote unlocking in Use Conditions 3 and 4 may be waived provided not more than 10 locks are necessary to be unlocked in order to move all occupants from one smoke compartment to an area of refuge within 3 min.

The opening of all necessary locks shall be accomplished with no more than two separate keys. This exception shall not be used for smoke barrier doors serving a smoke compartment containing more than 20 persons.

1024.2.9.5. All remote-release operated doors shall be provided with a redundant means of operation as follows:
- Power-operated sliding doors or power-operated locks shall be so constructed that in the event of power failure a manual mechanical means to release and open the doors is provided at each door.
- Mechanically operated sliding doors or mechanically operated locks shall be provided with a manual mechanical means at the door to release and open the door at the door.

1024.2.9.6. Emergency power shall be provided for all electrically power-operated sliding doors and power-operated locks. Automatic transfer from the normal power service shall be accomplished within 10 sec and operate under full load conditions for at least 1-1/2 hrs.

> EXCEPTION: This provision is not applicable for facilities with 10 locks or less complying with the exception in Section 1024.2.9.4.

1024.2.9.7. Doors remotely unlocked under emergency conditions shall not automatically relock when closed unless specified action is taken at the remote location to enable doors to relock.

1024.2.10 Stairs

1024.2.10.1. Spiral stairs meeting the requirements of Section 1007.8.2 are permitted for access to and between staff locations.

1024.2.10.2. Alternating-tread stairways meeting the requirements of Section 1007.8.4 are permitted for access to and between staff locations.

1024.2.10.3. Solid risers, intermediate handrails, latticework, or similar facilities required by Sections 1007.1.2 and 1007.5.8 which would interfere with visual supervision of residents are not required.

1024.2.11 Handrails and Guardrails

Handrails and guardrails shall be installed in accordance with Sections 1007.5 and 1015.

> **EXCEPTION**: In areas not accessible to the public in Group I Restrained, the clear distance between rails measured at right angles to the rails shall not exceed 21 in (533 mm).

1024.2.12 Illumination and Marking of Means of Egress

1024.2.12.1. Illumination shall be in accordance with Section 1016.

1024.2.12.2. Emergency lighting shall be provided in accordance with Section 1016.1.2.

1024.2.12.3. Exit markings shall be provided in areas accessible to the public in accordance with Section 1016.2.

> **EXCEPTION**: Exit signs may be omitted in sleeping room areas.

1025.0 MERCANTILE [309.0; 309.1]

1025.1 Single Exit

A single exit is permitted in Group M Occupancies when meeting the following conditions:

- One story maximum.
- The floor area does not exceed 2250 sq ft (209 m^2).
- The maximum distance of travel to the exit does not exceed 50 ft (15 m).

1025.2 Locks

A key locking device may be used from the egress side on the main exterior exit doors in Group M Occupancies subject to the following:

- There is a readily visible durable sign on or adjacent to the door stating: "This exit to remain unlocked when this building is occupied." The sign shall be in letters no less than 1 in (25.4 mm) high on a contrasting background.
- The locking device must be of a type that will be readily distinguishable as locked.
- The main exit door is a single door or one pair of doors.
- When unlocked, the door or both leaves of the pair must be free.

The use of the key locking device may be revoked by the building official for due cause.

1026.0 RESIDENTIAL [309.0; 309.1]
1026.1 Single Exit
1026.1.1. In Group R1 and R2 Occupancies one common exit is permitted provided all the following conditions are met:
- Maximum distance of travel to reach the exit from the entrance door to any dwelling unit shall not exceed 30 ft (9144 mm).
- Maximum number of dwelling units served by the exit shall not exceed four per floor.
- Maximum gross area of the dwelling units served by the exit shall not exceed 3500 sq ft (325 m^2) per floor.
- Maximum building height shall be one story above the level of exit discharge.

1026.1.2. A single exit is permitted in R3 Occupancies.

1026.2 Doors
Egress doors shall conform to the requirements of Section 1012.1.2, except doors within a dwelling or dwelling unit need not be of the side-swinging type unless such doors open onto common corridors or common balconies or are required exits.

1026.3 Locks
A night latch, dead bolt or security device may be used on exit doors from a dwelling unit or a hotel guest room or suite provided such devices are openable from the inside without the use of a key, tool, special knowledge, or effort and the device is mounted at a height not to exceed 48 in (1219 mm) above the finished floor.

1026.4 Guardrails
1026.4.1. Guardrails for dwellings and within individual dwelling units or guest rooms shall be a minimum of 36 in (9144 mm) high.

1026.4.2. For one- and two-family dwellings, only one intermediate rail located between 14 and 18 in (356 and 457 mm) above floor level shall be required between the top of the guardrail and the floor level of boat docks, piers, landings, decks on beach fronts, and dune walkovers, providing the floor or deck level is not more than 6 ft (1829 mm) above the mean high-water level or average grade of the beach, dune, or ground below. No guardrail shall be required on that portion of a boat dock used for docking a boat.

1026.4.3. A bottom rail or curb is not required on guardrails within dwellings or dwelling units.

1027.0 STORAGE [311.0; 311.1]
1027.1 Number of Exits
1027.1.1 Single Exits
A single exit is permitted in Group S Occupancies when meeting the following conditions:
- One story maximum.
- The floor area does not exceed 2500 sq ft (232 m²).
- The maximum distance of travel to the exit does not exceed 50 ft (15 m).

1027.1.2 Helistops
Exits and stairways from helistops shall comply with the provisions of this chapter, except that all landing areas located on buildings or structures shall have two or more exits. For landing platforms or roof areas less than 60 ft (18 m) long, or less than 2000 sq ft (186 m²) in area, the second exit may be a fire escape or ladder leading to the floor below.

1027.2 Travel Distance
For allowable increase in travel distance, see Section 1004.1.4.

1027.3 Doors
Egress doors shall conform to the requirements of Section 1012.1.2 except in automobile parking garages and storage areas with an occupant load of 10 or less.

1027.4 Locks
A key locking device may be used from the egress side on the main exterior exit doors in Group S occupancies subject to the following:
- There is a readily visible durable sign on or adjacent to the door stating: "This exit to remain unlocked when this building is occupied." The sign shall be in letters no less than 1 in (25.4 mm) high on a contrasting background.
- The locking device must be of a type that will be readily distinguishable as locked.
- The main exit door is a single door or one pair of doors.
- When unlocked, the door or both leaves of the pair must be free.
- The use of the key locking device may be revoked by the building official for due cause.

1027.5 Handrails and Guardrails
Handrails and guardrails shall be installed in accordance with Sections 1007.5 and 1015.

EXCEPTION: In areas not accessible to the public in Group S, the clear distance between rails measured at right angles to the rails shall not exceed 21 in (533 mm).

Chapter 10
Exterior Walls

SECTION 1400.0 EXTERIOR WALLS [1400.0; 1401.1]

The national building codes incorporate recognized tested and applied industry standards for use in judging the performance of exterior walls, materials, systems, and installations. This provides for a system that uses the equal treatment of both proven traditional materials and systems, and the introduction of new construction technologies. The national building codes, in this manner, provide for the efficient introduction of new materials into the traditional construction process. The purpose here is to assure a high level of public safety and construction technologies advancement.

1401.1 Scope
Provisions of this chapter shall govern the construction of exterior veneered walls, architectural trim, balconies, bay windows, and openings for fire department access.

1402.0 DEFINITIONS [1402.0; 1402.0]
This chapter contains no unique definitions. For general definitions, see Chapter 2. For definitions specific to masonry, see Section 2102.

1403.0 VENEERED WALLS [1403.0; 1405.0]
1403.1. Veneer refers to a facing of brick; tile; concrete; masonry units; metal, including metal coated with porcelain enamel; glass; wood; or similar material securely attached to a wall for the purpose of providing ornamentation, protection, or insulation but not so bonded as to exert a common reaction under load. [Figure 10-1 shows a cross-sectional view of a typical veneered wall.]

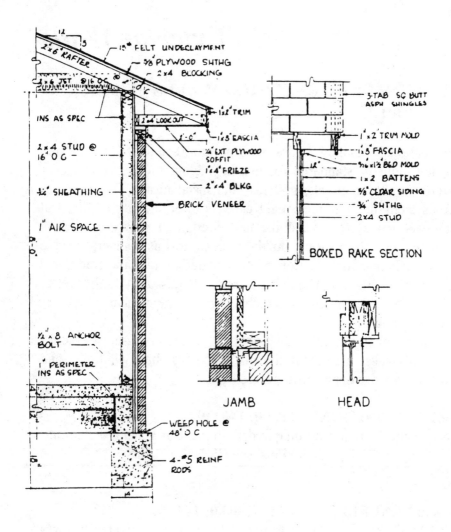

**Figure 10-1
Veneered Wall and Details**

1403.1.2. Veneer shall not be assumed to support any load other than its own weight, nor shall it be assumed to add to the strength of the wall.

1403.3.3. Veneered walls shall provide weather protection for the building at the walls.

1403.1.4. Flashing shall be provided as necessary to prevent the entrance of water at openings in or projections through veneered walls. Flashing shall be provided at intersections of veneered walls of different materials unless such materials provide a self-flashing joint and at other points subject to the entrance of water. Caulking shall be provided where such flashing is determined by the building official to be impractical.

1403.1.4.1. Flashing and weep holes as outlined in SBC section 2111.1.3 shall be located in the first course of masonry above finished ground level above the foundation wall or slab, and other points of support, including structural floors, shelf angles, and lintels when anchored veneers are designed in accordance with Sections 1403.2.4, 1403.2.5, and 1403.2.6.

1403.2 Anchored Masonry Veneer

1403.2.1. Anchored veneer is secured with approved mechanical fasteners to an approved backing. All masonry units, mortar, and metal accessories used in anchored veneer walls shall meet the physical requirements of this section. Anchored veneer units shall be not less than 1-5/8 in (41 mm) in actual thickness for solid masonry units and not less than 2-5/8 in (67 mm) in actual thickness for hollow masonry units. [The scaffold system used to construct anchored masonry veneer must also meet the Occupational Safety and Health Administration (OSHA) code requirements shown in Figure 10-2.]

1403.2.2 Support

1403.2.2.1. The weight of anchored veneer shall be vertically supported on footings, foundation walls, or other approved noncombustible structural supports. Wood foundations meeting the requirements of Section 1804.8 are permitted to vertically support anchored veneer.

1403.2.2.2. Anchored veneer supported laterally by wood frame shall be limited to a maximum height of 30 ft (9144 mm). Anchored veneer installed more than 30 ft (9144 mm) in height above the noncombustible foundation or support shall be laterally supported

242 Chapter Ten

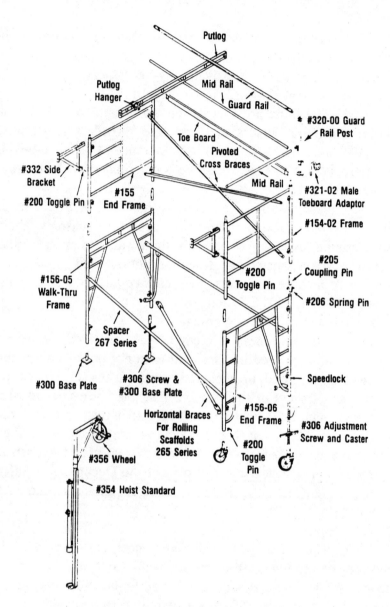

Figure 10-2
Wall Construction Scaffold

Exterior Walls 243

by noncombustible structural framing. The structural framing shall horizontally support the weight of veneer at least at the initial 30-ft (9144 mm) height and at each story height thereafter.

> **EXCEPTION**: These height restrictions may be increased for noncombustible structural framing when special design techniques, approved by the building official, are used in construction.

1403.2.3. Noncombustible lintels and noncombustible supports shall be provided over all openings where the anchored veneer is not self-supporting. The deflections of all structural lintels and horizontal supports for reinforced masonry required by Sections 1403.2.2 and 1403.2.3 shall not exceed 1/600 of the span or 0.3 in (7.62 mm).

1403.2.4. Masonry veneer anchored to wood framing shall be attached with corrosion-resistant corrugated sheet metal that is not less than 0.029 in (0.74 mm) by 7/8 in (22.2 mm) wide or corrosion-resistant ties of strand wire that are not less than no. 9 gauge (3.76 mm) wire with ends of the wire bent to a 90^0 (1.6 rad) angle to form a hook not less than 2 in (51 mm) long. The metal ties shall be embedded in the mortar joint a minimum of one-half the veneer thickness. Each metal tie shall support not more than 3 sq ft (0.28 m^2) of wall area with a maximum spacing of 16 in (406 mm) vertically and 32 in (813 mm) horizontally.

When anchored veneer is applied over wood frame, the studs shall be spaced a maximum of 24 in (610 mm) on center horizontally and be faced with sheathing materials as specified in SBC Table 2308.2.2B or insulation board on both sides. The minimum thickness of the sheathing material or insulation board installed on each side of the studs shall be 1/2 in (12.7 mm). In addition, a 1-in (25.4 mm) minimum air space shall be maintained between the anchored veneer and the exterior face of the sheathing material or insulation board. Moisture protection shall be provided as required by Section 2303.3.

1403.2.5. Masonry veneer anchored to corrosion-resistant steel framing shall be attached with corrosion-resistant ties of strand wire and shall be not less than 0.148-in (3.76 mm) (no. 9 gauge) wire with the ends of the wire bent to a 90^0 (1.6 rad) angle to form a hook not less than 2 in (51 mm) long. The wire ties shall be embedded in the mortar joint a minimum of one-half the veneer thickness. Each metal tie shall support not more than 3 sq ft (0.28 m²) of wall area with a maximum spacing of 16 in (406 mm) vertically and 32 in (813 mm) horizontally.

When anchored veneer is applied over steel framing, the studs shall be spaced a maximum of 24 in (610 mm) on center horizontally and be faced with sheathing materials as specified in SBC Table 2308.2.2B or insulation board on both sides. The minimum thickness of the sheathing material or insulation board installed on each side of the steel framing shall be 1/2 in (12.7 mm). A 1 in (25.4 mm) minimum air space shall be maintained between the anchored veneer and the exterior face of the sheathing material or insulation board. Moisture protection shall be provided as required by Section 2303.3.

1403.2.6. Masonry veneer anchored to masonry or concrete walls shall be attached with corrosion-resistant corrugated sheet metal that is not less than 0.029 in (0.74 mm) by 7/8 in (22.2 mm) wide or corrosion-resistant ties of strand wire that are not less than no. 9 gauge wire (3.76 mm) with ends of the wire bent to a 90° (1.6 rad) angle to form a hook not less than 2 in (51 mm) long. The metal ties shall be embedded in the mortar joint a minimum of one-half the veneer thickness. Each metal tie shall support not more than 3 sq ft (0.28 m²) of wall area with a maximum spacing of 16 in (406 mm) vertically and 32 in (813 mm) horizontally. A 1-in (25.4 mm) minimum air space shall be maintained between the anchored veneer and the supporting masonry or concrete walls.

1403.2.7. Stone veneer units not exceeding 10 in (254 mm) in thickness may be anchored directly to masonry, concrete, or stud construction by one of the following methods:

1. With concrete or masonry backing, anchor ties shall be not less than 0.1055-in (2.68 mm) corrosion-resistant wire, or approved equal, formed beyond the base of the backing. The legs of the loops shall be not less than 6 in (152 mm) in length bent at right angles and laid in the mortar joint and spaced so that the eyes or loops are 12 in (305 mm) maximum on center in both directions. There shall be provided not less than a 0.1055-in (2.68 mm) corrosion-resistant wire tie, or approved equal, threaded through the exposed loops for every 2 sq. ft. (0.2 m²) of stone veneer. This tie shall be a loop having legs not less than 15

in (381 mm) in length bent so that it will lie in the stone veneer mortar joint. The last 2 in (51 mm) of each wire leg shall have a right-angle bend. A layer of 1 in (25.4 mm) minimum thickness of cement grout shall be placed between the backing and the stone veneer.

2. With a stud backing, a 2-in by 2-in (51 by 51 mm) mesh made of 0.0625 in (1.59 mm) corrosion-resistant wire with two layers of waterproof paper backing shall be applied directly to wood studs spaced a maximum of 16 in (406 mm) on center. On studs the mesh shall be attached with 2-in (51 mm) long corrosion-resistant steel wire furring nails at 4 in (102 mm) on center providing a minimum 1-1/8 in (29 mm) penetration into each stud and with 8d common nails at 8 in (203 mm) on center into top and bottom plates. The corrosion-resistant wire mesh may be attached to steel studs with equivalent wire ties. There shall be not less than a 0.1055-in (268 mm) corrosion-resistant wire, or approved equal, looped through the mesh for every 2 sq ft (0.2 m^2) of stone veneer.

This tie shall be a loop having legs not less than 15 in (381 mm) in length, so bent that it will lie in the stone veneer mortar joint. The last 2 in (51 mm) of each wire leg shall have a right-angle bend. A 1-in layer of (25.4 mm) minimum thickness of cement grout shall be placed between the backing and the stone veneer.

1403.2.8. Any slab-type veneer units not exceeding 2 in (51 mm) in thickness may be anchored directly to any masonry, concrete, or stud construction. For veneer units of marble, travertine, granite, or other stone units of slab form, ties of corrosion-resistant dowels in drilled holes must be located in the middle third of the edge of the units spaced a maximum of 24 in (610 mm) apart around the periphery of each unit with not less than four ties per veneer unit. Units shall not exceed 20 sq ft (1.9 m^2) in area.

If the dowels are not tight-fitting, the holes may be drilled not more than 1/16 in (1.6 mm) larger in diameter than the dowel with the hole countersunk to a diameter and depth equal to twice the diameter of the dowel in order to provide a tight-fitting key of cement mortar at the dowel locations when the mortar in the joint has set. All veneer ties shall be corrosion-resistant metal capable of resisting tension or compression force equal to two times the weight of the attached veneer. If made of sheet metal, veneer ties shall be not smaller in area than 0.0336 X 1 in (0.853 X 25.4 mm) or, if made of wire, not smaller in diameter than 0.1483 in (3.76 mm).

1403.2.9. Anchored terra-cotta or ceramic units not less than 1-5/8 in (41 mm) thick may be anchored directly to masonry, concrete or stud construction. Tied terra-cotta or ceramic veneer units shall be not less than 1-5/8 in (41 mm) thick with projecting dovetail webs on the back surface spaced approximately 8 in (203 mm) on center. The facing shall be tied to the backing wall with corrosion-resistant metal anchors of not less than no. 8 gauge wire installed at the top of each piece in horizontal bed joints not less than 12 in (305 mm) nor more than 18 in (457 mm) on center; these anchors shall be secured to 1/4-in (6.4 mm) corrosion-resistant pencil rods which pass through the vertical aligned loop anchors in the backing wall.

The veneer ties shall have sufficient strength to support the full weight of the veneer in tension. The facing shall be set with not less than a 2-in (51 mm) space from the backing wall, and the space shall be filled solidly with portland cement grout and pea gravel. Immediately prior to setting, the backing wall and the facing shall be drenched with clean water and shall be distinctly damp when the grout is poured.

1403.3 Adhered Masonry Veneer

1403.3.1. Adhered veneer is secured and supported through the adhesion of an approved bonding material applied to an approved backing. All masonry units used in adhered veneer walls shall meet the physical requirements of SBC chapter 21. Adhered veneer units shall be less than 1-5/8 in (41 mm) thick and the units shall not be assumed to support any superimposed loads. With the exception of ceramic tile, adhered veneer and its backing shall be designed to provide a bond to the supporting element sufficient to withstand a shearing stress of 50 psi (345 kPa) after curing 28 days.

1403.3.2. Backing permitted for adhered veneer shall be continuous and may be of any material permitted by code. The backing shall have surfaces prepared to secure and support the imposed loads of the adhered veneer.

1403.3.3. Exterior adhered veneer shall not be attached to wood frame construction at a point more than 30 ft (9144 mm) in height

Exterior Walls

above the noncombustible foundation. The 30-ft (9144 mm) limit may be increased when special design techniques, approved by the building official, are used in construction.

1403.3.4. Adhered veneer units shall not exceed 36 in (9144 mm) in the greatest dimension nor more than 720 sq in (0.5 m^2) in total area and shall not weigh more than 15 psf (718 Pa) unless approved by the building official.

> **EXCEPTION**: Adhered veneer units weighing less than 3 lb/sq ft (144 Pa) shall not be limited in dimension or area.

1403.3.5. Adhered veneer units may be adhered directly to the backing by one of the following methods:

1. A paste of neat portland cement shall be brushed on all of the backing and the back of the veneer unit. Type S mortar then shall be applied to the full backing and the veneer unit. Sufficient mortar shall be used to create a slight excess to be forced out the edges of the units. The units shall be tapped into place so as to completely fill the space between all the units and the backing. The resulting thickness of mortar in back of the units shall be not less than 1/2 in (12.7 mm) nor more than 1-1/4 in (32 mm).

2. Units of masonry, stone, or terra-cotta, not over 1 in (25.4 mm) in thickness shall be restricted to 81 sq in (0.05 m^2) in area unless the back side of each unit is ground or box screeded to true up any deviation from plane. Those units not over 2 X 2 X 3/8 in (51 X 51 X 9.5 mm) in size may be adhered by means of portland cement. Backing may be of masonry, concrete, or portland cement plaster on metal lath. Metal lath shall be fastened to the supports in full accordance with the requirements of SBC chapter 25.

   Mortar as described in SBC Table 1403.3 shall be applied to the backing as a setting bed. The minimum setting bed shall be 3/8-in (9.5 mm) thick and a maximum thickness of 3/4 in (19.1 mm) thick. A paste of neat portland cement or half portland cement and half graded sand shall be applied to the back of the exterior veneer units and to the setting bed, and the veneer shall be pressed and tapped into place to provide complete coverage between the mortar bed and veneer unit. A neat portland cement grout shall be used to tuck-point the veneer.

1403.3.6. Adhered veneer units of ceramic tile shall be bonded to the backing as provided in SBC section 2104.10.

1403.3.7. Adhered veneer over wood frame shall be backed by solid sheathing covered with waterproof building paper except where the sheathing is water-repellent.

1403.4 Metal Veneers

1403.4.1. Metal veneers may be formed metal not less than 0.0149 in (0.38 mm) (28 gauge) thick. Aluminum siding shall conform to the American Aluminum Manufacturers Association (AAMA) 1402.

1403.4.2. Exterior metal veneer shall be securely attached to the supporting masonry or framing members with corrosion-resistant fastenings, metal ties, or other approved devices or methods. The spacing of the fastenings or ties shall not exceed 24 in (610 mm) either vertically or horizontally, but where units exceed 4 sq ft (0.4 m^2) in area there shall be not less than four attachments per unit. The metal attachments shall have a cross-sectional area not less than provided by no. 9 gauge wire (3.76 mm). Such attachments and their supports shall be capable of resisting a horizontal force equal to the wind loads specified in code, but in no case less than 20 psf (958 Pa).

1403.4.3. Metal supports for exterior metal veneer shall be protected by painting, galvanizing, or other equivalent coating or treatment. Wood studs, furring strips, or other wood supports for exterior metal veneer shall be approved pressure-treated wood or protected as required in Section 2303.3.

1403.4.4. All joints and edges in metal veneer that are exposed to the weather shall be caulked or painted with durable waterproofing material or shall be protected by other means to prevent penetration of moisture.

1403.4.5. Masonry backup shall not be required for metal veneer except as is necessary to meet the fire-resistance requirements of code.

1403.4.6. Metal veneers fastened to supporting elements which are not a part of the grounded metal framing of a building shall be made electrically continuous by contact or interconnection of individual units and shall be effectively grounded. The conductor used to ground the veneer shall have no greater resistance than the conductor used to ground the electrical system within the building.

Where a metal veneer is applied to a building with no electrical wiring system, grounding shall be required only if determined to be necessary by the building official.

1403.5 Glass Veneer

1403.5.1. The area of a single section of thin exterior structural glass veneer shall not exceed 10 sq ft (0.93 m²) where it is not more than 15 ft (4572 mm) above the level of the sidewalk or grade level directly below, and shall not exceed 6 sq ft (0.56 m²) where it is more than 15 ft (4572 mm) above that level.

1403.5.2. The length or height of any section of thin exterior structural glass veneer shall not exceed 48 in (1219 mm).

1403.5.3. The thickness of thin exterior structural glass veneer shall not be less than 11/32 in (8.7 mm).

1403.5.4. Thin exterior structural glass veneer shall be set only after backing is thoroughly dry and after an approved bond coat is applied uniformly over the entire surface of the backing so as to effectively seal the surface. Glass shall be set in place with an approved mastic cement in sufficient quantity so that at least 50% of the area of each glass unit is directly bonded to the backing by mastic not less than 1/4 in (6.4 mm) thick and not more than 5/8 in (15.9 mm) thick. Bond coat and mastic shall preferably be from the same manufacturer and shall bond firmly together.

1403.5.5. Where glass extends to the sidewalk surface, each section shall rest in an approved metal molding and be set at least 1/4 in (6.4 mm) above the highest point of the sidewalk. The space between the molding and the sidewalk shall be thoroughly caulked and made watertight.

1403.5.6 Joints

1403.5.6.1. Unless otherwise specifically approved by the building official, all abutting edges of thin exterior structural glass veneer shall be ground square. Mitered joints shall not be used except when specifically approved for wide angles.

1403.5.6.2. All joints shall be uniformly buttered with an approved jointing compound, and all horizontal joints shall be held to not less than 1/16 in (1.6 mm) by an approved nonrigid substance or device.

1403.5.6.3. Where thin exterior structural glass veneer abuts nonresilient material at the sides or top, expansion joints not less than 1/4 in (6.4 mm) wide shall be provided.

1403.5.7. When thin exterior structural glass veneer is installed above the level at the top of a bulkhead facing, or at a level more than 36 in (914 mm) above the sidewalk level, the mastic cement binding shall be supplemented with approved nonferrous metal shelf angles located in the horizontal joints in every course.

Such shelf angles shall be not less than 0.0478 in (1.2 mm) thick and not less than 2 in (51 mm) long and shall be spaced at approved intervals, with not less than two angles for each glass unit. Shelf angles shall be secured to the wall or backing with expansion bolts, toggle bolts, or other approved methods.

1403.5.8 Mechanical Fastenings

1403.5.8.1. All thin exterior structural glass veneer installed above the level of the heads of snow windows and all such veneer installed more than 12 ft (3658 mm) above sidewalk level shall, in addition to the mastic cement and shelf angles, be held in place by the use of fastenings at each vertical or horizontal edge or at the four corners of each glass unit.

1403.5.8.2. Fastenings shall be secured to the wall or backing with expansion bolts, toggle bolts, or other methods.

1403.5.8.3. Fastenings shall be so designed as to hold the glass veneer in a vertical plane independently of the mastic cement. Shelf angles providing both support and fastenings may be used.

1403.5.9. Exposed edges of thin exterior structural glass veneer shall be flashed with overlapping corrosion-resistant metal flashing and caulked with a waterproof compound in a manner to effectively prevent the entrance of moisture between the glass veneer and the backing.

1403.6 Wood

1403.6.1. Wood siding patterns known as rustic drop siding or shiplap shall have an average thickness in place of not less than 19/32 in (15.1 mm) and shall have a minimum thickness of not less than 3/8 in (9.5 mm). Bevel siding shall have a minimum thickness measured at the butt section of not less than 7/16 in (11.1 mm) and a tip thickness of not less than 3/16 in (4.8 mm). Siding of lesser dimensions may be used provided such wall covering is placed over sheathing which conforms to the provisions of Section 2308.2.

1403.6.2. Board siding applied vertically shall be nailed to horizontal nailing strips or blocking set 24 in (610 mm) on center.

The nails shall penetrate 1-1/2 in (38 mm) into studs, blocking, studs or blocking and sheathing combined, or nailing strips.

1403.6.3. Wood shakes and shingles shall be applied in accordance with the Construction Specifications Institute (CSI) Design and Application Manual for Exterior and Interior Walls. [UBC and BOCA each have separate specs for wood shakes under sections 1507.0 and 2303.1 and 1507.0 and 2303.0, respectively.]

1403.6.4. Wood structural panels shall be of the exterior type and shall have a thickness of 3/8 in (9.5 mm), except as provided in SBC Table 2308.1D. All wood structural panel joints shall be backed solidly with nailing pieces not less than 2 in (51 mm) wide, unless wood, wood structural panel, or particleboard sheathing is used or shall be otherwise made waterproof as required in Section 2303.3.

> **EXCEPTION**: The mounting framework is not required to be protected in accordance with Section 2303.3 when the joints are protected by a continuous wood batt, caulking, flashing, or vertical or horizontal shiplap.

1403.6.5. Fiberboard siding shall be of medium density not less than 1/2 in (12.7 mm) nominal thickness.

1403.6.6. Hardboard siding shall conform with the requirements of ANSI/AHA (American Hardwood Association) A135.4, A135.5, or A135.6 and shall be identified as to classification.

1403.6.7. Particleboard siding used for covering the exterior of outside walls shall be of the Exterior Type 2-M grades conforming to ANSI A208.1. Particleboard panel siding shall be installed in accordance with Tables 2306.1 and 1403.6. Nails shall be spaced not less than 3/8 in (9.5 mm) from edges and ends. Joints shall occur over framing members unless particleboard panel siding is applied over 5/8 in (15.9 mm) net wood sheathing or 15/32 in (11.9 mm) plywood or 1/2 in (12.7 mm) particleboard sheathing. The framework shall be protected as required in Section 2303.3.

> **EXCEPTION**: The mounting framework is not required to be protected in accordance with Section 2303.3 when the joints are protected with a continuous wood batt, caulking, flashing, or vertical or horizontal shiplap.

1403.6.8. Wood veneers on exterior wall panels of Types I, II, III, IV, and V construction shall comply with Sections 1403.6.8.1 and 1403.6.8.2.

1403.6.8.1. Wood veneers of not less than 1 in (25.4 mm) nominal thickness, 7/16-in (11.1 mm) exterior hardboard siding, or 3/8-in (9.5 mm) exterior type wood structural panels or particleboard may be used on exterior walls when all the following conditions are met:

- The wall to which the veneer is attached faces a street or a permanent open space of 30 ft (9144 mm) or more wide.
- The veneer does not exceed two stories in height, measured from grade, except when it is fire-retardant-treated for exterior use, it may be four stories in height.
- The veneer is attached to or furred from a noncombustible backing of the fire-resistance required by other provisions of this chapter.
- Where open or spaced wood veneers (without concealed spaces) are used, they shall not project more than 24 in (610 mm) from the building wall.

1403.6.8.2. Where the wood veneer is furred from the wall and forms a solid surface, the distance between the back of the veneer and the wall shall not exceed 1-5/8 in (41 mm) and the space thereby created shall be fireblocked in accordance with Section 2305 and arranged so that there will be no open space exceeding 100 sq ft (9.3 m^2). Where wood furring strips are used, they shall be of approved wood of natural decay resistance or pressure-treated wood.

1403.7 Asbestos Shingles

Asbestos shingles attached to sheathing other than wood, plywood, or 2-M-W particleboard shall be secured with approved mechanically bonding nails or by corrosion-resistant common nails on shingle nailing boards securely nailed to each stud with two 8d nails, except that asbestos shingles may be attached directly to fiberboard nail base sheathing with corrosion-resistant annular grooved nails. Asbestos shingles shall have a minimum thickness of 5/32 in (4 mm).

1403.8 Stucco

Stucco or exterior plaster shall conform to requirements of Section 2504. [These exterior wall surfacings include rough stucco as shown in Figure 10-3 and smooth texture stucco as shown in Figure 10-4.]

Exterior Walls 253

Figure 10-3
Rough Stucco Exterior Finish

254 Chapter Ten

**Figure 10-4
Smooth Stucco Exterior Finish**

1403.9 Rigid Vinyl

1403.9.1. Rigid vinyl siding shall conform with the requirements of ASTM D 3679 and is limited to Type VI construction.

1403.9.2. Provisions for exterior plastic veneers other than rigid vinyl are found in SBC section 2604.9.

1404.0 ARCHITECTURAL TRIM, BALCONIES, AND BAY WINDOWS [1402.0; 1404.0]

1404.1 Architectural Trim

1404.1.1. Architectural trim on buildings of Type I, II, and IV construction not more than three stories or 40 ft (12.2 m) high may be of Type VI construction and on all buildings of Type III, V, and VI construction may be of Type VI construction. Trim shall be secured to the wall with metal or other approved brackets or fasteners. When architectural trim is located along the top of exterior walls, it shall be completely backed by the exterior wall and shall not extend over the top of exterior walls.

1404.1.2. For projection over public property, see SBC section 3206.

1404.2 Balconies and Bay Windows

Balconies not used as required exits and bay windows shall conform to the type of construction required for the building to which they are attached, except that exterior fire-retardant-treated wood is permitted on buildings three stories or less for Type I and II exterior walls.

1404.3 Combustible Projections

Combustible projections from walls located where protection of openings is required shall be 1-hr fire-resistance or heavy timber construction. Projections shall not extend more than 12 in (305 mm) into the areas where openings are prohibited.

1405.0 FIRE DEPARTMENT ACCESS IN EXTERIOR WALLS [1406.0; 1406.0]

1405.1 General

Exterior walls shall have access openings for fire department use serving each story above grade on an accessible side of the building up to a height of 75 ft (22.9 m). Such access openings shall be a minimum of 32 in (813 mm) wide and 48 in (1219 mm) high and with the bottom of the opening not more than 32 in (813 mm) above the floor.

> **EXCEPTION**: Fire department access to high-piled combustible storage and high-rack storage systems shall be in accordance with chapter 36 of the Fire Prevention Code.

1405.2 Spacing

Openings shall be so spaced that there will be one opening in each 50 ft (15.2 m) of exterior wall on an accessible side of the building.

> **EXCEPTION**: Buildings equipped with an automatic sprinkler system throughout in accordance with NFiPA 13 shall have access panels as set forth for each 200 ft (61 m) of wall.

1405.3 Identification

Where complying access openings are not apparent, they shall have distinctive markings for identification.

1405.4 Obstruction

Access openings shall open into a fire aisle within the building and no shelving, loose or fixed; no containers or equipment of any description; nor any loose merchandise shall be placed so as to block aisleways.

Housewraps

Because both the builder and the customer are demanding more energy efficiency from homes today, the exterior wall component systems should include a vapor-air infiltration barrier known as a housewrap. These are becoming adopted by the national building codes because of their proven energy efficiency. Wrapping the house in a sheath of new technology fabric is extensive, but the accrued savings in annual heating and air-conditioning energy costs is going to make this a money-making investment to home buyers, if the builder just takes time to point it out to them.

Housewraps provide an air infiltration barrier and exfiltration retarder for the frame of the house. It basically keeps conditioned air from leaking out and external air from seeping in. It allows water vapor to escape, preventing moisture buildup that can rot walls. Plus, it covers construction gaps and seams left by joints and corners. In many areas around the country, an approved housewrap has become a code requirement. It is most definitely a

Exterior Walls 257

major selling point feature that knowledgeable home buyers are looking for from modern builders.

There are several brands of housewrap on the market currently. Price varies, but what is really important for the builder is the strength of the wrap. If the product tears easily, it will lose some or all of its barrier capabilities. As usual for the builder, quality of the product equates to quality of the job. Here are the specs you should call for in a quality housewrap used by the manufacturer:

- Trapezoidal tear strength (prior to exposure): 200%
- Tear strength after exposure (0 weeks exposure): 100%
- 8 weeks exposure: 90%
- 16 weeks exposure: 50%
- Air porosity (percentage of air stopped): >95%
- Moisture vapor transmission: >35 g/m^2 per 24 hr
- Ultraviolet resistance (exposure time before product must be covered): 120 days

By code requirements, homes are constructed to be energy-efficient. Homes built for areas of cold winters have standard size walls constructed of 2 in X 6 in studs with full exterior wall insulation. Roof systems use a minimum of 12-in blown cellulose insulation with a minimum R value of 38. Other new energy saving-techniques include seamless housewraps for corners and interior wall plates, which provide moisture control and roof insulation. The building process also allows for installation of the vapor barrier in the ceiling before the interior walls are in place. This way, the vapor barrier is virtually continuous and not broken by interior wall plates.

Chapter 11

Roof Structures

SECTION 1501.0 GENERAL [1500.1; 1501.1]

Requirements for parapet walls on a roof perimeter are that no masonry or concrete can be supported by wood members. Support for masonry or concrete in buildings over one story must meet either a 1-hr fire rating or meet the fire-rating requirements of the supporting wall, whichever is greater. Parapet walls must also have the same fire-resistive rating as the support wall and project a minimum of 30 in above the intersection point where the wall meets the roof; the top 18 in must be noncombustible construction on the roof side.

If the roof pitch exceeds 2:12, the top of the parapet wall must be at the same elevation as any portion of the roof within the distance where protection of wall openings is required. Provisions of this chapter shall govern the construction of roof structures and the materials, application, installation, wind resistance, and fire-resistance of roof coverings. Typical roof trusses are designed to meet the factors shown in Figure 11-1.

1502.0 DEFINITIONS [1502.0; 1502.0]

The following words and terms shall, for the purposes of this chapter and as stated elsewhere in code, have the meanings shown herein. Refer to Chapter 2 for general definitions.

PENTHOUSE - An enclosed structure above the roof of a building, other than a roof structure or bulkhead, occupying not more than one-third of the roof area.

ROOF COVERING SYSTEM - A system designed to provide weather protection and resistance to design loads. The system may consist of two components, the roof covering and the roof deck, or a single component serving as both the roof covering and the roof

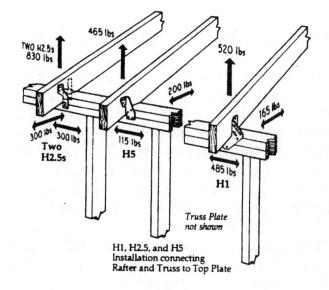

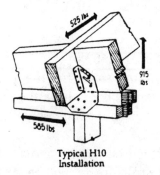

**Figure 11-1
Roof Truss Torsion Factors**

deck. The roof covering provides the weather protection, and the roof deck provides the structural support for the design loads.

ROOF STRUCTURE - An enclosed structure on or above the roof of any part of a building.

SCUPPER - An opening in a wall or parapet that allows water to drain from a roof.

1503.0 PENTHOUSE-TYPE ROOF STRUCTURES [1501; 1502]

1503.1 Height and Area

1503.1.1. A penthouse or other projection above the roof in structures of other than Type I construction shall not exceed 28 ft (8534 mm) above the roof when used as an enclosure for tanks or for elevators which run to the roof and, in all other cases, shall not extend more than 12 ft (3658 mm) above the roof.

1503.1.2. The aggregate area of all penthouses and other roof structures shall not exceed one-third the area of the supporting roof.

1503.1.3. A penthouse, bulkhead, or any other similar projection above the roof shall not be used for purposes other than shelter of mechanical equipment or shelter of vertical shaft openings in the roof. Penthouses or bulkheads used for purposes other than permitted by this section shall conform to the requirements of code for an additional story. [An example of concrete masonry unit (CMU) block wall bulkhead to roof assembly is shown in Figure 11-2.]

1503.2 Type of Construction

1503.2.1. Roof structures shall be constructed with walls, floors, and roof as required for the main portion of the building.

> **EXCEPTIONS**:
> 1. On buildings of Type I and II construction, the exterior walls and roofs of penthouses which are more than 5 ft (1524 mm) and less than 20 ft (6096 mm) from a common property line shall be of at least 1-hr noncombustible construction. Walls and roofs which are over 20 ft (6096 mm) from a common property line may be of noncombustible construction. All interior framing and walls shall be of noncombustible construction.

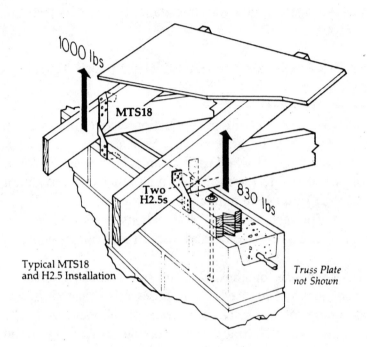

**Figure 11-2
CMU Block Bulkhead to Roof Connection**

2. On buildings of Type III, IV, and V construction, the exterior walls and roofs of penthouses which are more than 5 ft (1524 mm) and less than 20 ft (6096 mm) from a common property line shall be of at least 1-hr fire-resistive noncombustible construction. Walls which are over 20 ft (6096 mm) from a common property line may be of heavy timber construction or noncombustible construction. Roofs may be of wood frame construction. All interior framing and walls shall be of heavy timber construction or noncombustible construction.
3. Enclosures housing only mechanical equipment and located at least 20 ft (6096 mm) from adjacent property lines may be of unprotected noncombustible construction.
4. On one-story buildings, unroofed mechanical equipment screens, fences, or similar enclosures may be of combustible construction when located at least 20 ft (6096 mm) from adjacent property lines and when not exceeding 4 ft (1219 mm) in height above the roof surface.
5. Dormers shall be of the same type of construction as the roof on which they are placed or as the exterior walls of the building.

1503.2.2. The restrictions of this section shall not prohibit the placing of wood flagpoles or similar structures on the roof of any building.

1504.0 TANKS [307.1 & 312.4; 308.0]
1504.1 Tanks Exceeding 500 Gal (2 m^3)

1504.1.1. Tanks of more than 500 gal (2 m^3) capacity placed in or on a building shall be supported on masonry, reinforced concrete, or steel construction, except that portion of the supporting structure which is above the roof of the building may be of heavy timbers, provided that, when such construction is within the building, it shall be as required for Type I construction.

1504.1.1.1. Such tanks shall have in the bottom or on the side near the bottom, a pipe or outlet, fitted with a suitable quick opening valve for discharging the contents in an emergency through an adequate drain.

1504.1.1.2. Such tanks shall not be placed over nor near a line of stairs or an elevator shaft, unless there is a solid roof or floor underneath the tank.

1504.2 Unenclosed Roof Tanks
All unenclosed roof tanks shall have covers sloping toward the outer edges.

1505.0 COOLING TOWERS [N/A; 1500.0]
Cooling towers in excess of 250 sq ft (23.2 m²) in base area or in excess of 15 ft (4572 mm) high when located on buildings more than 50 ft (15.2 m) high shall be of noncombustible construction, except that drip boards may be of wood not less than 1 in (25.4 mm) nominal thickness and the enclosing framework may be of wood, if covered on the exterior of the tower with noncombustible material. Cooling towers shall not exceed one-third of the supporting roof area.

1506.0 OTHER ROOF STRUCTURES [1505.0; 1510.0]
1506.1 Type of Construction
1506.1.1 Minimum Type I or II Construction
Any tower, spire, dome, or cupola shall be of a type of construction not less in fire-resistance rating than required for the building to which it is attached except that any such tower, spire, dome, or cupola which exceeds 60 ft (18.3 m) in height above grade, and all construction upon which it is supported, shall be of Type I or II construction when the area at any horizontal section of such tower, spire, dome, or cupola exceeds 200 sq ft (18.6 m²) or when it is used for any purpose other than a belfry or an architectural embellishment.

1506.1.2 Minimum Noncombustible Construction
1506.1.2.1. Any tower, spire, dome, or cupola which exceeds 25 ft (7620 mm) in height above the highest point at which it comes in contact with the roof; which exceeds 200 sq ft (18.6 m²) in area at any horizontal section, or which is intended to be used for any purpose other than a belfry or architectural embellishment shall be constructed entirely of and supported by noncombustible materials.

Such structures shall be separated from the building below by construction having a fire-resistance rating of not less than 1-1/2 hr and, if access doors are provided, such doors shall be of an approved fire-resistant type.

1506.1.2.2. Structures, except aerial supports 12 ft (3658 mm) high or less, flagpoles, water tanks, and cooling towers, placed above the roof of any building more than 50 ft (15.2 m) in height, shall be of noncombustible material and shall be supported by construction of noncombustible material.

1506.2 Towers and Spires

1506.2.1. Towers and spires when enclosed shall have exterior walls as required for the building to which they are attached. The roof covering of spires shall be of the same class of roof covering as required for the main roof of the rest of the structure.

1507.0 PARAPET WALLS [709.4; 1505.0]

1507.1 Framing

Parapet walls shall be designed as provided in this section.

1507.2 Coping

All parapet walls shall be properly coped with noncombustible, weatherproof materials of a width no less than the thickness of the parapet wall.

1507.3 Flashing

Proper flashing shall be installed in such a manner as to prevent moisture entering the wall through the joints in the coping, through moisture permeable material, at intersections with the roof plane, or at parapet wall penetrations.

1507.4 Scuppers

1507.4.1. Where required for roof drainage, a scupper shall be placed level with the roof surface in a wall or parapet. The scupper shall be located as determined by the slope and the contributing area of the roof. The exterior facing or lining of a scupper, if metal, shall be the same as valley lining material required by Section 1509 for the particular type of covering specified for the building. For other materials follow manufacturer's specifications.

1507.4.2. A scupper shall be sized in accordance with chapter 11 of the Standard Plumbing Code [UPC; NPC].

1507.4.3. When other means of drainage of overflow water is not provided, overflow scuppers shall be placed in walls or parapets not less than 2 in (51 mm) nor more than 4 in (102 mm) above the roof deck and shall be located as close as practical to required vertical leaders or downspouts or wall and parapet scuppers. An overflow scupper shall be sized in accordance with chapter 11 of the Standard Plumbing Code [UPC; NPC].

1508.0 GUTTERS AND LEADERS [1506.0; 1506.1]
1508.1 Gutters and leaders placed on the outside of buildings other than one- or two-family dwellings, private garages, and buildings of Type VI construction, shall be of noncombustible material or a minimum Schedule 40 plastic pipe. See chapter 11 of the Standard Plumbing Code [UPC; NPC] for the sizing of vertical leaders and horizontal storm drains.

1509.0 ROOF COVERINGS [1507.0; 1507.1]
1509.1 General
The requirements set forth in this section shall be construed as minimum requirements and shall apply to the application and installation of roof covering materials specified herein, excluding pre-engineered steel buildings. Roof coverings shall be applied in accordance with this chapter and/or meet manufacturer's recommendations. It should also be noted that this chapter does not deal with the minimum design loads of roofing materials. Those requirements, with which all roofing systems must comply, are covered in Section 1606.
1509.1.2 Covering
1509.1.2.1. Roof coverings shall provide weather protection for the building at the roof.
1509.1.2.2. All roof coverings shall be applied to a solid or closely fitted deck, except where the roof covering is specifically designed to be applied to spaced supports.
1509.1.2.3. Low slope roofs shall be designed for a minimum 1/4 in/ft (20.8 mm/m) slope unless specific water-retaining roof materials are to be installed.
1509.1.3 Insulation
1509.1.3.1. The use of above-deck thermal insulation is permitted on top of both the roof deck and the roof membrane provided such insulation is covered with an approved covering applied directly thereto in accordance with the manufacturer's recommendations.

1509.1.3.2. A minimum of 1/2-in (12.7 mm) thick insulation shall be installed over metal decking when a roof covering is installed subject to the manufacturer's flute span table.

1509.1.4 Fasteners

1509.1.4.1. Nails, clips, or similar fastening devices shall be hot dipped galvanized, stainless-steel, nonferrous metal, or other suitable corrosion-resistant material.

1509.1.4.2. Fasteners for wood shingles and shakes shall conform with the requirements of SBC sections 1209.8.5.1 and 1209.8.6.1.

1509.1.4.3. The composition flashing shall be mechanically attached using suitable manufacturer-approved fasteners spaced a maximum of 8 in (203 mm) on center and 1 in (25.4 mm) minimum from the top edge of flashing sheet.

1509.1.4.4. Tin caps shall be not less than 1-5/8 in (41 mm) in diameter and a minimum of 0.0134 in (0.34 mm) thick sheet metal.

1509.1.4.5. Caphead nails shall be a minimum 1-in (25.4 mm) long annular threaded 12 gauge (2.66 mm) wire nail with a head not less than 1 in (25.4 mm) in diameter.

1509.1.5 Composition Flashing

Composition base flashing should extend a minimum of 8 in (203 mm) nominal above the roof line. When using such flashing, wood or fiber cants must be provided at any 90^0 (1.57 rad) angle created by rectangular curbs or projections. Wood nailers should be provided on all prefabricated curbs.

1509.1.6 Mechanical Units

For new construction, mechanical units mounted on pipe standards or curbs beneath which roofing materials will extend must be mounted to a height sufficiently above the roof to allow room to install the roof system and to make repairs beneath the unit.

Heavy loads, such as large mechanical units, shall not be rolled over the completed membrane as they may cause damage to the roof. A failure in horizontal shear between the membrane, insulation, or deck from these loads may result in future splitting of the roof.

1509.1.7 Flashing

Flashing shall be placed around openings and extensions of mechanical appliances or equipment through the roof and otherwise as necessary to provide adequate drainage.

1509.2 Fire Resistance Classification

1509.2.1 General

Roof coverings shall be divided into the classes defined below. All Class A, B, and C roof coverings required to be listed by this section shall be tested in accordance with ASTM E 108. In addition, fire-retardant-treated wood roof coverings shall be tested in accordance with ASTM D 2898. All roof coverings shall be installed in accordance with Section 1509 or other approved nationally recognized standards.

1509.2.2 Class A Roof Coverings

Class A roof coverings shall include brick, concrete, slate, tile, or assemblies listed and identified as Class A by an approved testing laboratory, inspection agency, or product evaluation organization.

1509.2.3 Class B Roof Coverings

Class B roof coverings shall include corrugated-steel sheets, galvanized-steel sheets, galvanized-steel shingles, sheet copper galvanized iron, or assemblies listed and identified as Class B by an approved testing laboratory, inspection agency, or product evaluation organization.

1509.2.4 Class C Roof Coverings

Class C roof coverings shall include the assemblies listed and identified as Class C by an approved testing laboratory, inspection agency, or product evaluation organization.

1509.2.5 Requirements for Roofs

Roofs on buildings shall have Class A, B, or Class C roof coverings, as specified herein. Unclassified wood shingles or shakes may be used as provided in Section 1509.8. Private detached garages, carports, and farm buildings as defined in SBC section 411.11 are not regulated by this section.

1509.3 Wind Loads and Wind Resistance

1509.3.1. Wind loads on roof decks and other structural members supporting roof coverings are specified in Section 1606.

1509.3.2. Roof systems with built-up, modified bitumen, fully adhered, or mechanically attached single-ply metal panels or other types of membrane roof coverings shall be designed to withstand the appropriate wind loads prescribed in Section 1606.

1509.3.3. Ballasted single-ply roof system coverings shall be designed in accordance with ANSI/RMA/SPRI RP-4.

1509.3.4. Asphalt shingles shall have self-seal strips or be interlocking and shall have the type and minimum number of fasteners recommended by the manufacturer.

1509.3.5. Self-seal asphalt strip shingles shall have a minimum of six fasteners per shingle when the roof is in one of the following categories:
- The basic wind speed is 90 miles per hour (mph) (40.2 m/s) or greater and the eave is 20 ft (6096 mm) or higher above grade.
- The basic wind speed is 90 mph (40.2 m/s) or greater and the Use Factor in Table 1606 (see Appendix) is 1.15.
- The basic wind speed is 100 mph (44.7 m/s) or greater.

1509.4 Asphalt Shingles

1509.4.1 General

1509.4.1.1. The installation of asphalt shingles used as a roof covering shall comply with the requirements of this section.

1509.4.1.2. Shingle application shall be as specified in the manufacturer's published application instructions. [A typical example appears in Figure 11-3.]

1509.4.2 Application

1509.4.2.1. For 2:12 up to 4:12 pitch, the underlayment shall be two layers of Type 15 asphalt saturated, nonperforated felt applied in the following manner. Apply a 19-in (483 mm) strip of Type 15 asphalt saturated, shingle underlayment felt parallel with and starting at the eaves, fastened sufficiently to hold in place.

Starting at the eave, apply 36 in (914 mm) wide sheets of underlayment overlapping successive sheets by 19 in (483 mm) and fastened sufficiently to hold in place. Where January mean temperatures are 30°F (-1°C) or less, the full width of the 19-in (483 mm) laps from the eave to a point 24 in (610 mm) from the inside of the exterior wall line of the building should be coated with asphalt-based roofing cement.

1509.4.2.2. For 4:12 to 20:12 pitch the underlayment shall be Type 15 asphalt saturated, nonperforated felt applied shingle fashion, parallel to and starting from the eave and lapped 2 in (51 mm), fastened only as necessary to hold in place. [Pitches on roof trusses are typically 4:12 as these are set by jig during manufacturing as shown in Figure 11-4.]

1509.4.2.3. Asphalt shingles shall be fastened along the rake. Asphalt shingles shall be fastened and cemented at all valleys, rakes,

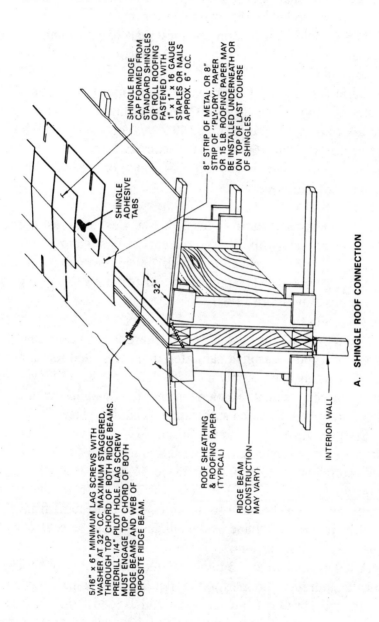

**Figure 11-3
Shingle Roof Connection**

Roof Structures 271

**Figure 11-4
Roof Truss Manufacturing**

penetrations, and vertical projections. Eaves must be cemented or the metal eave drip shall be installed under the felt.

1509.4.2.4. Fasteners shall penetrate through the roofing material and at least 3/4 in (19 mm) into or through the roof sheathing.

1509.4.2.5. When slopes exceed 20:12, special methods of fastening are required. Follow manufacturer's printed instructions.

1509.5 Mineral Fiber Shingles

1509.5.1. For a minimum slope of 3:12 up to 4-1/2:12, the underlayment shall be a minimum of one layer Type 30 asphalt saturated felt nailed and one layer of Type 15 asphalt saturated felt laid with hot asphalt or cold applied cement solid mopped.

1509.5.2. For slopes of 4-1/2:12 and greater, underlayment shall be one layer of Type 30 asphalt saturated felt.

1509.5.3. Application shall be in accordance with the recommendations of the manufacturer.

1509.6 Slate Shingles

1509.6.1. Slate shingles shall only be used on slopes of 4:12 or greater.

1509.6.2. Underlayment shall be one layer of Type 30 asphalt saturated felt.

1509.6.3. Maximum exposure shall be calculated using the formula: $E = (L - H)/2$ where L is the shingle length, and H is the shingle head lap [e.g., for an 18-in (457 mm) shingle with a 3-in (76 mm) head lap, maximum exposure is $(18 - 3)/2 = 7\text{-}1/2\text{-in}$ (191 mm)].

1509.6.4. Minimum head lap of the shingles shall be in accordance with Table 1509.6.

Table 1509.6
Slate Shingle Head Lap

Roof Slope	Minimum, in (mm)
4:12 to < 8:12	4 (102)
8:12 to < 20:12	3 (76)
20:12 and greater	2 (51)

1509.7 Concrete and Clay Roof Tile
1509.7.1 General

1509.7.1.1. Each roof tile shall have a permanent manufacturer's identification mark.

1509.7.1.2. The tile manufacturer's written application specifications shall be available and shall include but not be limited to the following:
- The tile's placement and spacing
- Amount, and placement of mortar
- Underlayment
- Slope requirement

1509.7.1.3. Roof tile shall be in accordance with the physical test requirements as follows:

1. Except for an overlapping lip, tile shall have a minimum thickness of not less than 1/2 in (12.7 mm) for barrel tile and 3/8 in (9.5 mm) for shingle tile. Barrel tile shall be test loaded by being supported on sand 2 in (51 mm) deep in a sandbox that is 4 in (102 mm) wider than the width of the tile.
2. Shingle tile shall be tested using cloth tubes parallel to the edge of the tile. Sand tubes shall be 2 in (51mm) in diameter loosely filled with dry 40/60 silicon and shall be placed under the edge of the tile with a center-to-center distance equal to the width of the tile. A test load shall be applied on a 3-in (76 mm) square steel plate being on a sandbag set at the center of the tile.
3. The breaking load of any individual shingle tile shall be not less than 200 lb (890 N) and the average breaking load of 5 shingle tiles shall not be less than 250 lb (1110 N). The average breaking load of 5 barrel tiles shall be not less than 300 lb (1330 N).
4. Roof tiles shall absorb not more than 12% of the dry weight of the tile during a 24-hr immersion test.
   Roof tiles shall meet or exceed the requirements of ASTM C 666 for freeze-thaw requirements.

1509.7.1.4. The substrate to which tile is to be installed shall be uniform, smooth, clean, and dry. Neither underlayment nor tiles shall be installed on wet, frozen, or icy surfaces.

1509.7.1.5. Underlayment materials used shall be in accordance with the following minimum standards.

1509.7.1.5.1. Single-ply system (mechanically fastened systems only):

1. Organic Type II (#30) having a minimum weight of 26 lb (12 kg) per roll
2. Organic Type 1 (#43 base sheet) having a minimum weight of 37 lb (17 kg) per roll
3. #90 mineral surface roll roofing having a minimum weight of 74 lb (34 kg) per roll

1509.7.1.5.2. Two-ply system (cemented and mechanically fastened systems):

1. Organic Type I (#30) having a minimum weight of 26 lb (12 kg) per roll
2. #90 mineral surface roll roofing having a minimum weight of 74 lb (34 kg) per roll

1509.7.1.5.3. Membrane materials shall comply with the following minimum standards.

1. <u>Organic</u>. Conforming to ASTM D 173, asphalt impregnated membrane, a minimum of 3 in (76 mm) wide
2. <u>Inorganic</u>. Conforming to ASTM D 1668, asphalt impregnated fiberglass membrane, a minimum of 3 in (76 mm) wide

1509.7.1.5.4. Fasteners shall be in accordance with the following:

1. Nails shall be of corrosion-resistant minimum hot dipped galvanized steel or approved equal of sufficient length to properly penetrate the deck a minimum of 3/4 in (19 mm) or through the thickness of the deck, whichever is less.
2. Caphead nails.
3. Staples shall be corrosion-resistant minimum with a 7/16-in (12 mm) crown.
4. Tin caps shall be in accordance with Section 1509.1.4.4.
5. Storm clips shall be applied in accordance with manufacturer's recommendations.

1509.7.1.5.5. Metal flashing shall be a minimum 26 gauge (0.455 mm) G 90 corrosion-resistant metal conforming to ASTM A 525 and A 90 or minimum 16 ounce (oz) (4.9 kg/m^2) copper conforming to ASTM B 370 or approved equal.

1509.7.1.5.6. Adhesive or sealant shall be in accordance with the following:
- Asphalt plastic roof cement shall conform to ASTM D 2822, Type II. Nonrunning, heavy body material composed of asphalt and other mineral ingredients.
- Cold applied liquid roof coating shall conform to ASTM D 3019, Type II.
- Structural bonding adhesive shall conform to ASTM C 557 or ASTM D 3498.
- Hot steep asphalt shall conform to ASTM D 312.

1509.7.1.5.7. Mortar components shall be in accordance with the following:
1. Cements:
   Blended cement shall conform to ASTM C 91, Type M.
   Portland cement shall conform to ASTM C 150, Type 1.
   Masonry cement shall conform to ASTM C 91, Type M.
2. Sand shall conform to ASTM C 144 and be uniformly graded, clean, and free from organic materials.

1509.7.1.5.8. Mortar mixes shall conform to Type M in accordance with Section 2104.7.

1509.7.1.5.9. Mortar flow $110 \pm 5\%$ shall conform to ASTM C 230 flow table.

1509.7.1.5.10. Eave closure shall be one of the following installed according to manufacturer's recommendations.
1. Prefabricated synthetic rubber conforming to ASTM D 1056.
2. Prefabricated minimum 26 gauge (0.455 mm) corrosion-resistant metal eave closure.
3. Mortar for mineral surface roll roofing.

1509.7.1.5.11. Lumber shall be in accordance with the following:
1. Fasteners which penetrate fire-retardant-treated wood and pressure-treated wood shall comply with SBC section 2306.3.
2. Sheathing shall comply with Section 2301.4.
3. Nailer boards shall be pressure-treated wood.

1509.7.2 Cement Applied Tile

1509.7.2.1 Subroof Application

Two-ply underlayment is required for pitches 2.5:12 and greater.

1509.7.2.1.1 Base Ply

Starting at the eave edge, one course of Type 30 roofing felt shall be applied horizontally along the roof line lapping the end joints a minimum of 4 in (102 mm), tin-capped and secured with nails, staples, or caphead nails a maximum of 12 in (305 mm) on center (o.c.) in field and at 12 in (305 mm) on all head laps with a weave pattern at all valleys. Each succeeding course shall be applied in the same manner allowing a minimum 2-in (51 mm) head lap. All hips and ridges shall be overlapped a minimum of 6 in (152 mm).

1509.7.2.1.2. Drip-edge metal shall be nailed or stapled along and directly on top of the #30 felt at the eave, fastened 6 in (152 mm) o.c. and 1/2 in (12.7 mm) from the top flange. All joints shall be lapped a minimum of 2 in (51 mm) continuing from the eave up the rake or gable in the same manner.

1509.7.2.1.3. Valleys shall be nailed within 1 in (25.4 mm) of the metal edges, a maximum of 6 in (152 mm) o.c. with joints lapped a minimum of 6 in (152 mm). Plastic roof cement shall be applied between the laps. Valleys shall meet the following requirements.

1. Standard valley shall be a minimum of 16 in (407 mm) in width.
2. Preformed closed valley shall be a minimum of 16 in (407 mm) in width with a center diverter.
3. Preformed open valley shall be a minimum of 16 in (407 mm) in width with twin center diverters.

1509.7.2.1.4 Flashing and Counterflashing at Wall Abutments

An "L"-shaped metal flashing shall be installed flush to the base of walls over the #30 felt and nailed within 1 in (25.4 mm) of the metal edges. Joints shall be lapped a minimum of 4 in (102 mm) and plastic roof cement applied between the laps. Work shall start at the lower portion and work upward to ensure watertightness. The top edge of the vertical flange shall be sealed, covering all nail penetrations with plastic roof cement and membrane.

If counterflashing is to be installed, the top flange of the base flashing shall be lapped a minimum of 3 in (76 mm). The metal shall be nailed within 1 in (25.4 mm) of the metal edge a minimum of 6 in (152 mm) o.c. or set into regrets and thoroughly caulked. Joints shall be lapped a minimum of 3 in (76 mm) and plastic roof cement applied between laps.

1509.7.2.1.5. Standard skylights, chimneys, etc. shall be installed in accordance with regular flashing installation procedures.

1509.7.2.1.6. Plastic roof cement shall be applied around the base of the protrusion and on the bottom side of the metal flanges sealing the unit base flashing to the deck. All sides of the base flashing shall be nailed within 1 in (25.4 mm) of the edge to secure the base flush to the deck.

1509.7.2.1.7 Top Ply

Starting at the eave edge, #90 roll roofing shall be applied horizontally along the roof line over the base ply lapping the end joints a minimum of 6 in (152 mm). Hot asphalt shall be applied between the plies of roofing felt so that no felt touches felt. Cold applied or plastic roof cement shall be permitted to be substituted for hot asphalt for pitches above 6:12. Felt shall be back-nailed using tin caps and roofing nails or caphead nails a maximum of 12 in (305 mm) o.c. and a minimum 1 in (25.4 mm) from the top edge of the felt.

Each succeeding course shall be applied in the same manner, allowing a minimum 4-in (102 mm) head lap. Hip and ridges shall be overlapped a minimum of 6 in (152 mm) with a weave pattern in valleys or trimmed a maximum of 4 in (102 mm) past nail penetrations. When preformed valley metal is used, the edge of the #90 felt shall be sealed with plastic roof cement and membrane. All metal flashings and roof protrusions shall be hot mopped or cold applied. Felt overhanging at eaves and gables shall be trimmed. Fishmouths shall be cut and sealed with plastic roof cement and membrane. Plastic roof cement or hot asphalt shall be applied along the edge of the felt wherever it meets wall bases.

1509.7.2.2 Tile Installation

1509.7.2.2.1. Clay tile to be set in mortar shall be wetted prior to setting in mortar bed.

1509.7.2.2.2. Eave treatment shall be one of the following:

1. <u>Prefabricated eave closure</u>. Closure strips shall be installed along the eave according to the manufacturer's recommendation.
2. <u>Metal eave closure</u>. Closure strips shall be installed along the eave according to the manufacturer's recommendation.
3. <u>Thickbutt tile</u>. Thickbutt tile shall be installed along the eave according to the manufacturer's recommendation.

4. <u>Mortar eave closure</u>. Starting at the lower left-hand corner (facing down roof), the first course of tile shall be installed with the eave edge of tile elevated with mortar, then tuck-pointed, and provided with a weep hole flush with the deck to allow for drainage and ventilation.
5. <u>Fascia</u>. For raised fascia or wood starter strip, #30 felt shall be installed in accordance with Section 1509.7.2.1. Fascia board shall be installed approximately 1-1/2 in (38 mm) above the roof deck or a 2 X 2 wood starter strip installed at the roof edge. A tapered cant strip shall be installed behind the fascia or starter strip to support metal flashing and mortar. Antiponding metal flashing of sufficient width to ensure positive drainage over the fascia or starter strip shall be installed. The top edge of the flange shall be nailed to the roof.

1509.7.2.2.3. The first course of tile shall be installed starting at the lower left-hand corner (facing down roof). All tile shall overhang the drip edge evenly along the entire course and at the same height when using mortar to elevate the first course.

1509.7.2.2.4 Low Profile, High Profile, and Flat Tile
Tile shall be set in a bed of Type M mortar. A full 10-in (254 mm) minimum length mason trowel full of mortar [weighing approximately 4 to 5 lbs (1.8 to 2.3 kg) dry weight] shall be installed vertically under the pan or flat portion of tile. When the tile has more than one pan or flat portion, the mortar shall be placed under the pan closest to the underlock of the previously installed tile. For flat tile, mortar shall be placed adjacent to the underlock of the previously installed tile.

Mortar shall be placed from the head of the tile in the previous course to within 2 to 4 in (51 to 162 mm) of the head of the tile being set. Mortar shall not be placed under the lugs, the head of the tile, nor onto the underlock of the adjacent tile. A half-starter or finisher tile shall be used for proper staggering of the tile courses when using the staggered or cross bond method of installation. Tile shall be set in stepped course fashion or in a horizontal and/or vertical fashion when utilizing the straight bond method. Succeeding courses of field tile shall be laid in the same manner. The bed of mortar shall make contact with the head of the lower course of tile and the underside of the tile being set.

1509.7.2.2.5 Two-piece barrel tile
A 10-in (254 mm) mason's trowel full of mortar [weighing approximately 4 to 5 lb (1.8 to 2.3 kg) dry weight] shall be applied vertically over the chalk line and under the center of each pan with the narrow end facing down the roof. Mortar shall be placed so as to make a two-tile bond between the pan being set and the pan below in the previous course. A bed of mortar shall be placed along the inside edges of the pans and the covers set with the wide end facing down the roof. Mortar shall be pointed to the next acceptable straightedge finish, with good strong contact along the edges. Succeeding courses of field tile shall be laid in the same manner. The bed of mortar shall make contact with the head of the lower course of tile and the underside of the tile being set.

1509.7.2.2.6. Fastening shall be in accordance with the following:
1. <u>Steep roof pitch installations</u>. For pitches 4:12 and above, the eave course shall be fastened with one nail in addition to the mortar. For pitches from 6:12 up to and including 7:12, every third tile in every fifth course shall be fastened with one nail in addition to the mortar. Plastic cement shall be applied to seal all nail penetrations. For pitches above 7:12, every tile shall be nailed in addition to the application of mortar.
2. <u>Elevation requirements</u>. Tile installed 55 ft (17 m) above grade or greater shall be fastened with a minimum of two hot dipped galvanized nails in addition to the mortar.

1509.7.2.2.7. Hip and ridge tiles shall be set in a continuous bed of mortar lapping the tile a minimum of 1 in (25.4 mm). Mortar should be pointed and finished to match the tile surface. Hip starter tiles shall be one of the following:
1. Prefabricated hip starter
2. Mitered tile as hip starter to match eave lines
3. Standard hip tile as starter

1509.7.2.2.8. Pressure-treated nailer boards should be installed per the manufacturer's recommendations where required for steep roof pitch installation, for use with hip or ridge tile, and/or two-piece barrel tile.

1509.7.2.2.9. Rake treatment shall be one of the following methods:

1. _Rake tile_. The first rake tile shall be installed to the exposed length of the first course of field tile. The factory finish end of the tile should be installed toward the eave. The rake tile shall be nailed with a minimum of two hot dipped galvanized nails of sufficient length to penetrate the framing a minimum of 3/4 in (19 mm). Each succeeding rake tile shall be abutted to the nose of field tile above, maintaining a constant head lap and tuck-pointed with mortar to match the tile surface along the inside edge.
2. _Hush Finish_. A mortar bed shall be placed along the roof edge and the field tile set in mortar flush with the edge and pointed smooth to a straight edge finish.

1509.7.2.2.10. At wall abutments, tile shall be installed adjacent to the wall and voids filled with mortar and finished to match the tile surface.

1509.7.2.2.11. Valleys shall be one of the following:
1. _Closed Valley_. Tile shall be mitered to meet at the center of the valley or mitered to form a straight border on either side of the water diverter.
2. _Open Valley_. Tile shall be mitered to form a straight border on either side of the two water diverters. A bed of mortar shall be placed a minimum of 2 in (51 mm) from the valley center. Tile shall be mitered to form a straight border and tuck-pointed to match the tile surface. A minimum of one two-by-four shall be placed on edge down the center of the valley. A continuous bed of mortar shall be applied along the edge of the two-by-four. The tile shall be installed a minimum of 1 in to a minimum of 4 in (25 to 102 mm) from the two-by-four. A line of mortar shall be consistent through the valley. The mortar shall be smoothed and formed to match the tile contour. After initial set, the two-by-four shall be removed and the mortar tuck-pointed to match the tile surface.

1509.7.2.2.12. Tile shall be installed to accommodate all roof penetrations. Voids shall be filled with mortar and tuck-pointed to match the tile surface.

1509.7.2.2.13. Acrylic sealer shall be applied to exposed mortar in accordance with the manufacturer's recommendation.

1509.7.2.2.14. In a cemented application, an average nonadherence of no more than one tile in ten is allowable.

1509.7.3 Mechanically Fastened Tile - Subroof Application Subroof application shall be in accordance with one of the methods described in Sections 1509.7.3.1, 1509.7.3.2, and 1509.7.3.3.

1509.7.3.1. Method 1: Single-ply underlayment, minimum 4:12 pitch (not sealed). This method utilizes preformed metal flashing, minimum 2-in (51 mm) paper head lap, 6-in (152 mm) side lap, and minimum 3-in (76 mm) tile head lap. Paper laps or nail penetrations of tile need not be sealed when using this method.

1509.7.3.1.1. Drip-edge metal shall be nailed or stapled along and directly on top of sheathing 6 in (152 mm) o.c. and 1/2 in (12.7 mm) from the top flange. All joints shall be lapped a minimum of 3 in (76 mm). Fastening of the drip edge may be decreased when additional fastenings will be used during application of underlayment and/or eave closure.

1509.7.3.1.2 Underlayment
A 36-in (914 mm) side strip of underlayment (sweat sheet) shall be applied down the center of the valley, tin-capped, and fastened with nails, staples, or caphead nails a maximum of 24 in (610 mm) o.c. along the edge of the sheet. Starting at the eave edge, one course of underlayment should be applied horizontally along the roof line with end joints lapped a minimum of 6 in (152 mm), tin-capped, and fastened with nails, staples, or caphead nails a maximum 36 in (914 mm) o.c. along the top edge of the sheet.

Each succeeding course shall be applied in the same manner allowing a minimum 2-in (51 mm) head lap, tin-capped, and fastened with nails, staples, or caphead nails approximately 12 in (305 mm) o.c. at the head lap with a weave pattern at all valleys. All hips and ridges shall be overlapped a minimum of 6 in (152 mm). Underlayment should extend a maximum 4 in (102 mm) up abutting walls and other protrusions. Underlayment shall be fastened a minimum of 12 in (305 mm) o.c. at the eave.

1509.7.3.1.3. The rake treatment shall be one of the following:
1. Underlayment wrapped gable. Not recommended for flush finish. Trim tile shall be installed. Underlayment shall extend beyond the rake and be folded down onto the fascia on a barge board, tin-capped, and fastened with nails, staples, or caphead nails a maximum 6 in (152 mm) o.c.

2. **Metal drip edged gable.** Drip-edge metal shall be nailed or stapled along and directly on top of underlayment 6 in (152 mm) o.c. and 1/2 in (12.7 mm) in from the top flange. All joints shall be lapped a minimum of 3 in (76 mm). Metal shall continue from the eave up the rake.

1509.7.3.1.4. Valleys shall be in accordance with the following:
1. **Preformed closed valley.** Valleys shall be a minimum 16 in (407 mm) wide with a minimum 2-1/2-in (64 mm) high center diverter and with minimum 1-in (25.4 mm) metal edge returns. All joints shall be lapped a minimum of 6 in (152 mm). A coating or separator sheet for corrosion resistance shall be installed when using any ferrous metals.
2. **Preformed open valley.** Valleys shall be a minimum 16 in (407 mm) wide with a minimum 2-1/2-in (64 mm) high twin center diverter and with minimum 1-in (25.4 mm) metal edge returns. All joints shall be lapped a minimum of 6 in (152 mm). A coating or separator sheet for corrosion resistance shall be installed when using any ferrous metals.
3. Valleys shall be secured with clips fabricated from the same material 24 in (610 mm) o.c. One-in (25.4 mm) metal edge returns shall be clipped to either deck or batten strip with roofing nails through a metal strap. Metal shall be trimmed and a lead soaker installed at all valley or ridge junctions. Lead should be turned up 1 in (25.4 mm) to create water diverter.
4. Valleys terminating onto the roof plane shall be installed in accordance with regular valley flashing installation procedures. An 18- X 18-in (457 X 457 mm) lead soaker skirt shall be applied underneath the eave end of the valley to carry water off of the valley back onto the field tile.

1509.7.3.1.5 Flashing and Counterflashing at Wall Abutments
A preformed metal wall tray shall be installed flush to the base of walls over underlayment. Work shall be started at the lower portion to ensure watertightness. Flashing shall be secured with clips on the horizontal metal flange 24 in (610 mm) o.c. A 1-in (25.4 mm) metal edge diverter shall be clipped to the deck or batten strip with a roofing nail through a metal strap. The vertical metal flange shall be nailed within 1 in (25.4 mm) of the metal edge. Joints shall be

lapped a minimum of 4 in (102 mm), and a coating or separator sheet for corrosion resistance applied when using ferrous metals.

The entire edge of the vertical metal flange shall be sealed. Where counterflashing is installed, the top flange of base flashing shall be lapped a minimum of 3 in (76 mm). Metal shall be nailed within 1 in (25.4 mm) of the metal edge a minimum of 6 in (152 mm) o.c. or set into regrets and thoroughly caulked. Joints shall be lapped a minimum of 3 in (76 mm) and plastic roof cement or sealant applied between laps. All head or apron flashing shall be installed on top of the tile. The deck flange shall conform to the pitch of the roof and extend a minimum of 4 in (102 mm) onto the field tile. Flashing shall be installed to either channel water to the eave under the tile or redirect water back on top of the field tile.

1509.7.3.1.6 Curb-Mounted Skylights, Hood Vents, Turbines

A cricket shall be installed on the ridge side of any curb greater than 48 in (1219 mm) wide. A minimum 12-in (305 mm) width of lead shall be installed at the eave end of curb, trimmed as necessary to ensure water shedding capabilities on top of the field tile. It shall be secured with roofing nails 6 in (152 mm) o.c., with nails covered by the skylight or hood vent flange. Lead shall be continued on both sides of the curb working up toward the ridge, trimmed as necessary to ensure water shedding capabilities onto the field tile. It shall be secured with roofing nails 6 in (152 mm) o.c.

The ridge end of the curb shall be installed with a minimum 24 in (610 mm) width of lead extending over the course of tile abutting the top of the curb, and under the second course of tile at the top of the curb or lead shall be extended under both courses of the tile at the top of the curb. Lead shall be folded to create a 1-in (25.4 mm) water diverter at the top and sides of the 24 in (610 mm) lead saddle. All nail penetrations, lead or skylight, or hood vent joints shall be sealed with approved sealant or caulk.

1509.7.3.1.7. Prefabricated curbed skylights shall be installed in accordance with the skylight manufacturer's recommendations for nail-on tile system skylights.

1509.7.3.1.8. Chimneys and Wall Abutments Terminating Onto Roof Plane

A minimum 12-in (305 mm) width of lead shall be installed at the eave end of the protrusion, trimmed as necessary to ensure water shedding capabilities on top of the field tile. It shall be secured with roofing nails a maximum 6 in (152 mm) o.c., 1 in (25.4 mm) from the vertical flange. A wall pan flashing with a minimum 1-in (25.4 mm) water diverter along sides of the chimney or wall abutments shall be installed, trimmed, and folded at the ridge end of side flashing. The wall pan flashing shall terminate a minimum of 8 in (203 mm) from the eave end of the protrusion. Prior to securement, a lead saddle to carry water from wall pan flashing on top of the field tile shall be applied. The pan flashing shall be sealed to the lead saddle and secured with roofing nails a maximum 6 in (152 mm) o.c., 1 in (25.4 mm) from the vertical flange.

A metal saddle shall be installed at the ridge end of the chimney using preformed corrosion-resistant metal or lead, trimmed and folded to ensure water shedding capabilities onto the side pan flashing. It shall be secured with roofing nails a maximum 6 in (152 mm) o.c., 1 in (25.4 mm) from the vertical flange. The vertical flange shall be sealed with plastic cement or sealant. On flat tile installations, rigid corrosion-resistant metal shall be permitted to be substituted for lead flashing at the chimney-to-wall abutment and eave end.

1509.7.3.1.9. Pipe stacks shall be sealed with plastic cement. An 18-in (457 mm) skirt lead stack shall be installed over the last field tile previously installed. The lead shall extend under the course of tile above the pipe stack course and be sealed with approved sealant or caulk. On flat tile installations, rigid corrosion-resistant metal shall be permitted to be substituted for lead.

1509.7.3.2. Method 2: Single-ply underlayment, minimum 4:12 pitch (sealed). This system utilizes standard metal flashing, minimum #43 base sheet with a sealed minimum 2-in (51 mm) paper head lap, sealed 6-in (152 mm) side lap, and minimum 2-in (51 mm) tile head lap. All tile nail penetrations shall be sealed with plastic cement.

1509.7.3.2.1. Drip-edge metal shall be nailed or stapled along and directly on top of sheathing, fastened a minimum 6 in (152 mm) o.c. and 1/2 in (12.7 mm) from the top flange. All joints shall be lapped a minimum of 3 in (76 mm) and sealed along the entire length of the top edge of the eave drip with plastic cement. Securement of the drip edge may be decreased when additional securement will be used during application of the underlayment and/or metal or rubber eave closure.

1509.7.3.2.2 Underlayment
A 36-in (914 mm) wide strip of underlayment (sweat sheet) shall be applied down the center of the valley, tin-capped, and secured with nails, staples, or caphead nails a maximum 24 in (610 mm) o.c. along the edge of the sheet. Starting at the eave edge, one course of underlayment shall be applied horizontally along the roof line with the end joints lapped a minimum of 6 in (152 mm). End laps shall be sealed with plastic cement. The underlayment shall be tin-capped and secured with nails, staples, or caphead nails a maximum 12 in (305 mm) o.c. along the top edge of the sheet.

The entire length of the top edge of underlayment shall be sealed, covering all tin-caps or caphead nails with plastic cement. Each succeeding course shall be applied in the same manner allowing a minimum 2-in (51 mm) head lap, ensuring a weave pattern at all valleys. Hips and ridges shall be overlapped a minimum of 6 in (152 mm). Underlayment shall be trimmed at all wall bases.

1509.7.3.2.3 Rake Treatment
Drip-edge metal should be nailed or stapled along and directly on top of the underlayment, fastened 6 in (152 mm) o.c. and 1/2 in (12.7 mm) from the top flange. All joints shall be lapped a minimum of 3 in (76 mm). Metal shall continue from the eave up to the rake or gable. The entire edge of the metal flange shall be sealed, covering all nail penetrations with plastic cement and membrane.

1509.7.3.2.4. Valleys shall be nailed within 1 in (25.4 mm) of the metal edge a maximum 6 in (152 mm) o.c. Metal joints shall be lapped a minimum of 6 in (152 mm) with plastic roof cement applied between laps. The entire edge of the metal flange shall be sealed, covering all nail penetrations with plastic cement and membrane. Valleys shall be one of the following:
1. Standard. Material shall be a minimum of 16 in (407 mm) in width.
2. Preformed closed valley. Material shall be a minimum of 16 in (407 mm) in width with a minimum 2-1/2-in (64 mm) high center diverter.
3. Preformed open valley. Material shall be a minimum of 16 in (407 mm) in width with minimum 2-1/2-in (64 mm) high twin center diverters.

1509.7.3.2.5 Flashing and Counterflashing at Wall Abutments
"L"-shaped metal flashing shall be installed flush to the base of walls over the underlayment, nailed within 1 in (25.4 mm) of the metal edge. The horizontal flange shall be nailed a maximum 6 in (152 mm) o.c. Joints shall be lapped a minimum 4 in (102 mm) and plastic roof cement applied between laps. Flashing shall start at the lower portion and work up the roof to ensure water-shedding capabilities of all metal laps. The entire edge of the metal flange shall be sealed, covering all nail penetrations with plastic roof cement and membrane. If counterflashing is installed, the top flange of base flashing shall be lapped a minimum of 3 in (76 mm).

Metal shall be nailed within 1 in (25.4 mm) of the metal edge a minimum of 6 in (152 mm) o.c. or set into regrets (secured properly) and thoroughly caulked. Joints shall be lapped a minimum of 3 in (76 mm) with plastic roof cement applied between laps. All head and apron flashing shall be installed on top of the underlayment, conforming to the pitch of the roof and extending a minimum of 4 in (102 mm) onto the deck. The metal edge shall be sealed with plastic cement and membrane.

1509.7.3.2.6. Standard skylights, chimneys, etc., shall be installed in accordance with regular flashing installation procedures.

1509.7.3.2.7. For pipes, turbines, vents, etc., plastic roof cement shall be applied around the base of protrusion and on the bottom side of metal flanges, sealing the unit base flashing to the underlayment. All sides of base flashing shall be nailed and secured within 1 in (25.4 mm) of the edge. The base shall be flush to the deck. The edge of the metal flanges shall be sealed, covering all nail penetrations with plastic roof cement and membrane.

1509.7.3.3. Method 3: Two-ply underlayment - pitches 3:12 and greater. This system utilizes standard metal flashing, minimum #30 dry-in sheet, tin-capped, hot mopped minimum 74-lb (34 kg) mineral surfaced roll roofing, and a minimum 2-in (51 mm) headlap. Plastic cement shall be applied at all file nail penetrations except for pitches 6:12 and above.

1509.7.3.3.1 Base Ply
Starting at the eave edge, one course of #30 roofing felt shall be applied horizontally along the roof line, lapping end joints a minimum of 4 in (102 mm). #30 felt should be secured with nails, staples, or caphead nails, a maximum of 12 in (305 mm) o.c. in the

field with 12-in (305 mm) head laps, ensuring a weave pattern at the valleys. Each succeeding course shall be applied in the same manner allowing a minimum 2-in (51 mm) head lap. All hips and ridges shall be overlapped a minimum of 6 in (152 mm).

1509.7.3.3.2. Drip edge metal shall be nailed and stapled along and directly on top of the #30 felt at the eave, fastened a maximum 6 in (152 mm) o.c. and 1/2 in (12.7 mm) from the top flange. All joints shall be lapped a minimum of 3 in (76 mm). Metal shall continue from the eave up to the rake or gable in the same manner.

1509.7.3.3.3 Valleys

Metal shall be nailed within 1 in (25.4 mm) of the metal edges, a maximum of 6 in (152 mm) o.c. Metal joints shall be lapped a minimum of 6 in (152 mm). Plastic roof cement shall be applied between the laps. Valleys shall be one of the following:

1. <u>Standard</u>. Material shall be a minimum of 16 in (407 mm) in width.
2. <u>Preformed closed valley</u>. Material shall be a minimum of 16 in (407 mm) in width with a minimum 2-1/2-in (64 mm) high center diverter.
3. <u>Preformed open valley</u>. Material shall be a minimum of 16 in (407 mm) in width with minimum 2-1/2-in (64 mm) high twin center diverters.

1509.7.3.3.4 Flashing and Counterflashing at Wall Abutments

"L"-shaped metal shall be installed flush to the base of walls over the #30 felt and nailed within 1 in (25.4 mm) of the metal edges. Joints shall be lapped a minimum of 4 in (102 mm) with plastic roof cement applied between the laps, starting at the lower portion and working up to ensure watertightness. The top edge of the vertical flange shall be sealed, covering all nail penetrations with plastic roof cement and membrane. If counterflashing is installed, the top flange of base flashing shall be lapped a minimum of 3 in (76 mm).

Metal shall be nailed within 1 in (25.4 mm) of metal edge a minimum of 6 in (152 mm) o.c. or set into reglets (secured properly) and thoroughly caulked. Joints shall be lapped a minimum of 3 in (76 mm), and plastic roof cement shall be applied between laps. All head or apron flashing shall be installed on top of the mineral surface roll roofing. The deck flange should conform to the pitch of the roof and extend a minimum of 4 in (102 mm) onto the deck.

1509.7.3.3.5. Standard skylights, chimneys, and other projections shall be installed in accordance with regular flashing installation procedures.

1509.7.3.3.6 Pipes, Turbines, Vents, and Other Penetrations
Plastic roof cement shall be applied around the base of the protrusion and on the bottom side of the metal flanges, sealing the unit base flashing to the deck. All sides of base flashing shall be nailed and secured within 1 in of the edge, making certain the base is flush to the deck.

1509.7.3.3.7 Top Ply
Starting at the eave edge, #90 roll roofing shall be applied horizontally along the roof line over the base ply, lapping the end joints a minimum of 6 in (152 mm). Hot asphalt shall be applied between the plies of roofing felt so that no felt touches felt. For pitches above 6:12, cold applied or plastic roof cement shall be permitted. The ply shall then be back-nailed, using tin-cap and roofing nails or caphead nails, a maximum of 12 in (305 mm) o.c., and a minimum 1 in (25.4 mm) from the top edge of the felt.

Each succeeding course shall be applied in the same manner, allowing a minimum 4-in (102 mm) head lap. Hips and ridges shall be lapped a minimum of 6 in (152 mm), ensuring a weave pattern in the valley or trim, a maximum of 4 in (102 mm) past nail penetrations. When preformed valley metal is used, the edge of #90 felt shall be sealed with plastic roof cement and membrane. All metal flashing and roof protrusions shall be hot mopped or cold applied in order to ensure weathertightness. Any felt overhanging at the eave and gable shall be trimmed. Fishmouths shall be cut and sealed with plastic roof cement and membrane. Plastic roof cement or hot asphalt shall be applied along the edge of the felt wherever it meets the wall bases.

Fire-Rated Wall Cement Board Roof Panels

Noncombustible structural fiber-reinforced cement board is a safe and cost-effective alternative to plywood problems. It reduces fire sprinkler requirements and allows lower insurance costs and fire-resistant design. The builder can gain the durability of portland cement on sloped roofs without pouring concrete. Select the metal deck section in combination with cement board thickness for load-carrying capacity over given framing spacing. Limit the deflection of metal deck under design loads to 1:240 maximum deflection for cement board. Use a metal deck in all roof and floor assemblies where spans are greater than 24 in o.c. Noncombustible structural fiber-reinforced cement board allows for construction of totally noncombustible roof designs. In panelized roof construction, it offers the durability of concrete without the time delays and equipment expense related to pouring wet concrete.

The production of panelized construction is done on 4 X 8 jig tables on which the modular manufacturing plant puts together the closed wall panels, as shown in Figure 11-5. Steel in the panels typically ranges from 25 to 14 gauge. Headers are back-to-back C-channels. Self-tapping screws are driven by coil-fed electric or pneumatic tools. The reason panelized construction makes use of more steel orientation than wood is because closed wood panels erected in a wall sometimes tend to grow due to swelling of the wood. In a 35-ft-long wall, as much as 1/2 to 3/4 in of expansion due to swelling from one end of the wall to the other can happen in areas of high-temperature swings. This leads to significant problems with both the electrical and plumbing systems. By contrast, in a 35-foot-long steel-framed wall, also closed panel, the expansion is only about 3/16" from one end to the other. The most common cause of the higher expansion in the wood is the use of green lumber, a common supply problem in the southwest, in addition to the area's hot-cold climate conditions. Climate has a slight effect on shrinking or expanding steel-framed walls.

Roof and wall panels can be designed to include electrical and fire sprinkler or plumbing lines preinstalled, complete with interior drywall that is taped, textured, and painted. The relatively small 4 X 8 panel sizes represent the standard modulus of construction, and in many areas, where the homes are to be shipped, mechanical lifting equipment is not available, but there seems to be plenty of manual labor. Two workers can quite easily handle 4 X 8 panels in construction. Also, because the framework is doubled at each side of the 4 X 8 panes, the builder ends up with a stronger dwelling. On some homes, such as three-story designs, 4 X 8 shear panels made of 1/4-in rolled steel are placed in strategic areas to give the light-gauge steel-framed homes "engineering validity."

290 Chapter Eleven

**Figure 11-5
Panelized Construction**

Panelized roof construction includes fiberglass and tile panel roofing combining the popular look of Spanish tile with the best features of metal panels, including lightweight, durability, and all-weather performance. Proven wind resistance of 230+ mph is specified for the best-quality stuff. Systems are typically made up of 3-ft-wide panels in lengths up to 20 ft, with a full range of accessories designed for builder-friendly installation. These are available in a wide range of traditional tile styles and designer colors. Some even look just like granite slate. You cannot tell the difference standing in the driveway looking at them. Cool stuff indeed, and no need for an expensive 2 X 6 rafter system to carry the heavy load of the real thing. Panelized walls using exterior steel doors combine the beauty of wood with the security of 22 gauge steel. A special primer coat accepts most stains to give the look and feel of natural wood. Recessed hinge systems allow for alignment adjustments of exterior doors in their frames.

Chapter

12

Structural Loads

SECTION 1600.0 GENERAL [1600.0; 1600.1]

This section clearly describes the function of the national building codes in relation to structural design, construction, and installation work. By the word <u>installation</u>, it is made clear that the code regulation applies to systems in their manner of installation as well as in the components used. Effective provision of structural loads and selection of component materials for particular applications involve selective engineering skills of the designer, above routine adherence to code requirements Designers must know the physical characteristics, limitations, and advantages of the many materials they specify for use in modern construction. Safety is not automatically made a characteristic of a system by simply observing code compliance.

Although each of the national building codes continually strive to attain that criterion, safety must be initially designed into any construction system. The provisions of this chapter govern the structural design of buildings, structures, and portions thereof. The national building codes contain provisions considered necessary for structural safety, but they do not always provide adequately detailed information of a design nature, other than for safety purposes In such case, designers are encouraged to contact the local building official having jurisdiction. His or her interpretation will mirror the intent of the code-making councils

All buildings and structures must be designed and constructed to sustain, within the stress limitations specified in the code, all dead loads and other loads, especially in living areas built over other living areas as shown in Figure 12-1. Design loads are computed from four different factors which determine either the uniform or concentrated loads, the combination of which a structure must bear safely. These are defined as follows:

294 Chapter Twelve

**Figure 12-1
Multistory Construction**

- **Dead load.** The vertical load due to the weight of all permanent structural and nonstructural components of a building, such as walls, floors, roofs, and fixed service equipment.
- **Live load.** The load superimposed by the use and occupancy of the building not including the wind load, earthquake load, or dead load.
- **Load duration.** The period of continuous application of a given load, or the aggregate of periods of intermittent application of the same load.
- **Unit live load.** A concentrated load localized in one specific area and not distributed evenly over the general floor or load-bearing structure area. Locating and planning for heavy equipment and machinery falls under this design load definition.

Buildings and structures must also be designed and constructed for roof loads to resist snow loads (if applicable to the region) and wind effects that are assumed to come from any horizontal direction. No reduction in wind pressure is permitted by the codes for the shielding effect of adjacent structures Each code has a wind load formula with pressure coefficients as well as a snow load formula for its relative region that is used to calculate the minimum roof live load construction.

Designers must also contend with three other load factors: *uplift loads*, which are negative loads imposed by winds on roofs; *hydrostatic uplift*, which is upward pressure caused by over-optimum groundwater pressure; and *earthquake loads* All these load factors are covered by the individual national building codes with different tables relative to the regions served by the separate codes

1601.2 Structural Safety

1601.2.1. Every building and structure shall be of sufficient strength to support the loads and forces encountered, or combinations thereof, without exceeding in any of its structural elements the stresses prescribed elsewhere in code.

1601.2.2. Buildings and structural systems shall possess general structural integrity to reduce the hazards associated with progressive collapse to levels consistent with good engineering practice. The structural system shall be able to sustain local damage or failure with the overall structure remaining stable. Compliance with the applicable provisions of ASCE 7 shall be considered as meeting the requirements of this section.

1601.3 Restrictions on Loading
It shall be unlawful to place, or cause or permit to be placed, on any floor or roof of a building or other structure a load greater than is permitted by these requirements

1601.4 Occupancy Permits for Changed Loading
Plans for other than residential buildings filed with the building official with applications for permits shall show on each drawing the live loads per square foot of area covered, for which the building is designed; and occupancy permits for buildings hereafter erected shall not be issued until the floor load signs, required by Section 106.4, have been installed.

1601.5 Items not Specifically Covered
Loads and forces for occupancies or uses not covered in this chapter shall be subject to the approval of the building official.

1602.0 DEFINITIONS [1601.0; 1601.1]
The following term shall, for the purposes of this chapter and as stated elsewhere in code, have the meaning shown herein. Refer to Chapter 2 for general definitions

CRANE LOAD - The dead, live, and impact loads and forces resulting from the operation of permanent cranes

1603.0 DEAD LOADS [1602.0; 1602.0]
1603.1 Weights of Materials and Construction
In estimating dead loads for purposes of design, the actual weights of materials and constructions shall be used, provided that in the absence of definite information, values satisfactory to the building official may be assumed.

1603.2 Provision for Partitions
The actual weight of all permanent partitions shall be included in the dead load. Where partitions are likely to be used, although not definitely located, or where they are likely to be shifted, 20 psf (958 Pa) shall be added to the dead load in the areas supporting them, except in the case of light partitioning.

1603.3 Weight of Fixed Service Equipment

In estimating dead loads for the purpose of design, the weight of fixed service equipment, such as plumbing stacks and risers; electrical feeders; and heating, ventilating and air-conditioning systems, shall be included whenever such equipment is supported by structural elements

1604.0 LIVE LOADS [1603.0; 1603.0]
1604.1 Uniform Floor Live Loads

The live loads assumed for purposes of design shall be the greatest loads that probably will be produced by the intended uses and occupancies, provided that the minimum live loads to be considered as uniformly distributed shall be as given in Table 1604.1.

1604.2 Reduction of Uniform Floor Live Load

Floor live loads in Table 1604.1 may be reduced in accordance with the following provisions Such reductions shall apply to slab systems designed for flexure in more than one direction, beams, girders, columns, piers, walls, and foundations

1. A reduction shall not be permitted in Group A Occupancies
2. A reduction shall not be permitted when the live load exceeds 100 psf (4.8 kPa) except that the design live load for columns may be reduced 20%.
3. For live loads not exceeding 100 psf (4.8 kPa), the design live load for any structural member supporting 150 sq ft (14 m^2) or more may be reduced at the rate of 0.08% per sq ft of the area supported. Such reduction shall not exceed 40% for horizontal members, 60% for vertical members, nor R as determined by the following formula:

$$R = 23.1 (1 + D/L)$$

where R = reduction, %
D = dead load per square foot of area supported
L = live load per square foot of area supported

Table 1604.1

Minimum Uniformly Distributed Live Loads

Occupancy or Use	Live Load, psf
Apartments (see Residential)	
Armories and drill rooms	150
Assembly halls and other places of assembly:	
Fixed seats	50
Movable seats	100
Balcony and decks (exterior) same as occupancy but not less than	60
On one- and two-family dwellings	40
Bowling alleys, poolrooms, and similar recreational areas	75
Corridors:	
First floor	100
Other floors, same as occupancy served except as indicated	100
Dance halls and ballrooms	100
Dining rooms and restaurants	100
Dwellings (see Residential)	
Fire escapes	100
On multi- or single-family residential buildings only	40
Garages (passenger cars only)	50
For trucks and buses use AASHTO[1] lane loads	
Grandstands (see Reviewing Stands and Bleachers)	
Gymnasiums, main floors, and balconies	100
Hospitals:	
Operating rooms, laboratories	60
Private rooms	40
Wards	40
Corridors, above first floor	80
Hotels (see Residential)	
Libraries:	
Reading rooms	60
Stack rooms (books and shelving at 65 lb/cu ft)	125

Corridors, above first floor	80
Manufacturing:	
Light	100
Heavy	150
Marquees	75
Office buildings:	
Offices	50
Lobbies	100
Corridors, above first floor	80
File and computer rooms require heavier loads based upon anticipated occupancy	
Penal institutions:	
Cell blocks	40
Corridors	100
Residential:	
Multifamily houses:	
Private apartments	40
Public rooms	100
Corridors	80
Dwellings:	
Sleeping rooms	30
Attics with storage	30
Attics without storage	10
All other rooms	40
Hotels:	
Guest rooms	40
Public rooms	100
Corridors serving public rooms	100
Corridors	80
Reviewing stands and bleachers[2]	100
Schools:	
Classrooms	40
Corridors	80
Sidewalks, vehicular driveways and yards, subject to trucking	200
Skating rinks	100
Stairs and exitways	100
Storage warehouse:	
Light	125

Heavy	250
Stores:	
Retail:	
First floor, rooms	75
Upper floors	75
Wholesale	100
Theaters:	
Aisles, corridors and lobbies	100
Orchestra floors	50
Balconies	50
Stages and platforms	125
Catwalks	40
Followspot, projection, and control rooms	50
Yards and terraces, pedestrians	100

1 psf = 47.8803 Pa

Notes:
1. AASHTO = American Association of State Highway and Transportation Officials
2. For detailed recommendations, see NFiPA 102.

1604.3 Concentrated Floor Live Loads

In the design of floors, probable concentrated loads shall be considered. Where such loads may occur, the supporting beams, girders, and slabs shall be designed to carry either the concentrated loads or the live load described in Section 1604.1, whichever produces the greater stresses Concentrated loads shall be equal to the machinery, vehicle, equipment, or apparatus anticipated, but they shall be not less than the loads specified in Table 1604.3.

Table 1604.3
Minimum Concentrated Loads[1]

Location	Load, lb
Elevator machine room grating (on area of 4 sq in)	300
Finish light floor plate construction (on area of 1 sq in)	200
Garages	See Note 2
Office floors	2000
Scuttles, skylight ribs, and accessible ceilings	200
Stair treads (on area of 4 sq in at center of tread)	300

1 lb = 4.4482 N
1 sq in = 645.16 mm^2
1 sq ft = 0.0929 m^2

Notes:
1. Load distributed uniformly over an area of 2-1/2 sq ft unless noted otherwise.
2. Floors in garages or portions of buildings used for the storage of motor vehicles shall be designed for the uniformly distributed live loads of Table 1604.1 or the following concentrated loads:
   a: For passenger cars accommodating not more than nine passengers, 2000 lb acting on an area of 20 sq in
   b: Mechanical parking structures without slab or deck, passenger cars only, 1500 lb per wheel
   c: For truck or buses, with a maximum wheel load on an area of 20 sq in

1604.4 Distribution of Live Loads

Where structural members are arranged so as to create continuity, the distribution of the live loads, such as on adjacent spans or alternate spans, which would cause maximum design conditions shall be upset, except that roof live loads shall be distributed uniformly as provided in Section 1604.6.

EXCEPTION: The distribution of live loads on reinforced concrete structures shall be in accordance with ACI 318.

1604.5 Interior Wall Loads

Interior walls, permanent partitions, and temporary partitions shall be designed to resist all loads to which they are subjected but not less than 5 psf (240 Pa) applied perpendicular to the walls, except for decorative screen walls

1604.6 Roof Live Loads

1604.6.1. The design of roof live loads shall take into account the effects of occupancy and water but shall be not less than the minimum roof live loads as set forth in Table 1604.6.

**Table 1604.6
Minimum Roof Live Load
(Pounds per square foot of horizontal projection)**

Roof Slope	Tributary Loaded Area for Any Structural Member, sq ft		
	0 to 200	201 to 600	Over 600
Flat or rise less than 4 in/ft per ft			
Arch or dome with rise less than 1/8 of span; rise 4 in/ft to less than 12 in/ft	20	16	12
Arch or dome with rise 1/8 of span to less than 3/8 of span; rise 12 in/ft and greater	16	14	12
Arch or dome with rise 3/8 of span or greater	12	12	12
Awnings except cloth covered	5	5	5
Greenhouses, lath houses, and agricultural buildings	10	10	10

1 in/ft = 83.33 mm/m
1 psf = 47.8803 Pa
1 sq ft = 0.0929 m^2

1604.6.2. Rain loads shall be designed in accordance with all the following. Roof drainage systems shall be designed in accordance with chapter 11 of the plumbing code and shall be designed to preclude instability from ponding loads

Each portion of a roof shall be designed to sustain the load of all rainwater that could accumulate on it if the primary drainage system for that portion is blocked. In determining the load that could result should the primary drainage system be blocked, the load due to the depth of water (i.e., head) needed to cause the water to flow out of the secondary drainage system at the rate required by chapter 11 of the plumbing code shall be included. Pending instability shall be considered in this situation. If the overflow drainage provisions contain drain lines, such lines shall be independent of any primary drain lines

Roofs equipped with controlled drainage provisions shall be equipped with a secondary drainage system at a higher elevation which prevents ponding on the roof above the design water depth. Such roofs shall be designed to sustain all rainwater loads on them to the elevation of the secondary drainage system, plus the load due to the depth of water (i.e., head) needed to cause the water to flow out of the secondary drainage system. Ponding instability shall be considered in this situation.

1604.6.3. Roofs designed as future floors for the parking of automobiles or for other occupancy loadings shall comply with the provisions of Sections 1604.1 and 1604.3.

1604.6.4. Snow loads shall comply with the provisions of Section 1605.

1604.6.5. Wind loads shall comply with the provisions of Section 1606.

1604.7 Impact Loads

1604.7.1. For structures carrying live loads which induce unusual impact, the assumed live load shall be increased sufficiently to provide for same. If not otherwise specified, the increase shall be
- For supports of elevators: 100%
- For cab-operated traveling crane support girders and their connections: 25%*
- For pendant-operated traveling crane support girders and their connections: 10%*

- For supports of light machinery, shaft or motor driven, not less than 20%
- For supports of reciprocating machinery or power-driven units, not less than 50%
- For hangers supporting floors and balconies: 33%

---

* Live loads on crane support girders shall be taken as the maximum crane wheel loads

1604.7.2. The lateral force on crane runways to provide for the effect of moving crane trolleys shall, if not otherwise specified, be 20% of the sum of the weights of the lifted load and of the crane trolley exclusive of other parts of the crane.

The force shall be assumed to be applied at the top of the rails acting in either direction normal to the runway rails and shall be distributed with due regard for lateral stiffness of the structure supporting these rails The longitudinal force shall, if not otherwise specified, be taken as 10% of the maximum wheel loads of the crane applied at the top of the rail.

1604.8 Supports for Walkway
Where walkways are to be installed above ceilings, supports shall be designed to carry a load of 200 lb (890 N) occupying a any space 2-1/2 sq ft (0.23 m²), so placed as to produce maximum stresses in the affected members

1604.9 Sidewalks
Sidewalks shall be designed to carry either a uniformly distributed load of 200 psf (9.6 kPa) or a concentrated load of 8000 lb (35.6 kN) on a space 2-1/2 sq ft (0.58 m²) and placed in any position, whichever will produce the greater stresses This does not apply to sidewalks on grade.

1605.0 SNOW LOADS [1605.0; 1605.0]
Design roof snow loads shall be calculated in compliance with provisions in section 7, Snow Loads, of ASCE 7.

1606.0 WIND LOADS [1606.0; 1606.0]
1606.1 Applications
All buildings, structures, and parts thereof shall be designed to withstand the appropriate wind loads prescribed herein. Decreases

in wind loads shall not be made for the effect of shielding by other structures Wind pressures shall be assumed to act normal to the surfaces considered.

1606.1.1 Determination of Wind Forces

Wind forces on every building or structure shall be determined by the provisions of ASCE 7.

EXCEPTIONS:
1. Provisions of Section 1606.2 shall be permitted for buildings 60 ft (18.3 m) high or less
2. Wind tunnel tests together with applicable subsections of Section 1606.2.
3. Provisions of section SSTD 10-93 shall be permitted for applicable Group R2 and R3 buildings
4. ANSI/NAAMM FP 1001 Specification for Design Loads of Metal Flagpoles

1606.1.2 Limitations

1606.1.2.1. Mixing of provisions from ASCE 7 and Section 1606.2 shall not be permitted.

1606.1.2.2. Provisions of Section 1606.2 do not apply to buildings or structures having unusual geometric shapes, site locations, or response characteristics for which channeling effects or buffeting in the wake of upwind obstructions may warrant special consideration. For these cases, wind loads shall be based on wind tunnel tests or nationally recognized data.

1606.1.3 Minimum Wind Loads

No part (component, cladding, or fastener) of a building or structure shall be designed for less than 10 psf (479 Pa).

1606.2 Buildings 60 Ft High or Less

1606.2.1 Scope

Procedures in Section 1606.2 shall be used for determining and applying wind pressures in the design of buildings with flat, single sloped, hipped, and gable-shaped roofs whose mean roof heights exceed neither 60 ft (18.3 m) nor the least horizontal dimension of the building.

1606.2.2 Wind Pressures

1606.2.2.1 Structural members, cladding, fasteners, and systems providing for the structural integrity of the building shall be designed for the velocity pressures from Table 1606.2A (see Appendix). [Coefficients include a gust factor and thus do not correspond to coefficients used in many other sources] Mixing of coefficients and dynamic pressures from different sources shall not be permitted.

1606.2.2.2 Members that act as both part of the main wind force resisting system and as components and cladding shall be designed for separate load cases

1606.2.3 Special Definitions for Section 1606.2

For the purpose of this section, certain special terms are defined as follows:

COMPONENTS and CLADDING - Elements that are either directly loaded by the wind or receive wind loads originating at relatively close locations and that transfer these loads to the main wind force resisting system.

EFFECTIVE WIND AREA for COMPONENTS and CLADDING - The tributary area used to determine pressure coefficients of the element considered but need not be less than one-third the square of the span.

ENCLOSED BUILDING - A building that encloses a space and does not have openings that qualify it as a partially enclosed or open building.

MAIN WIND FORCE RESISTING SYSTEM (MWFRS) - An assemblage of major structural elements assigned to provide support for secondary members and cladding. The system primarily receives wind loading from more than one surface.

OPEN BUILDING - A building having all walls at least 80% open.

OPENINGS - Windows, doors, skylights, or other apertures in the building envelope (roof and exterior wall surfaces) that are not designed as components and cladding.

PARTIALLY ENCLOSED BUILDING - A building which complies with all the following conditions:
- The total area of openings in a wall, or wall and roof, that receives positive external pressure exceeds the sum of the areas of openings in the balance of the building envelope (walls and roof)
- The total area of openings in a wall, or wall and roof, that receives positive external pressure exceeds 5% of the area of that wall, or wall and roof

- The openings in the balance of the building envelope do not exceed 20%

The three conditions can be expressed by the following equations:

$A_o \geq A_{oi}$ and $A_o > 0.05 A_g$ and $A_{oi} / A_{gi} \leq 0.20$, where:

A_o = Total area of openings in a wall, or wall and roof, that receives positive external pressure, in sq. ft.
A_g = Gross area of that wall, or wall and roof, in which A_o is identified, in sq ft
A_{oi} = Sum of the areas of openings in the building envelope (walls and roof) not including A_o, in sq. ft.
A_{gi} = Sum of gross surface areas of the building envelope (walls and roof) not including A_g, in sq. ft.

1606.2.4 Main Wind Force Resisting System (MWFRS)
Pressure coefficients GC_p for all wind loading actions arising from combining loads acting simultaneously on more than one surface shall be determined by Section 1606.2.4.

1606.2.4.1 End Zones
The width of X end zones shall be twice the value of Z determined in Section 1606.2.5.1. For framed buildings whose end bay spacings are greater than or equal to X the difference in end zone and interior zone loading can be allocated entirely to the end frame.

1606.2.4.2 Applicability of Coefficients
Pressure coefficient GC_p for the MWFRS shall be taken from Tables 1606.2B or 1606.2C (see Appendix) and applied with consideration for the torsional effect in each individual load case, as specified in Section 1606.2.2. Where more than one load case exists, buildings shall be designed for all load cases

1606.2.4.3 Overhang Coefficients
The pressure coefficients GC_p to be used for the effects of roof overhangs on MWFRS for each of the load cases and windward and leeward surfaces shall be as indicated in Table 1606.2D (see Appendix). Roof overhang members shall be designed in accordance with Section 1606.2.5.

1606.2.5 Components and Cladding
Pressure coefficients GC_p for wind loading actions on components and cladding shall be determined from Tables 1606.2C and 1606.2D (see Appendix), based on effective wind area.

1606.2.5.1 Edge Strips
The width of the edge strips "Z" for walls and roofs shall be determined by the smaller of 10% of the least horizontal dimension of the building or 40% of the eave height, but not less than the larger of 4% of the least horizontal dimension of the building or 3 ft (914 mm).

1606.2.5.2 Walls
Pressure coefficients GC_p for components and cladding of walls shall be taken from Table 1606.2C according to their effective wind areas and applied to the corresponding regions of the building as shown in Table 1606.2C. Coefficients taken from this table may be reduced 10% if the angle of the roof is no greater than 10 degrees from horizontal.

1606.2.5.3 Gable and Hip Roofs
Pressure coefficients GC_p for components and cladding or roofs shall be taken from Table 1606.2D. Pressure coefficients for roof angles greater than $45°$ (0.785 rad) shall be based on nationally recognized data.

1606.2.5.4 Monoslope Roofs
Pressure coefficients GC_p for monoslope roofs shall be taken from Table 1606.2D for roof slopes between 0 and $3°$ (9 and 0.052 rad), and from Tables 1606.2F and 1606.2G for roof slopes greater than $3°$ (0.052 rad).

1606.2.6 Simplified Wind Loads for Buildings with Vertically Spanning Walls
Wind loads for enclosed buildings with vertically spanning walls whose MWFRS consists of floor and roof diaphragms that are laterally supported by shear walls, braced frames or moment frames shall be permitted to be designed in accordance with this section.

Structural members, cladding, fasteners, and systems providing for the structural integrity of the building shall be designed for the velocity pressures from Table 1606.2A, multiplied by the appropriate pressure coefficient use factor from Table 1606 (see Appendix).

1606.2.6.1 Main Wind Force Resisting System

All elements and connections of the MWFRS shall be designed for vertical and horizontal loads based on the combined leeward and windward wall and roof coefficients GC_p as given in Table 1606.2E. The design wind load shall be applied nonconcurrently to each major axis of the structure. The width of the end zone shall be determined from Section 1606.2.4.1.

Wall elements subjected to wind loads which also support roof framing members shall be considered part of the MWFRS These elements shall be designed for the interaction of vertical and horizontal wind loads or have independent resistance mechanisms for vertical and horizontal loads. The horizontal load shall be based on $GC_p = \pm 0.95$ for end zones and $GC_p = \pm 0.70$ for interior zones

1606.2.6.2 Components and Cladding

All wall and roof framing including cladding and connections of these elements shall be designed using pressure coefficients GC_p determined from Tables 1606.2C and 1606.2D.

1606.3 Roof Systems

1606.3.1 Roof Deck

The roof deck shall be designed to withstand the wind pressures determined under Section 1606.2 for buildings 60 ft (18.3 m) or less in height or ASCE 7 for buildings of any height.

1606.3.2 Roof Coverings

Roof coverings shall comply with Section 1606.3.1. Rigid tile roof coverings that are air-permeable and installed over a roof deck shall be permitted to be designed in accordance with Section 1606.3.3.

1606.3.3 Rigid Tile

Wind loads on rigid tile roof coverings shall be determined as the lifting moment M_a. The lifting moment shall be determined in accordance with the following formula:

$$M_a = q_h C_L bLL_a(1.0 - GC_p)$$

where

M_a = aerodynamic uplift moment acting to raise tail of tile (ft-lb)

q_h = wind velocity pressure determined from Table 1606.2A (psf)
C_L = lift coefficient determined from SBC Table 1606.3.3

b = exposed width of roof tile (ft)
L = length of roof tile (ft)
L_a = moment arm from axis of rotation to point of uplift on roof tile (ft)

The point of uplift shall be taken at 0.76L from the head of the tile and the middle of the exposed width. For roof tiles with nails or screws (with or without a tail clip), the axis of rotation shall be taken as the head of the tile for direct deck applications and as the top edge of the batten for battened applications For roof tiles fastened only by a nail or screw along the side of the tile, the axis of rotation shall be determined by testing.

For roof tiles installed with battens and fastened only by a clip near the tail of the tile, the moment arm shall be determined about the top edge of the batten with consideration given for the point of rotation of the tiles based on straight bond or broken bond and the tile profile. The roof coefficient (GC_p) for each applicable zone is determined from SBC Table 1606.2E. The roof coefficient is not to be adjusted for internal pressure.

Loose laid or mechanically fastened concrete and clay roof tiles complying with the following limitations shall be designed to withstand the wind loads prescribed in this section, i.e.,

- The roof tiles shall be either loose laid on battens or mechanically fastened.
- The roof tiles shall be installed on solid sheathing which has been designed as components and cladding in accordance with Section 1606.2.3.3.
- An underlayment shall be installed in accordance with Section 1509.7.
- The tile shall be single-lapped interlocking with a minimum head lap of not less than 2 in (51 mm).
- The length of the tile shall be between 1.0 and 1.75 ft (305 and 533 mm).
- The exposed width of the tile shall be between 0.73 and 1.25 ft (223 and 381 mm).
- The maximum thickness of the tail of the roof tile shall not exceed 1.3 in (33 mm).

1606.4 Stability

Anchorage shall be provided to resist excess overturning, uplift, and sliding forces The overturning moment due to wind load shall not exceed two-thirds of the dead load stabilizing moment unless the building or structure is anchored so as to resist the excess moment.

Table 1606
Use Factors for Buildings and Other Structures

Nature of Occupancy	Use Factor
All buildings and structures except those listed below	1.0
Buildings and structures where the occupant load is 300 or more in any one room	1.15
Buildings and structures designated as essential facilities, including, but not limited to:	
Hospital and other medical facilities having surgery or emergency treatment areas	1.15
Fire or rescue and police stations	1.15
Primary communication facilities and disaster operation centers	1.15
Power stations and other utilities required in an emergency	1.15
Buildings and structures that represent a legal hazard to human life in the event of failure, such as agricultural buildings, certain temporary facilities, and minor storage facilities	0.9

Table 1606.2A
Velocity Pressure $(q)^1$ (psf)

Mean Roof Height H^3(ft)	Fastest Mile Wind Speed, $(V)^2$ in mph				
	70	80	90	100	110
0-15	10.0	13.1	16.6	20.4	24.7
16	10.2	13.3	16.9	20.8	25.2
17	10.4	13.6	17.2	21.2	25.6
18	10.5	13.8	17.4	21.5	26.1
19	10.7	14.0	17.7	21.9	26.5
20	10.9	14.2	18.0	22.2	26.8
22	11.2	14.6	18.5	22.8	27.6
24	11.5	15.0	18.9	23.4	28.3
26	11.7	15.3	19.4	23.9	28.9
28	12.0	15.6	19.8	24.4	29.6
30	12.2	15.9	20.2	24.9	30.1
33	12.5	16.4	20.7	25.6	31.0
35	12.8	16.7	21.1	26.0	31.5
40	13.3	17.3	21.9	27.0	32.7
45	13.7	17.9	22.7	28.0	33.8
50	14.1	18.4	23.3	28.8	34.9
55	14.5	19.0	24.0	29.6	35.8
60	14.9	19.4	24.6	30.4	36.7

1 mph = 0.447 m/s
1 psf = 47.8803 Pa
1 ft = 0.305 m

Notes:
    1. A single value for velocity pressure (q) is used for the entire building is q = 0.00256 V^2 ($\underline{H}$) 2/7, where
       V = Fastest mile wind speed in miles per hour determined from Table 1606.

Structural Loads 313

H = Mean height of roof above ground or 15 ft whichever is greater. Eave height may be substituted for mean roof height if roof angle "a" is not more than 10^0.
2. Plus and minus signs signify pressures acting toward and away from the surfaces, respectively.
3. Each component shall be designed for maximum positive and negative pressures

1607.0 EARTHQUAKE LOADS [1607.0; 1607.0]
1607.1 General
1607.1.1. Every building and structure, and portion thereof, shall be designed and constructed to resist the effects of earthquake motions determined in accordance with Section 1607. Additions and change of occupancy to existing buildings and structures shall be designed and constructed to resist the effects of earthquake motions determined in accordance with Section l607.

[Same section in both UBC and BOCA.] Special structures, including but not limited to vehicular bridges, transmission towers, industrial towers and equipment, piers and wharves, and hydraulic structures shall be designed for earthquake loads using a properly substantiated analysis

EXCEPTIONS:
- Buildings of detached one- and two-family dwellings (Group R3) that are located in seismic map areas having an effective peak-velocity-related acceleration value, A_V, according to Section 1607.1.5, less than 0.15 are exempt from the requirements of Section 1607.
- Agricultural storage buildings which are intended only for incidental human occupancy are exempt from the requirements of Section 1607.
- Buildings or structures located where the seismic coefficient representing the effective peak velocity-related acceleration, A_V, is less than 0.05 need comply with Section 1607.3.6.1.
- Buildings of detached one- and two-family dwellings (Group R3) with a building height not more than 35 ft (10.7 m) or two stories, which have seismic-load-resisting systems which are entirely of wood frame construction in accordance with the requirements of SBC chapter 23 and are located in seismic map areas having an effective peak

velocity-related acceleration, A_V, equal to or greater than 0.15, need only comply with Section 1607.3.6.1.
- Buildings assigned to Seismic Performance Category B, according to Sections 1607.1.5 and 1607.1.8, which have seismic-load-resisting systems which are entirely of light frame wood construction in accordance with the provisions of SBC chapter 23, need only comply with Section 1607.3.6.1.

1607.1.2 Required Design Data

Where earthquake loads are applicable, the following design data shall be indicated on the design drawings:
- The peak-velocity-related acceleration, A_V, according to Section 1607.1.5.
- The peak acceleration, A_a, according to Section 1607.1.5.
- The seismic hazard exposure group according to Section 1607.1.6.
- The seismic performance category according to Section 1607.1.8.
- The soil profile type according to Table 1607.3.1.
- The basic structural system and seismic resisting system according to Table 1607.3.3 (see Appendix).
- The response modification factor, R, and the deflection amplification factor, C_d, according to Table 1607.3.3.
- The analysis procedure utilized in accordance with Section 1607.4 or 1607.5 as applicable.

1607.1.3 Additions to Existing Buildings

An addition which is structurally independent from an existing building shall be designed and constructed in accordance with the seismic requirements for new buildings An addition which is not structurally independent from an existing building shall be designed and constructed such that the entire building conforms to the seismic requirements for new buildings unless the following three provisions are complied with:
- The addition complies with the seismic requirements for new buildings
- The addition shall not increase the seismic forces in any structural element of the existing building by more than 5% unless the increased forces on the element are still in compliance with these provisions

- The addition shall not decrease the seismic resistance of any structural element of the existing building below that required for new buildings

1607.1.4 Change of Occupancy
When a change of occupancy results in an existing building being reclassified to a higher seismic hazard exposure group, the building shall conform to the seismic requirements for new buildings

Upgrading the building for the seismic requirements of this section is not required for buildings located in seismic map areas having an effective peak-velocity-related acceleration, A_V, value of less than 0.15 when the change of use results in a building being reclassified from Seismic Hazard Exposure Group I to Seismic Hazard Exposure Group II.

1607.1.5 Seismic Ground Acceleration Maps
The effective peak-velocity-related acceleration, A_V, and the effective peak acceleration, A_a, shall be determined from SBC Tables 1607.1.5A and 1607.1.5B, respectively. Interpolation shall be permitted in the determination of the effective peak velocity-related acceleration, A_V, and the effective peak acceleration, A_a. Where site-specific ground motions are used or required, they shall be developed with 90% probability of ground motion not being exceeded in 50 years

1607.1.6 Seismic Hazard Exposure Groups
All buildings shall be assigned to one of the following seismic hazard exposure groups in Table 1607.1.6 (see Appendix A).

1607.1.6.1 Mixed Use
Where a building is occupied for two or more uses, not included in the same seismic hazard exposure group, the building shall be assigned the classification of the highest seismic hazard exposure group occupancy.

1607.1.7 Group III Building Operational Access
Where operational access to a Seismic Hazard Exposure Group III building is required through an adjacent building, the adjacent building shall conform to the requirements for Group III buildings

Where operational access is less than 10 ft (3048 mm) from the interior property line or another building on the same lot, protection from potential falling debris from adjacent property shall be provided by the owner of the Seismic Hazard Exposure Group III building.

1607.1.8 Seismic Performance Category
Buildings shall be assigned a Seismic Performance Category in accordance with SBC Table 1607.1.8 (see Appendix).

1607.1.9 Site Limitation for Seismic Performance Category E
A building assigned to Category E shall not be sited where there is the potential for an active fault to cause rupture of the ground surface under the building.

1607.2 Definitions

Acceleration
Effective Peak - Coefficient A_a according to Section 1607.1.5, for determining the prescribed seismic forces
Effective Peak Velocity Related - Coefficient A_a, for determining the prescribed seismic forces given in Section 1607.1.5.
Base - The level at which the horizontal seismic ground motions are considered to be imparted to the building.
Base Shear - Total design lateral force or shear at the base of the building.
Bearing Wall System - A structural system with bearing walls providing support for all, or major portions of, the vertical loads Shear walls or braced frames provide seismic force resistance.
Design Earthquake - The earthquake at the site under consideration that produces ground motions having a 90% probability of not being exceeded in 50 years
Designated Seismic Systems - The seismic-load-resisting system and those architectural, electrical, and mechanical systems and their components that require special performance characteristics
Diaphragm - A horizontal, or nearly horizontal, portion of the seismic-load-resisting system, which is designed to transmit seismic forces to the vertical elements of the seismic-load-resisting system.

Frame
Braced - An essentially vertical truss, or its equivalent, of the concentric or eccentric type that is provided in a bearing wall, building frame or dual system to resist seismic forces
Concentrically Braced Frame (CBF) - A diagonally braced steel frame in which at least one end of each brace frames into a beam a short distance from a beam-column joint or from another diagonal brace. These short beam segments are called link beams

Diagonal Brace - A member of an CBF placed diagonally in the bay of the frame.

Lateral Support Members - Secondary members designed to prevent lateral or torsional buckling of beams in an CBF.

Link Beam - The horizontal beam in an CBF which has a length of the clear distance between the diagonal braces or between the diagonal brace and the column face.

Link Beam End Web Stiffeners - Vertical web stiffeners placed on the sides of the web at the diagonal brace ends of the link beam.

Link Beam Intermediate Web Stiffener - Vertical web stiffeners placed within the link beam.

Link Beam Rotation Angle - The angle between the beam outside of the link beam and the link beam occurring at a total story drift of the deflection amplification factor, C_d, times the elastic drift at the prescribed design forces The rotation angle is permitted to be computed assuming the CBF bay is deformed as a rigid, ideally plastic mechanism.

Intermediate Moment Frame - A frame in which members and joints are capable of resisting forces by flexure as well as along the axis of the members Intermediate moment frames of reinforced concrete shall conform to SBC section 1912.3.2.

Ordinary Moment Frame - A frame in which members and joints are capable of resisting forces by flexure as well as along the axis of the members

Special Moment Frame - A frame in which members and joints are capable of resisting forces by flexure as well as along the axis of the members Special moment frames shall conform to the applicable requirements of SBC sections 1912 or 2212.

Space Frame - A structural system composed of interconnected members, other than bearing walls, that is capable of supporting vertical loads and, if so designed, resisting the seismic forces

Frame System

Building - A structural system with an essentially complete space frame providing support for vertical loads Seismic force resistance is provided by shear walls or braced frames

Dual - A structural system with an essentially complete space frame providing support for vertical loads A moment-resisting frame shall be provided that shall be capable of resisting at least 25% of the prescribed seismic forces

The total seismic force resistance is provided by the combination of the moment-resisting frame together with shear walls or braced frames in proportion to their relative rigidities

High Temperature Energy Source - A fluid, gas, or vapor whose temperature exceeds 220^0F (104^0C).

Inverted Pendulum-Type Structures - Structures which have a large portion of their mass concentrated near the top and thus have essentially one degree of freedom in horizontal translation. The structures are usually T-shaped with a single column supporting the beams or slab at the top.

Light Framed Wall - A wall with wood or steel studs

Moment Resisting - A structural system with an essentially complete space frame providing support for vertical loads Seismic force resistance is provided by special, intermediate, or ordinary moment frames capable of resisting the total prescribed forces

P-Delta Effect - The secondary effect on shears and moments of frame members due to the action of the vertical loads induced by displacement of the building frame resulting from lateral forces

Resilient Stable Mounting System - A system incorporating helical springs, air cushions, rubber-in-shear mounts, fiber-in-shear mounts, or other comparable approved systems The force displacement ratios are equal in the horizontal and vertical directions

Restraining Device - A device used to limit the vertical or horizontal movement of the mounting system due to earthquake motions

Elastic Restraining Device - A fixed restraining device that incorporates an elastic element to reduce the seismic forces transmitted to the structure due to impact from the resilient mounting system.

Fixed Restraining Device - A nonyielding or rigid type of restraining device.

Seismic Activated - That part of the structural system that has been considered in the design to provide the required resistance to the seismic forces prescribed herein.

Shear Wall - A wall, bearing or nonbearing, designed to resist seismic forces, from other than its own mass, acting in the place of the wall.
Story Drift Ratio - The story drift divided by the story height.
Story Shear - The summation of design lateral forces at levels above the story under consideration.

1607.3 Structural Design Requirements
The seismic analysis and design procedures used in the design of buildings and their structural components shall be in accordance with the requirements of this section. The design seismic forces, and their distribution over the height of the building, shall be in accordance with the procedures in Sections 1607.4 or 1607.5. The corresponding internal forces in the structural components of the building shall be determined by the use of a linearly elastic model.

Structural concepts other than those in this section shall be permitted when evidence is submitted showing that equivalent ductility and energy dissipation are provided. [These must be designed by a state licensed professional engineer (P.E.) or civil engineer.] Other procedures used to establish the seismic forces and the distribution of such forces shall be permitted if the corresponding internal forces and deformations in the structural components are determined using a model consistent with the approved procedure. Individual structural members shall be designed for the shears, axial forces, and moments determined in accordance with this section.

Connections shall be designed to develop the strength of the connected members or the analysis forces, whichever is less The design story drift of the building shall not exceed the allowable story drift requirements of Section 1607.3.7 when the building is subjected to the design seismic forces All structural components of the building that transmit seismic force shall be connected, with adequate strength and stiffness, through a continuous path to the final point of resistance. The foundation shall be designed to accommodate the forces developed and the movements imparted to the building by the design ground motions The foundation design shall consider the dynamic nature of the seismic forces and the design ground motions

1607.3.1 Site Coefficient

The value of the site coefficient, S, shall be determined from Table 1607.3.1. In locations where the soil properties are not known in sufficient detail to determine the soil profile type or where the soil profile does not fit any of the four types indicated in Table 1607.3.1, a site coefficient, S, of 2.0 shall be used.

**Table 1607.3.1
Site Coefficient**

Soil Profile Type	Description	Site Coefficient S
S_1	A soil profile with either: rock of any characteristic, either shale-like or crystalline in nature, which has a shear wave velocity greater than 2500 ft/s or still soil conditions where the soil types overlaying rock or stable deposits of sands, gravels or stiff clays	1.0
S_2	A soil profile with deep cohesionless or stiff clay conditions, where the soil depth exceeds 200 ft and soil types overlaying rock are stable deposits of sands, gravels, or stiff clays	1.2
S_3	A soil profile containing 20 to 40 ft in thickness of soft to medium-stiff clays with or without intervening layers of cohesionless soils	1.5
S_4	A soil profile characterized by a shear wave velocity of less than 500 ft/s containing more than 40 ft of soft clay.	2.0

1607.3.2 Soil and Structure Interaction
The design base shear, story shears, overturning moments, and deflections determined by the requirements of Sections 1607.4 or 1607.5 are permitted to be modified in accordance with approved procedures which account for the effects of soil and structure interaction.

1607.3.3 Structural Framing Systems
The basic structural framing systems to be used are indicated in Table 1607.3.3 (see Appendix). Each type is subdivided by the types of vertical structural elements to be used to resist the design lateral forces. The structural system used shall be in accordance with the seismic performance category and height limitations indicated in Table 1607.3.3.

The appropriate response modification factor, R, and the deflection amplification factor, C_d, indicated in Table 1607.3.3 shall be used in determining the base shear and the design story drift. Structural framing and seismic resisting systems which are not contained in Table 1607.3.3 shall be permitted if analysis and test data are submitted that establish the dynamic characteristics and demonstrate the lateral force resistance and energy absorption capacity to be equivalent to the structural systems listed in Table 1607.3.3 for equivalent response modification factor, R, values

Table 1607.3.3
Structural Systems

Seismic resisting system factor	Response modification factor - R	Deflection amplification factor - C_d	Structural system limitations and building height limitations[2] (ft) Seismic performance category			
			A&B	C	D[4]	E[5]
BEARING WALL SYSTEM						
Light framed walls w/shear panels	6-1/2	4	NL	NL	160	100

Reinforced concrete shear walls	4-1/2	4	NL	NL	160	100
Reinforced masonry shear walls	3-1/2	3	NL	NL	160	100
Concentrically braced frames	4	3-1/2	NL	NL	160	100
Unreinforced masonry shear walls	1-1/4	1-1/4	NL	Note 3	NP	NP
Plain concrete shear walls	1-1/2	1-1/2	NL	Note 3	NP	NP

BUILDING FRAME SYSTEM

Eccentrically braced frames (moment resisting connections at columns away from link beam)	8	4	NL	NL	160	100
Eccentrically braced frames (non-moment resisting connections at columns away from link beam)	7	9	NL	NL	160	100
Light framed walls with shear panels	7	4-1/4	NL	NL	160	100
Concentrically braced frames	5	4-1/2	NL	NL	160	100
Reinforced concrete shear walls	5-1/2	5	NL	NL	160	100
Reinforced masonry shear walls	4-1/2	4	NL	NL	160	100
Unreinforced masonry shear walls	1-1/2	1-1/2	NL	Note3	NP	NP
Plain Concrete shear walls	2	2	NL	Note3	NP	NP

MOMENT RESISTING FRAME SYSTEM

Special moment frame of steel	8	5-1/2	NL	NL	NL	NL
Special moment frames of reinforced concrete	6	5-1/2	NL	NL	NL	NL
Intermediale moment frames of	5	4-1/2	NL	NL	NP	NP

reinforced concrete					
Ordinary moment frames of steel	4-1/2	4	NL	NL	160	100
Ordinary moment frames of reinforced concrete	3	2-1/2	NL	NP	NP	NP

1607.3.3.1 Dual System

For a dual system, the moment frame shall be capable of resisting at least 25% of the design seismic forces The total seismic force resistance is to be provided by the combination of the moment frame and the seismic-load resisting elements in proportion to their rigidities

[To counter compression, tension, or shear forces, these systems are combined with the following subsections to produce moment-resisting structural design loads]

1607.3.3.2 Combinations of Framing Systems

Different structural framing systems are permitted along the two orthogonal areas of the building. Combinations of framing systems shall comply with the requirements of this section.

1607.3.3.2.1 Combination Framing Factor R

The response modification factor, R, in the direction under consideration at any story shall not exceed the lowest response modification factor R obtained from Table 1607.3.3 (see Appendix) for the seismic resisting system in the same direction considered above the story.

> **EXCEPTION**: Supported structural systems with a weight equal to or less than 10% of the weight of the building need not comply with this requirement.

1607.3.3.2.2 Combination Framing Detailing Requirements

The detailing requirements of Section 1607.3.6 required by the higher response modification factor R shall be used for structural components common to systems having different response modification factors

1607.3.3.3 Seismic Performance Categories A, B, and C

The structural framing system for buildings assigned to Seismic Performance Categories A, B, and C shall comply with the building height and structural system limitations in Table 1607.3.3.

1607.3.3.4 Seismic Performance Category D
The structural framing system for buildings assigned to Seismic Performance Category D shall comply with Section 1607.3.3.3 and the additional provisions of this section.

1607.3.3.4.1 Limited Building Height
A building having a structural system of steel or cast-in-place concrete braced frames or shear walls is limited to a height of 240 ft (73.1 m) where the braced frames or shear walls in one plane are so arranged as to resist no more than the following proportion of the seismic design force in each direction, including torsional effects:

- Sixty percent when the braced frame or shear walls are arranged only on the perimeter
- Forty percent when some of the braced frames or shear walls are arranged on the perimeter
- Thirty percent for other arrangements

1607.3.3.4.2 Interaction Effects
Moment-resisting frames that are enclosed or adjoined by more rigid elements not considered to be part of the seismic resisting system shall be designed so that the action or failure of the enclosing or adjoining elements will not impair the vertical load and seismic-force-resisting capability of the frame. The design shall provide for the effect of these rigid elements on the structural system at building deformations corresponding to the design story drift as determined in Section 1607.4.5.

1607.3.3.4.3 Deformational Compatibility
Every structural component not included in the seismic-force-resisting system in the direction under consideration shall be designed to be adequate for the vertical load-carrying capacity and the induced moments resulting from the design story drift as determined in accordance with Section 1607.4.5.

1607.3.3.4.4 Special Moment Frames
A special moment frame that is used but not required by Table 1607.3.3 (see Appendix) is permitted to be discontinuous and supported by a more rigid system with a lower response modification factor, R, provided the requirements of Section 1607.3.6.2.4 are met. Where a special moment frame is required by Table 1607.3.3, the frame shall be continuous to the foundation.

1607.3.3.5 Seismic Performance Category E
The framing systems of buildings assigned to Category E shall conform to the requirements of Section 1607.3.3.4 for Category D and to the additional requirements and limitations of this section. The building height limitation in Section 1607.3.3.4.1 is reduced to 160 ft (48.8 m) for buildings assigned to Seismic Performance Category E.

1607.3.4 Building Configuration
Buildings shall be classified as regular or irregular based on the plan and vertical configuration.

1607.3.4.1 Plan Irregularity
Buildings having one or more of the features listed in Table 1607.3.4.1 (see Appendix) shall be designed as having plan irregularity and shall comply with the requirements in the referenced code sections of that table.

1607.3.4.2 Vertical Irregularity
Buildings having one or more of the features listed in Table 1607.3.4.2 (see Appendix) shall be designated as having vertical irregularity and shall comply with the requirements in the referenced code sections of that table.

   EXCEPTIONS:
   - Structural irregularities of Type 1 or 2 in Table 1607.3.4.2 do not apply where the building story drift ratio is less than 130% of the story drift ratio of the next story above.
   - Torsional effects need not be considered in the calculation of story drifts The story drift ratio relationship for the top two stories of the building are not required to be evaluated.
   - Irregularity Types 1 and 2 of Table 1607.3.4.2 are not required to be considered for one- and two-story buildings

1607.3.5 Analysis Procedures
A structural analysis shall be made for all buildings in accordance with the requirements of this section. An alternative generally accepted procedure, including the use of an approved site-specific response spectrum, is permitted to be used, if approved by the building official. The limitations on the base shear in Section 1607.5 apply to dynamic modal analysis

1607.3.5.1 Seismic Performance Category A
Regular or irregular buildings assigned to Category A are not required to be analyzed for seismic forces for the building as a whole. The requirements of Section 1607.3.6.1 apply.

1607.3.5.2 Seismic Performance Categories B and C
Regular or irregular buildings assigned to Category B or C shall be analyzed in accordance with the procedures in Section 1607.4.

1607.3.5.3 Seismic Performance Categories D and E
Buildings assigned to Categories D and E shall be analyzed in accordance with the referenced sections in Table 1607.3.5.3 (see Appendix).

1607.3.6. Design, detailing requirements, and structural component load effects The design and detailing of structural components of the seismic-load-resisting system shall comply with the requirements of this section. Foundation design shall conform to the applicable requirements of SBC chapter 18.

1607.3.6.1 Seismic Performance Category A
The design and detailing of buildings assigned to Seismic Performance Category A shall comply with the requirements of this section.

1607.3.6.1.1 Ties and Continuity
Except for connections exempted by Section 1607.6, all parts of the building that transmit seismic force shall be interconnected to form a continuous path to the building's seismic-load-resisting system. Any smaller portion of the building shall be tied to the remainder of the building with elements having a strength capable of transmitting the seismic force, F_p, determined in accordance with Section 1607.6, but not less than one-third of the effective peak velocity-related acceleration, A_v, times the weight of the smaller portion, W_c, or 5% of the portion's weight, whichever is greater.

For a building which is exempt from a full seismic analysis by Section 1607.1 and is only required to comply with Section 1607.3.6.1, the building's main wind force resisting system according to Section 1606 shall be deemed to be the seismic-load-resisting system. A positive connection for resisting a horizontal force acting parallel to the member shall be provided for each beam, girder, or truss to its support. The connection shall have a minimum strength of 5% of the dead plus live load reaction.

Structural Loads 327

1607.3.6.1.2 Concrete or Masonry Wall Anchorage
Concrete and masonry walls shall be anchored to the roof and all floors that provide lateral support for the wall. The anchorage shall provide a direct connection between the walls and the roof or floor construction. The use of toe nailing or nails subject to withdrawal forces is not permitted. Wood ledgers shall not be subjected to cross-grain bending or cross-grain tension.

The connections shall be capable of resisting a lateral seismic force, F_p, in accordance with either Section 1607.3.6.2.8 or 1607.6, for bearing and nonbearing walls, respectively, but not less than 1000 times the effective peak-velocity-related acceleration, A_v, pounds per linear foot of wall. Walls shall be designed to resist bending between anchors where the anchor spacing exceeds 4 ft (1219 mm).

1607.3.6.2 Seismic Performance Category B
Buildings assigned to Category B shall conform to the requirements of Section 1607.3.6.1 for Category A and the requirements of this section.

1607.3.6.2.1 Materials
The materials and the systems composed of those materials shall conform to the requirements of code.

1607.3.6.2.2 Openings
Where openings occur in shear walls, diaphragms, or other plate-type elements, the edges of the openings shall be designed to transfer the stresses into the structure. The edge chord shall extend into the body of the wall or diaphragm a distance sufficient to develop the stress of the chord member.

1607.3.6.2.3 Orthogonal Effects
The design seismic forces shall be applied separately in each of two orthogonal directions

1607.3.6.2.4 Discontinuities in Vertical Systems
Buildings with discontinuity in lateral capacity, vertical irregularity Type 5 as defined in Table 1607.3.4.2 (see Appendix), shall not be over two stories or 30 ft (9144 mm) in height where the "weak" story has a calculated strength of less than 65% of the story above.

> **EXCEPTION**: Where the weak story is capable of resisting a total seismic force equal to 75% of the deflection amplification factor, C_d, times the design force prescribed in Section 1607.4.

1607.3.6.2.5 Nonredundant Systems
The building design shall comply with Section 1607.1.2 for progressive collapse.

1607.3.6.2.6 Collector Elements
Collector elements shall be provided that are capable of transferring the seismic forces originating in other portions of the building to the element providing the resistance to those forces

1607.3.6.2.7 Diaphragms
The deflection in the plane of the diaphragm, as determined by engineering analysis, shall not exceed the permissible deflection of the attached elements Permissible deflection shall be that deflection which will permit the attached element to maintain its structural integrity under the individual loading and continue to support the prescribed loads

Floor and roof diaphragms shall be designed to resist the following seismic forces: A minimum force equal to 50% the effective peak-velocity-related acceleration A_v, times the weight of the diaphragm and other elements of the building attached thereto plus the portion of the seismic shear force at that level, V_x, required to be transferred to the components of the vertical seismic-load-resisting system because of offsets or changes in stiffness of the vertical components above and below the diaphragm.

Diaphragms shall provide for both the shear and bending stresses resulting from these forces Diaphragms shall have ties or struts to distribute the wall anchorage forces into the diaphragm. Diaphragm connections shall be positive, mechanical type connections

1607.3.6.2.8 Bearing Walls
Exterior and interior bearing walls and their anchorage shall be designed for a force of the effective peak-velocity-related acceleration, A_v, times the weight of the wall, normal to the surface, with a minimum force of 10% of the weight of the wall. Interconnection of wall elements and connections to supporting framing systems shall have sufficient ductility, rotational capacity, or sufficient strength to resist shrinkage, thermal changes, and differential foundation settlement when combined with seismic forces

1607.3.6.2.9 Inverted Pendulum-type Structures
Supporting columns or piers of inverted pendulum-type structures shall be designed for the bending moment calculated at the base determined using the procedures given in Section 1607.4 and shall vary uniformly to a moment at the top equal to one-half the calculated bending moment at the base.

1607.3.6.3 Seismic Performance Category C
Buildings assigned to Category C shall conform to the requirements of Section 1607.3.6.2 for Category B and the requirements of this section.

1607.3.6.3.1 Plane Irregularity
Buildings that have plane structural irregularity Type 5 in Table 1607.3.4.1 (see Appendix) shall be analyzed for the critical load effect due to direction of application of seismic forces. Alternatively, the building shall be analyzed in any two orthogonal directions

Structural elements and foundations shall be designed for 100% of the forces for one direction plus a simultaneous load of 30% of the forces for the perpendicular direction.

1607.3.6.4 Seismic Performance Categories D and E
Buildings assigned to Category D or E shall conform to the requirements of Section 1607.3.6.3 for Category C and to the requirements of this section.

1607.3.6.4.1 Orthogonal Load Effects
Buildings shall be designed for 100% of the seismic forces for one direction plus a simultaneous load of 30% of the seismic forces for the perpendicular direction. The load combination requiring the maximum structural component strength shall be used.

> **EXCEPTION**: Diaphragms and components of the seismic resisting system utilized in only one of the two orthogonal directions are not required to be designed for the combined load effects

1607.3.6.4.2 Plan or Vertical Irregularities
For buildings having a plan irregularity of Type 1, 2, 3, or 4 in Table 1607.3.4.1 or a vertical irregularity of Type 4 in Table 1607.3.4.2, the design forces determined from Section 1607.4 shall be increased 25% for connections of diaphragms to vertical elements and to collectors and for connections of collectors to the vertical elements

1607.3.6.4.3 Vertical Seismic Loads

The vertical component of earthquake ground motion shall be considered in the design of horizontal cantilever and horizontal prestressed components. Horizontal prestressed components shall be designed for load combinations in accordance with Section 1609.2.

Horizontal cantilever structural components are to be designed for a net upward force of 0.2 times the dead load in addition to the applicable load combinations of Section 1609.1.

1607.3.7 Deflection and Drift Limits

The design story drift as determined in Section 1607.4.5 or 1607.5.8 shall not exceed the allowable story drift from SBC Table 1607.3.7 for any story. For structures with significant torsional deflections, the maximum drift shall include torsional effects

The total deflection of a building due to seismic design forces shall not encroach on an interior property line. All portions of the building shall be designed and constructed to act as an integral unit in resisting seismic forces unless separated structurally by a distance sufficient to avoid contact which would damage the structural system of the building under total deflection as determined by Section 1607.4.5.1.

1607.4 Equivalent Lateral Force Procedure

This section provides required standards for the equivalent lateral force procedure of seismic analysis of buildings For purposes of analysis, the building is considered to be fixed at the base. See Section 1607.3.5 for limitations on the use of this procedure.

1607.4.1 Seismic Base Shear

The seismic base shear, V, in a given direction, shall be determined in accordance with the following formula:

$$V = C_s W$$

where

C_s = is the seismic design coefficient determined in accordance with Section 1607.4.1.1 and

W = is the total dead load and applicable portions of other loads listed below. (In Group S Occupancies, a minimum of 25% of the floor live load shall be applicable.)

Structural Loads 331

EXCEPTION: Floor live load in parking garages is not applicable:
- Where an allowance for partition load is included in the floor load design, the actual partition weight or a minimum weight of 10 psf (1436 Pa) of floor area, whichever is greater, shall be applicable.
- Where total operating weight of permanent equipment is less than design strength.
- In areas where the ground snow load is equal to or greater than 30 psf (1436 Pa) of short duration and approved, a snow load reduction of 80% is permitted. Where the ground snow load is less than 30 psf (1436 Pa), the snow load is not required to be included.

1607.4.1.1 Calculation of Seismic Coefficient, C_s

When the fundamental period of the building is computed, the seismic design coefficient, C_s, shall be determined in accordance with the following formula:

$$C_s = \frac{1.2 A_v S}{R T^{2/3}}$$

where

A_v = Coefficient representing effective peak velocity-related acceleration from Section 1607.1.5
S = The coefficient for the soil profile characteristics of the site in Table 1607.3.1 (see Appendix)
R = The response modification factor in Table 1607.3.3
T = The fundamental period of the building determined in Section 1607.4.1.2

A soil-structure interaction reduction is permitted when determined from a generally accepted procedure approved by the building official. Alternatively, the seismic design coefficient, C_s, need not be greater than the following equation:

$$C_s = \frac{2.5 A_a}{R}$$

where

A_a = is the seismic coefficient representing the effective peak acceleration as determined in Section 1607.1.5

R = is the response modification factor in Section 1607.3.3.

1607.4.1.2 Period Determination

The fundamental period of the building, T, in the direction under consideration, shall be established using the structural properties and deformational characteristics of the resisting elements in a properly substantiated analysis

The fundamental period, T, shall not exceed the product of the coefficient for upper limit on calculated period, C_a, from Table 1607.4.1.2, and the approximate fundamental period, T_a. Alternatively, the fundamental period, T, shall be determined from the appropriate requirements of Section 1607.4.1.2.1.

1607.4.1.2.1 Approximate Fundamental Period, T_a

The approximate fundamental period, T_a, in seconds, shall be determined from the following formula:

$$T_a = C_T h_n 3/4$$

where:

h_n = height from base to highest level of building (ft)

C_T = 0.035 for moment-resisting frame systems of steel which provide 100% of required lateral force resistance, where frame is not enclosed or adjoined by more rigid components

C_T = 0.03 for moment-resisting frame systems of concrete which provide 100% of the required lateral force resistance, where the frame is not enclosed or adjoined by more rigid components, and

C_T = 0.03 for building frame systems with an eccentrically braced steel frame or dual systems with an eccentrically braced frame

C_T = 0.02 for seismic-load-resisting systems with shear walls, shear panels, or concentrically braced frames and all other building systems

Alternatively, the approximate fundamental period, T_a, in seconds, shall be determined from the following formula for buildings in which the lateral force-resisting system consists of concrete or steel moment-resisting frames which provide 100% of the required lateral force resistance and where such frames are not enclosed or adjoined by more rigid components tending to prevent the frames from deflecting when subjected to seismic forces Such buildings shall not exceed 12 stories in height and shall have a story height of not less than 10 ft (3048 mm).

$$T_a = 01.N$$

where

N is the number of stories

1607.4.2 Vertical Distribution of Seismic Forces

The lateral force, F_x, induced at any level, shall be determined from the following formulas:

$$F_x = C_{vx} V$$

$$Cvx = \frac{w_x h_x k}{n}$$

where

C_{vx} = vertical distribution factor.
V = total design lateral force or shear at base of building
w_i and w_x = portion of total gravity load of building, W, located or assigned to level i or x, respectively
h_i and h_x = height in feet from base to level i or x, respectively
k = an exponent related to the building period as follows:

For buildings having a period of 0.5 s or less, $k = 1$.
For buildings having a period of 2.5 s or more, $k = 2$.
For buildings having a period between 0.5 and 2.5 s, k shall be 2 or shall be determined by linear interpolation between 1 and 2.

1607.4.3 Horizontal Shear Distribution

The seismic design story shear in any story shall be determined from the portion of the seismic base shear, induced at a level determined by the adopted code. The seismic design story shear must be distributed to the various vertical elements of the seismic-load-resisting system in the story under consideration based on the relative lateral stiffness of the vertical resisting elements and the diaphragm.

1607.4.3.1 Torsion

The design shall include the torsion moment, M_t, resulting from the location of the building masses plus the accidental torsion moments, M_{ta}, caused by assumed displacement of the mass each way from its actual location by a distance equal to 5% of the dimension of the building perpendicular to the direction of the applied forces For buildings of Seismic Performance Categories C, D, and E, where Type 1 torsional irregularity exists as defined in Table 1607.3.4.1 (see Appendix), the effects shall be accounted for by increasing the accidental torsion at each level by a torsional amplification factor, A_x, determined in Table 1607.3.4.1. [The torsional amplification factor, A_x, is not required to exceed 3.0.]

1607.4.4 Overturning

The building shall be designed to resist overturning effects caused by the seismic forces determined in Section 1607.4.2. At any story, the increment of overturning moment in the story under consideration shall be distributed to the various vertical resisting elements in the same proportion as the distribution of the horizontal shears to those elements

The foundations of buildings, except inverted pendulum structures, shall be designed for the foundation overturning design moment, at the foundation-soil interface determined using the equation from the adopted code for the overturning moment with an overturning moment reduction factor for varying building heights

1607.4.5 Drift Determination and P-Delta Effects

Story drifts and, where required, member forces and moments due to P-delta effects, shall be determined in accordance with this section.

1607.4.5.1 Story Drift Determination

The design story drift shall be computed as the difference of the deflections at the top and bottom of the story under consideration. The deflections of level x at the center of the mass shall be determined in accordance with the deflection amplification factor in Table 1607.3.3 (see Appendix). [The deflections are determined by an elastic analysis]

The elastic analysis of the seismic-load-resisting system shall be made using the required seismic design forces of Section 1607.4.2. For determining compliance with the story drift limitation of Section 1607.3.7, the deflections of level x at the center of mass shall be calculated as required in this section. For purposes of this drift analysis only, it is permissible to use the computed fundamental period, T, of the building without the upper bound limitation specified in Section 1607.4.1.2 when determining drift level seismic design forces Where applicable, the design story drift shall be increased by the incremental factor relating to the P-delta effects as determined in Section 1607.4.5.2.

1607.4.5.2 P-Delta Effects

P-delta effects on story shears and moments, the resulting member forces and moments, and the story drifts induced by these effects are not required to be considered when the stability coefficient is determined in Section 1607.4.5.2.

1607.5 Modal Analysis Procedure

This section provides required standards for the modal analysis procedure of seismic analysis of buildings See Section 1607.3.5 for limitations on the use of this procedure.

1607.5.1 General

The symbols used in this method of analysis have the same meaning as those for similar terms used in Section 1607.4, with the subscript m denoting quantities in the m_{th} mode.

1607.5.2 Modeling

The building shall be modeled as a system of masses lumped at the floor levels with each mass having lateral displacement with one degree of freedom in the direction under consideration.

1607.5.3 Modes

The analysis shall include, for each of two mutually perpendicular axes, at least the lowest three modes of vibration or all modes of vibration with periods greater than 0.4 s, whichever is greater. The number of modes shall equal the number of stories for buildings less than three stories in height.

1607.5.4 Periods

The required periods and mode shapes of the building in the direction under consideration shall be calculated by established methods of structural analysis for the fixed base condition using the masses and elastic stiffness of the seismic-load-resisting system.

1607.5.5 Modal Base Shear

The portion of the base shear contributed by the m_{th} method, V_m, shall be as determined in Section 1607.4.5.2.

EXCEPTIONS:
- The limiting value of the modal seismic design coefficient, C_{sm}, is not applicable to Category D and E buildings with a period of 0.7 s or greater located on type S_4 soils
- For buildings on soil profile characteristics S_3 or S_4, the modal seismic design coefficient, C_{sm}, for modes other than the fundamental mode that have periods less than 0.3 s is permitted to be determined by the following formula:

$$C_{sm} = \frac{A_a(1.0 + 5.0\,T_m)}{R}$$

- For buildings where any modal period of vibration, T_m, exceeds 4.0 s, the modal seismic design coefficient, C_{sm}, for that mode is permitted to be determined by the following formula:

$$C_{sm} = \frac{3\,A_v\,S}{R\,T_m^{4/3}}$$

where

A_a = seismic coefficient representing the effective peak acceleration as determined in Section 1607.1.5

A_v = seismic coefficient representing the effective peak velocity-related acceleration as determined in Section 1607.1.5

R = response modification factor determined from Table 1607.3.3 (see Appendix)

T_m = modal period of vibration, in seconds, of m_{th} mode of buildings

$\quad$ S $\quad$ = coefficient for the soil profile characteristics of site as determined by Table 1607.3.1

1607.5.6 Modal Forces, Deflections, and Drifts
The modal force, F_{xm}, at each level shall be determined by the following formulas:

$$F_{xm} = C_{vxm} V_m$$

where

$\quad C_{vxm}$ = vertical distribution factor in the m_{th} mode
$\quad V_m$ = total design lateral force or shear at the base in the m_{th} mode
$\quad w_i$ and w_x = portion of the total gravity load of the building, W, located or assigned to level i or x.

$\quad$ The modal drift in a story shall be computed as the difference of the deflections at the top and bottom of the story under consideration.

1607.5.7 Modal Story Shears and Moments
The story shears, story overturning moments, and the shear forces and overturning moments in walls and braced frames at each level, due to the seismic forces determined from the appropriate equation in Section 1607.5.6, shall be computed for each mode by linear static methods

1607.5.8 Design Values
The design values for the modal base shear, V_t, for each of the story shear, moment, and drift quantities, and for the deflection at each level shall be determined by combining their modal values, obtained from Sections 1607.5.6 and 1607.5.7. The combination shall be determined by taking the square root of the sum of the squares of each of the modal values or by the complete quadratic combination technique.

$\quad$ The base shear, V, using the equivalent lateral force procedure in Section 1607.4 shall be calculated using a fundamental period of the building, T, in seconds, of 1.2 times the coefficient for the upper limit on the calculated period, C_a, times the approximate fundamental period of the building, T_a. Where the design value for the modal base shear, V_t, is less than the calculated base shear, V, using the equivalent lateral force procedure, the design story shears, moments, drifts, and floor deflections shall be multiplied by the following modification factor:

$$\frac{V}{V_t}$$

Chapter

13

Foundations and Retaining Walls

SECTION 1800.0 FOUNDATIONS [1800.0; 1800.1]

The foundation must resist vertical loads from the weight of the home, plus temporary extra roof loading, and it must resist side loads imposed on the home by wind blowing against the walls. Design data describing the roof and wind loads, which the home has been designed to resist, must be called out in the original design plans. Load zone maps of the United States showing roof load, wind load, and thermal zones are also used in calculations. Each of the national building codes has individual regulations governing foundation installation in your regional area which may have specific wind and roof load design requirements that vary from the indications on the standard load zone maps.

The piers used must be strong enough to transmit the vertical load, which includes the weight of the home, its furnishings, and temporary roof loading, to the foundation surface below. Recommended types of piers and footing sizes are described in the adopted local building code. Check with the local building official for any requirements for foundation construction of your project due to ground conditions. In areas where the ground is subject to freezing, the pier pads and footings must extend below the frost line established by local jurisdiction. The vertical loads imposed on the piers at the various locations under the home are shown on the foundation plan drawing under specifications by the engineer.

The foundation system must also resist the lifting, sliding, and overturning force resulting from side winds. A method frequently used is to install shear anchor connectors and tie-down straps in addition to the piers. The modern home is designed with provision for tie-down straps to resist side and uplift forces. Each connector or tie-down strap must have a working load capacity of at least 3150 lb and a total load capacity of at least 4725 lb. Because of local sheltered conditions, authorities may permit installation of the home without tie-downs. However, tie-downs as described are the minimum code required connectors necessary if the home is to withstand its design wind load without damage.

The foundation under-floor area must be ventilated to minimize the accumulation of moisture beneath the home. The ventilation is provided by openings with a net area of at least 1 sq ft for each 150 sq ft of under-floor area. The required area of openings should be approximately equally distributed along the length of at least two opposite sides with openings located close to corners to provide cross ventilation. It is recommended that a layer of 6-mil black polyethylene plastic or similar material be used to fully cover the ground under the home to form a water vapor retarder unless it can be demonstrated that the soil will remain dry.
1801.1 Provisions of this section shall govern the design, construction, and resistance to water intrusion of foundations for buildings and structures.

1802.0 DEFINITIONS [1802.1; 1802.2]
The following terms shall, for the purposes of this chapter and as stated elsewhere in code, have the meaning shown herein. Refer to Chapter 2 for general definitions.

<u>**Curb Level**</u> - Referring to a building, means the elevation at that point of the street grade that is opposite the center of the wall nearest to and facing the street line.

1803.0 EXCAVATIONS [1803.0; 1803.1]
1803.1 General
1803.1.1. When excavating for buildings or excavations accessory thereto, such excavations shall be made safe to prevent any danger to life and property.
1803.1.2. Permanent excavations shall have retaining walls of sufficient strength made of steel, masonry, or reinforced concrete to retain embankments, together with any surcharged loads.
1803.1.3. Excavations for any purpose shall not extend within 1 ft (305 mm) of the angle of repose or natural slope of the soil under

any footing or foundation, unless such footing or foundation is first properly underpinned or protected against settlement.

1803.2 Support of Adjoining Buildings and Structures

1803.2.1 Notice to Adjoining Structures

Notice to the owner of adjoining buildings or structures shall be served in writing by the one causing the excavation to be made at least 10 days before an excavation is commenced. The notice shall state the depth and location of the proposed excavation.

1803.2.2 Excavation 10 Ft (3048 mm) or Less

When an excavation extends not more than 10 ft (3048 mm) below the established curb grade nearest the point of excavation under consideration, the owner of the adjoining structure or building shall be afforded the necessary license to enter the premises where the excavation is to be made and, at his or her own expense, shall provide the necessary underpinning or protection.

1803.2.3 Excavation Greater Than 10 Ft (3048 mm)

1803.2.3.1. When an excavation extends more than 10 ft (3048 mm) below the established curb grade nearest the point of excavation under consideration, the one causing the excavation to be made, if given the necessary license to enter the adjoining premises, shall provide at his or her own expense one of the following:

1. Underpinning and protection required by that part of the excavation which extends to a depth greater than 10 ft (3048 mm) below the established curb grade nearest the point of excavation under consideration, whether or not the existing footings or foundations extend to the depth of 10 ft (3048 mm) or more below curb grade.
2. Shoring and bracing of the sides of the excavation required to prevent any soil movement into the excavation. If permanent lateral support is provided, the method used must satisfy requirements of the building official.

1803.2.3.2. If the necessary license is not afforded the person causing the excavation to be made, it shall be the duty of the owner failing to afford such license to provide the required underpinning or protection for which purpose he or she shall be afforded the necessary license to enter the premises where such excavation is to be made.

1803.2.4 Unestablished Curb Grade

If there is not an established curb grade, the depth of excavation shall be referred to the level of the ground at the point under consideration.

1803.2.5 Difference in Adjacent Curb Grades

1803.2.5.1. If an existing building or structure requiring underpinning or protection is so located that its curb grade or level is at a higher level than the level to which the excavation is properly referred, then such part of the required underpinning or protection that is necessary due to the difference in these levels shall be made and maintained at the joint expense of the owner of the building or structure and the person causing the excavation to be made.

1803.2.5.2. For the purpose of determining such part of the underpinning or protection that is necessary due to such difference in levels, the level to which a building more than 5 ft (1524 mm) back of the street line is properly referred shall be considered to be the level of the natural ground surface adjoining the building or structure.

1803.2.6 Party Walls

A party wall, which is in good condition and otherwise suitable for continued use, shall be underpinned or protected as required at the expense of the person causing the excavation to be made.

1803.2.7 Adjoining Structure Protection

Where the necessary license has been given to the person making an excavation to enter any adjoining structure for the purpose of underpinning or protecting it, the person receiving such license shall provide for such adjoining structure adequate protection against injury due to the elements resulting from such entry.

1803.2.8 Backfill

Only approved granular materials shall be used for backfill. It shall be properly compacted in order to prevent lateral displacements of the soil of the adjoining property after the removal of the shores or braces.

1804.0 FOOTINGS AND FOUNDATIONS [1804.0; 1804.1]

1804.1.1. Foundations shall be built on undisturbed soil or properly compacted fill material. Foundations shall be constructed of materials described in this section.

1804.1.2. Pile foundations shall be designed and constructed in accordance with Section 1805.

1804.1.3. The bottom of foundations shall extend below the depth of frost penetration shown in Section 1804.1 but no less than 12 in (305 mm) below finish grade.

1804.1.4. Temporary buildings and buildings not exceeding one story in height and 400 sq ft (37 m²) in area shall be exempt from these requirements.

1804.1.5. Excavations for foundations shall be backfilled with soil which is free of organic material, construction debris, and large rocks.

1804.1.6. Where water impacts the ground from a roof valley, downspout, scupper, or other rainwater collection or diversion device, provisions shall be made to prevent soil erosion and direct the water away from the foundation.

1804.1.7. Finish grade shall be sloped away from the foundation for drainage.

1804.1.8. The area under footings, foundations, and concrete slabs on grade shall have all vegetation, stumps, roots, and foreign materials removed prior to their construction. Fill material shall be free of vegetation and foreign material.

1804.2 Soils Investigation

1804.2.1 Plain Concrete, Masonry, or Timber Footings

Footings shall be so designed that the allowable bearing capacity of the soil is not exceeded. If structural plain concrete, masonry, or timber footings are used, they shall rest on undisturbed or minimum 90% compacted soil of uniform density and thickness.

1804.2.2 Questionable Soil

Where the bearing capacity of the soil is not definitely known or is in question, the building official may require load tests or other adequate proof as to the permissible safe bearing capacity at that particular location. To determine the safe bearing capacity of soil, the soil shall be tested at such locations and levels as conditions warrant, by loading an area not less than 4 sq ft (0.37 m²) to not less than twice the maximum bearing capacity desired for use.

Such double load shall be sustained by the soil for a period of not less than 48 hr with no additional settlement taking place, in order that such desired bearing capacity may be used. Examination of subsoil conditions shall be made at the expense of the owner, when deemed necessary by the building official.

1804.2.3 Natural Solid Ground or Piles

Foundations shall be built upon natural solid ground. Where solid natural ground does not occur at the foundation depth, such foundations shall be extended down to natural solid ground or piles shall be used. Foundations may be built upon mechanically compacted earth or fill material subject to approval by the building official upon submittal of evidence that proposed load will be adequately supported.

1804.2.4 Differential Settlement

Where footings are supported by soils of widely different bearing capacity, the allowable bearing values of the more yielding soil shall be reduced or special provisions shall be made in the design to prevent serious differential settlements.

1804.2.5 Shifting or Moving Soils

When it is definitely known that the top or subsoils are of a shifting or moving character, all footings shall be carried to a sufficient depth to ensure stability. The excavation around piers shall be backfilled with soils or materials which are not subject to such expansion or contraction.

1804.2.6 Groundwater Table Investigation

A subsurface soil investigation shall be performed to determine the possibility of the groundwater table rising above the proposed elevation of the lowest floor when such floor is located below the finished ground level adjacent to the foundation for more than 75% of the perimeter of the building.

> **EXCEPTION**: A subsurface soil investigation shall not be required when either of the following conditions is satisfied:
> 1. Waterproofing is provided in accordance with Section 1814.2.
> 2. Satisfactory data from adjacent areas is available which demonstrates that groundwater has not been a problem.

1804.3 Expansive Soils

Footings or foundations for buildings and structures founded on expansive soils shall be designed in accordance with this section. As an alternative to special design, the soil may be removed in

Foundations and Retaining Walls 345

accordance with Section 1804.3.4 or stabilized in accordance with Section 1804.3.5.

1804.3.2 Soil Tests

In areas likely to have expansive soil, the building official may require soil tests to determine if such soils do exist. Soils meeting all four of the following provisions shall be considered expansive, except that tests to show compliance with items 1, 2, and 3 shall not be required if the test prescribed in item 4 is conducted:

1. Plasticity index (PI) of #15 or greater, determined in accordance with ASTM D 4318.
2. More than 10% of the soil particles pass a #200 sieve, determined in accordance with ASTM D 422.
3. More than 10% of the soil particles are less than 5 micrometers in size (m), determined in accordance with ASTM D 422.
4. Expansion index greater than 20, determined in accordance with ASTM standard for expansive soil tests.

1804.3.3 Foundations

Footings or foundations placed on or within the active zone of expansive soils shall be designed to resist differential volume changes and to prevent structural damage to the supported structure. Deflection and racking of the supported structure shall be limited to that which will not interfere with the usability and serviceability of the structure.

1804.3.3.2. Foundations placed below where volume change occurs or below expansive soil shall comply with the following provisions:
- Foundations extending into or penetrating expansive soils shall be designed to prevent uplift of the supported structure.
- Foundations penetrating expansive soils shall be designed to resist forces exerted on the foundation due to soil volume changes or be isolated from the expansive soil.

1804.3.3.3. Slab-on-ground, mat, or raft foundations on expansive soils shall be designed and constructed in accordance with concrete reference specifications from WRI/ CRSI Design of Slab-on-Ground Foundations or PTI Design and Construction of Post-Tensioned Slabs-On-Ground.

> **EXCEPTION**: Slab-on-ground systems which have performed adequately in soil conditions similar to those

encountered at the building site may be used if approved by the building official.

1804.3.4 Removal of Expansive Soil

The expansive soil may be removed to a depth sufficient to assure a constant moisture content in the remaining soil. Fill material shall not contain expansive soils and shall be placed in accordance with the provisions of Section 1804.2.3.

> **EXCEPTION**: Expansive soil need not be removed to the depth of constant moisture, provided the confining pressure in the expansive soil created by the fill and supported structure exceeds the swell pressure.

1804.3.5 Stabilization

Stabilization of the active zone of expansive soils may be used when approved by the building official. Soils may be stabilized by chemical, dewatering, presaturation, or equivalent techniques.

1804.4 Footing Design

The base area of the footings of all buildings shall be designed in the following manner: The area of the footing which has the largest percentage of live load to total load shall be determined by dividing the total load by the allowable soil load. From the area thus obtained, the dead load soil pressure of such footing is determined, and the areas of all other footings of the building shall be determined on the basis of their respective dead loads only and such dead load soil pressure.

In no case shall the load per square foot under any portion of any footing, due to the combined dead, live, wind, and any other loads, exceed the safe sustaining power of the soil upon which the footing rests. The total reduced live load occurring in the column immediately above the footing shall be the live load used in the above computation.

1804.4.2. Footings shall be proportioned to sustain the applied loads and induced reactions without exceeding the allowable stresses specified in code.

1804.5 Concrete Footings

1804.5.1 Compressive Strength

Concrete in footings shall have a specified compressive strength of not less than 2500 psi (17,238 kPa) at 28 days.

1804.5.3 Footing Seismic Ties

Individual spread footings, bearing on Soil Profile Type S_2, S_3, or S_4, by Section 1607.3.1, and supporting buildings assigned to Seismic Performance Category D or E, by Section 1607.1.8, shall be interconnected by ties. All ties shall be capable of resisting, in tension or compression, a force equal to 25% of the effective peak velocity-related acceleration, A_v, times the column dead plus live load. [There are different types of footing ties, manufactured according to the connectors angle of deflection. The most common types are shown in Figures 13-1 through 13-11.]

Individual tie beams are not required when it is demonstrated that equivalent restraint will be provided by structural members within slabs on grade; reinforced concrete slabs on grade; or confinement by competent rock, hard cohesive soils, very dense granular soils, or other approved means.

1804.5.4 Pier Foundation Seismic Ties

Pier foundations shall be interconnected by ties for buildings assigned to Seismic Performance Category C, D, or E, by Section 1607.1.8. All ties shall be capable of resisting, in tension or compression, a force equal to 25% of the effective peak-velocity-related acceleration, A_v, times the column dead plus live load. Individual tie beams are not required when it is demonstrated that equivalent restraint will be provided by structural members within slabs on grade; reinforced concrete slabs on grade; or confinement by competent rock, hard cohesive soils, very dense granular soils, or other approved means.

1804.6 Foundation Walls

Foundation walls shall be designed and constructed in accordance with accepted engineering practice. Provisions of Section 1804.6.2 may be used without additional engineering design.

1804.6.2 Concrete and Masonry

1804.6.2.1. Foundation walls shall be not less in thickness than the walls immediately above them. Where the height of unbalanced fill (height of finished grade above basement floor or inside grade) and the height between lateral supports does not exceed 8 ft (2438 mm), and where the equivalent fluid weight of unbalanced fill does not exceed 30 lb/cu ft (481 kg/m³), the minimum thickness of foundation walls shall be that shown in SBC Table 1804.6.

Maximum depths of unbalanced fill permitted in Table

348 Chapter Thirteen

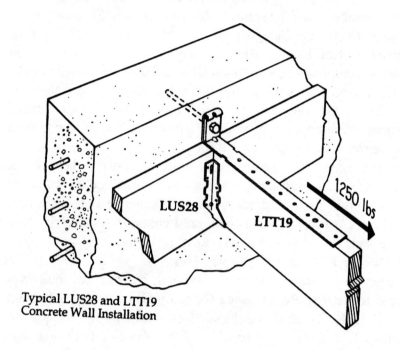

**Figure 13-1
Bolted Footing Seismic Tie**

Foundations and Retaining Walls 349

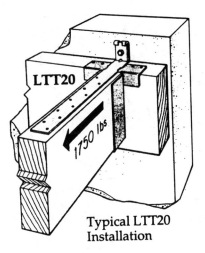

**Figure 13-2
Ledger Seismic Tie**

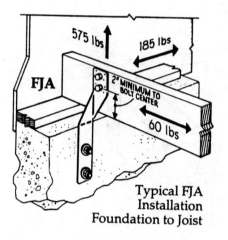

Figure 13-3
Three-Way Seismic Tie

Foundations and Retaining Walls 351

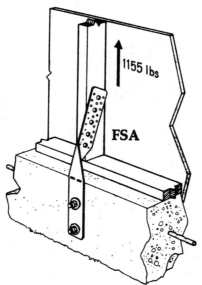

Typical FSA Installation
Foundation to Stud

**Figure 13-4
Seismic Strap Tie**

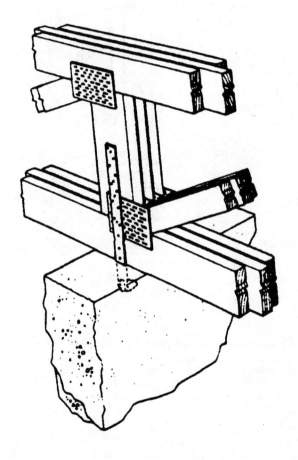

**Figure 13-5
Truss Seismic Tie**

Foundations and Retaining Walls 353

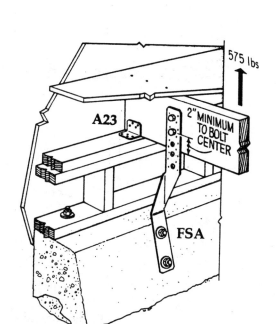

Typical FSA Installation
Foundation to Joist

**Figure 13-6
Joist to Stemwall Seismic Tie**

354 Chapter Thirteen

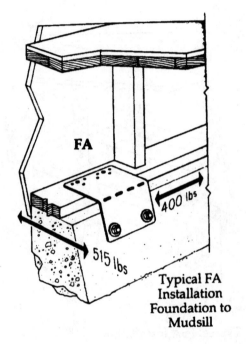

Figure 13-7
Two-Way Seismic Tie

Foundations and Retaining Walls 355

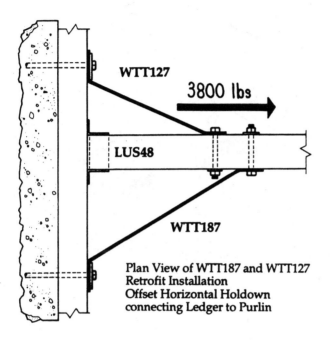

**Figure 13-8
Doubled Seismic Ties**

356 Chapter Thirteen

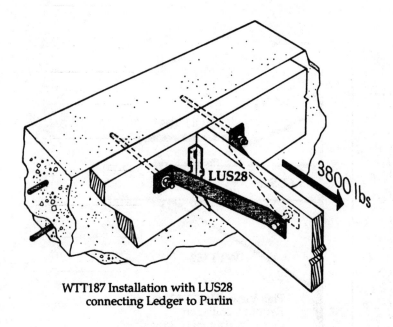

WTT187 Installation with LUS28
connecting Ledger to Purlin

**Figure 13-9
Doubled Seismic Ties on Joist**

Foundations and Retaining Walls 357

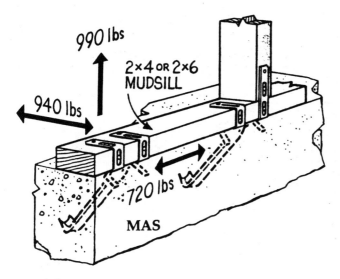

Typical MAS
Installation

**Figure 13-10
Sole Plate Seismic Tie**

358 Chapter Thirteen

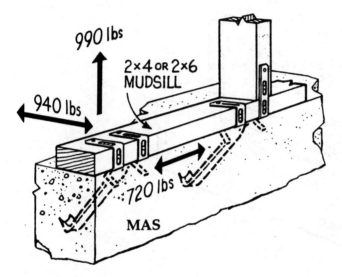

Typical MAS
Installation

**Figure 13-11
Three-Way Plate Seismic Tie**

1804.6 may be increased with the approval of the building official when local soil conditions warrant such an increase. Foundation walls shall be designed in accordance with Section 1901.1.2, where the height of unbalanced fill, the height between lateral supports, or the equivalent fluid weight of unbalanced fill exceeds that listed in this paragraph.

1804.6.2.2. Foundation walls 8 in (203 mm) thick, except as provided for in Section 1804.6.2.3 and conforming to the provisions of Section 1804.6.1, may be used as foundations for dwellings with walls of brick veneer on frame walls or 10-in (254 mm) cavity walls, provided that the dwelling is not more than two stories high and the total height of the wall, including the gable, is not more than 28 ft (8534 mm).

Foundation walls that are 8 in (203 mm) thick and that support brick veneer or cavity walls shall be corbeled with solid units to provide a bearing the full thickness of the wall above. The total projection shall not exceed 2 in (51 mm) with individual corbels projecting not more than one-third the thickness of the unit nor one-half the height of the unit. The top corbel course shall not be higher than the bottom of floor joists and shall be a full header course.

1804.6.2.3. Foundation walls of cast-in-place concrete when supporting one-story basementless structures may be 6 in (152 mm) thick if the total height of the foundation wall and the wall supported is within the allowable height permitted by code for 6-in (152 mm) thick walls.

1804.6.2.4. Pier and wall foundations shall be permitted to be used to support Type VI construction in dwellings not more than two stories in height, provided the following requirements are met:
- The wall shall be supported on a continuous concrete footing placed integrally with the exterior pier footings.
- The minimum actual thickness of the wall shall be not less than 3-5/8 in (92 mm) and integrally bonded into the piers.
- The maximum height of a 4-in (102 mm) thick wall shall not exceed 4 ft (1219 mm).
- Anchorage shall be provided in accordance with Section 1606.4.

- The unbalanced fill for 4-in (102 mm) thick walls shall not exceed 24 in (610 mm) for solid masonry or 12 in (305 mm) for hollow masonry.

(See Figure 13-12)

1804.6.2.5. Curtain walls between piers and nonbearing perimeter walls shall be permitted for frame construction and masonry veneer frame construction in dwellings not more than two stories in height, subject to the following limitations:

1. The minimum thickness of the curtain wall shall be 4 in nominal bonded into the piers and supported on a continuous concrete footing.
2. Masonry bearing piers shall comply with SBC section 2103. Pier spacing shall be governed by the beam or girder designed in accordance with Section 2307.2, maximum spacing of 8 ft (2438 mm) o. c. Piers shall provide a true and even bearing surface.
3. Unbalanced fill placed against a 4-in (102 mm) curtain wall shall not exceed 24 in (610 mm) for solid masonry or 16 in (406 mm) for hollow masonry.

1804.6.3 Openings
1804.6.3.1 Ventilation

Crawl spaces under buildings without basements shall be ventilated by approved mechanical means or by openings in foundation walls. Openings shall be arranged to provide cross ventilation and shall be covered with corrosion-resistant wire mesh of not less than 1/4 in (6.4 mm) nor more than 1/2 in (12.7 mm) in any dimension. Openings in foundation walls shall be not less than the following:

- Where wood floor systems are used, such openings shall have a net area of not less than 1 sq ft (0.093 m^2) for each 150 sq ft (14 m^2) of crawl space.
- Where other than wood floor systems are used, such openings shall be not less than 1-1/2 sq ft (0.14 m^2) of net opening for each 15 linear feet (4572 mm) or major fraction thereof of exterior wall.
- Where asphalt saturated felt weighing 55 lb (2.7 kg/m^2) per square, lapped at least 2 in (51 mm) at joints, or 4-mil (0.102 mm) polyethylene lapped at least 4 in (102 mm) at joints, or other approved vapor retarder is installed over the ground surface, the required net area of openings may be reduced to

Foundations and Retaining Walls 361

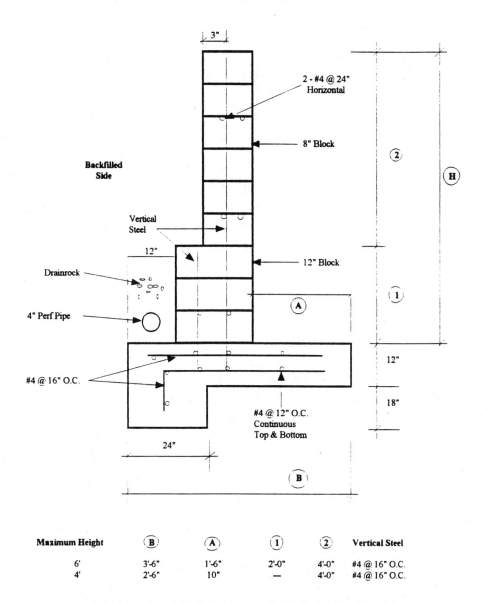

Maximum Height	B	A	1	2	Vertical Steel
6'	3'-6"	1'-6"	2'-0"	4'-0"	#4 @ 16" O.C.
4'	2'-6"	10"	—	4'-0"	#4 @ 16" O.C.

Material Specs: Concrete Block, Grade N - grouted solid. / Concrete: Fc = 2500 psi.
Rebar: Grade 40 - All lap splices minimum 24".

**Figure 13-12
Hollow Block Retaining Wall**

- 10% of that required above. There shall be one ventilation opening within 3 ft (914 mm) of each corner, and these shall be of equal size totaling a minimum of 50% of required openings.
- An operable vent louver shall be permitted only where an approved vapor barrier is installed over the ground surface.
- Where combustion equipment is installed within a crawl space, air for combustion shall be provided in accordance with Sections 705.1.1.4, and chapter 7 of the Standard Mechanical Code [UMC; NMC].

1804.6.3.2 Access

Usable crawl spaces under buildings without basements shall be provided with a minimum of one access opening not less than 18 X 24 in (457 X 610 mm). Access openings shall be readily accessible and provided with a door or device that may be easily removed or operated. For access to mechanical equipment installed in underfloor areas see section 304.5 of the Standard Mechanical Code [UMC; NMC].

1804.6.4 Masonry with Type VI Construction

Foundation walls of hollow masonry supporting Type V1 construction shall be capped with 4 in (102 mm) of solid masonry or concrete or shall have cavities of the top course filled with concrete or grout unless a sill plate of 2-in (51 mm) nominal thickness bears on both face shells.

1804.7 Timber Footings

Footings of wood may be used if they are entirely below permanent water level, or if they are pressure impregnated with an approved preservative in accordance with the standards listed in SBC chapter 35.

1804.8 Wood Foundation Systems

The foundation system may be of wood when the engineering design is based upon the bearing capacity of the soil (see Section 1804.2) and the design and construction complies with the provisions of NFiPA Technical Report No. 7.

1804.9 Seismic Provisions

1804.9.1 Seismic Investigation

Foundations of buildings assigned to Seismic Performance Category D or E, by Section 1607.1.8, shall conform to the requirements of this section. A soil investigation report, which determines the potential hazards due to slope instability, liquefaction, and surface rupture, due to faulting or lateral spreading and the determination of lateral pressures on below-ground building walls and retaining walls, shall be submitted to the building official.

1804.9.2 Soil Bearing

For the load combinations, including seismic as specified in Section 1609, the soil bearing strength shall be sufficient to resist loads, at acceptable strains, considering both the short duration of loading and the dynamic properties of the soil.

1804.9.3 Soil Seismic Lateral Pressure

Lateral soil pressure on below-ground surface building walls and retaining walls due to earthquake motions shall be included in the design of buildings assigned to Seismic Performance Category D or E, by Section 1607.1.8.

1805.0 PILES [1805.0; 1805.0]

1805.1 Investigation

Pile foundations shall be designed and installed on the basis of a foundation investigation and report which shall include borings, test pits or other subsurface exploration at locations and depths sufficient to determine the position and adequacy of the bearing soils except where sufficient data upon which to base the design and installation is available. The investigation and report shall include but not be limited to the following:

- Recommended pile types and installed capacities
- Driving criteria
- Installation and field inspection procedures
- Pile load test requirements
- Durability of pile materials
- Designation of bearing stratum or strata

1805.2 Special Types of Piles

The use of types of piles not specifically mentioned herein may be permitted, subject to the approval of the building official, upon the submission of acceptable test data, calculations and other information relating to the structural properties and load capacity of such piles. The allowable stresses shall not in any case exceed the limitations specified herein.

1805.3 Protection of Pile Materials
Where boring records or site conditions indicate possible deleterious action on pile materials because of soil constituents, changing water levels or other factors, the pile materials shall be adequately protected by materials, methods, or processes approved by the building official. Protective materials shall be applied to the piles so as not to be rendered ineffective by driving.

1805.4 Lateral Support
Any soil other than fluid soil shall be deemed to afford sufficient lateral support to the pile to prevent buckling and to permit the design of the pile in accordance with accepted engineering practice and the applicable provisions of code.

1805.4.2 Unbraced Piles
All piles standing unbraced in air, water, or soils not capable of providing lateral support shall be designed as columns in accordance with the provisions of code.

1805.4.3 Pile Bending Seismic Design
Piling for buildings assigned to Seismic Performance Category D or E, by Section 1607.1.8, shall be designed for the maximum imposed curvatures resulting from seismic forces on freestanding piles when the piles are located in loose granular soils or in soil profile type S_3 or S_4, by Section 1607.3.1. The piles shall be designed and detailed in accordance with Sections 1912.6 for a length equal to 120% of the flexural length. The flexural length shall be the distance from the point of fixity to the pile cap.

1805.5 Group Action
In cohesive soils, the compressive load capacity of a group of friction piles shall be analyzed by a rational method approved by the building official, and where such analysis indicates, the individual allowable pile load shall be reduced accordingly.

1805.6 Stability
All piles shall be braced to provide lateral stability in all directions. Three or more piles connected by a rigid cap shall be considered as being braced provided that the piles are located in radial directions from the centroid of the group not less than $60°$ (1 rad) apart. A two-pile group in a rigid cap shall be considered to be braced along the axis connecting the two piles. Methods used to brace piles shall be subject to the approval of the building official.

1805.6.2. Piles supporting walls shall be driven alternately in lines spaced at least 1 ft (305 mm) apart and located symmetrically under the center of gravity of the wall load carried, unless effective measures are taken to provide for eccentricity and lateral forces, or the wall piles are adequately braced to provide for lateral stability.

A single row of piles without lateral bracing may be used for one- and two-family dwellings and lightweight construction not exceeding two stories or 3.5 ft (1 m) in height provided the centers of the piles are located within the width of the foundation wall.

1805.7 Structural Integrity

Piles shall be installed in such a manner and sequence as to prevent distortion or damage to piles being installed or already in place to the extent that such distortion or damage affects the structural integrity of the piles.

1805.8 Spacing

The minimum center-to-center spacing of piles shall be not less than twice the average diameter of a round pile, nor less than 1-3/4 times the diagonal dimension of a rectangular pile. When driven to or penetrating into rocks, the spacing shall be not less than 24 in (610 mm).

When receiving principal support from end bearing on materials other than rock, or through frictional resistance, the spacing shall be not less than 30 in (762 mm) except for piles having enlarged bases formed either by compacting concrete or by driving a precast base. The minimum center-to-center spacing shall be 54 in (1372 mm). The spacing of piles shall be such that the average load on the supporting strata will not exceed the safe bearing value of those strata as determined by test borings or other approved methods.

1805.9 Splices

Splices shall be constructed so as to provide and maintain true alignment and position of the component parts of the pile during installation and subsequent thereto and shall be of adequate strength to transmit the vertical and lateral loads and moments occurring at the location of the splice during driving and under service loading. Splices shall develop not less than 50% of the least value of the pile in bending.

In addition, all pile splices occurring in the upper 10 ft (3048 mm) of the embedded portion of the pile shall be capable of resisting at allowable working stresses the moment and shear that would result from an assumed eccentricity of the pile load of 3 in (76 mm) or the pile shall be braced in accordance with Section 1805.6 to other piles that do not have splices in the upper 10 ft (3048 mm) of embedment.

1805.10 Pile Caps

1805.10.1 All Pile Caps

Pile caps shall be of reinforced concrete. The soil immediately below the pile cap shall not be considered as carrying any vertical load. The tops of all piles shall be embedded not less than 3 in (76 mm) into pile caps, and the caps shall extend at least 4 in (102 mm) beyond the edges of all piles. The tops of all piles shall be cut back to sound material before capping.

1805.10.2 Pile Cap Seismic Connection

1805.10.2.1. All concrete piles shall be connected to the pile cap for buildings assigned to Seismic Performance Category C, D, or E, by Section 1607.1.8. The connection shall consist of embedment of the pile reinforcement in the pile cap for a distance equal to the development length as specified in ACI 318.

1805.10.2.2. Field-placed dowels anchored in the plastic concrete piles are acceptable. The development length to be provided is the full development length for compression without reduction in length for excess area. Where seismic confinement reinforcement at the top of the pile is required, alternative measures for laterally confining concrete and maintaining toughness and ductilelike behavior at the top of the pile shall be permitted provided consideration is given to forcing the hinge to occur in the confined region.

1805.10.2.3. Where a minimum length for reinforcement or the extent of closely spaced confinement reinforcement is specified at the top of the pile, provisions shall be made so that those specified lengths or extents are maintained after pile cutoff.

1805.10.3 Pile Foundation Seismic Ties

Piles or pile caps shall be interconnected by ties for buildings assigned to Seismic Performance Category C, D, or E, by Section 1607.1.8. All ties shall be capable of resisting, in tension or compression, a force equal to 25% of the effective peak-velocity-related acceleration, A_v, times the column dead plus live load. Individual tie beams are not required when it is demonstrated that equivalent restraint will be provided by structural members within slabs on grade or reinforced concrete slabs on grade or confinement by competent rock, cohesive soils, very dense granular soils, or other approved means.

1805.11 Preexcavation
The use of jetting, auguring, or other methods of preexcavation shall be subject to the approval of the building official. When permitted, preexcavation shall be carried out in the same manner as used for piles subject to load tests and in such a manner that will not impair the carrying capacity of the piles already in place or damage adjacent structures. Pile tips shall be driven below the pre-excavated depth until the required resistance or penetration is obtained.

1805.12 Inspection
A qualified inspector approved by the building official shall be present when pile foundations are being installed and during tests. The inspector shall make and submit to the building official detailed records of the installation of each pile and the results of load tests. Records shall include the cutoff and tip elevation of each pile relative to a permanent reference.

1805.13 Identification
All pile materials shall be identified for conformity to the specified grade with this identity maintained continuously from the point of manufacture to the point of installation or shall be tested by an approved agency to determine conformity to the specified grade. The approved agency shall furnish an affidavit of compliance to the building official.

Insulating Concrete Forms (ICFs)
Insulating concrete forms, or ICFs as they are known in field construction terms, are new technology code compliance form-pour systems made of lightweight polystyrene interlocking hollow blocks made of plastic foam that are stacked to form the walls, then filled with steel reinforcing bar (rebar) followed by flowable concrete. Rebar is placed vertically and horizontally into the cavities which are then filled with concrete. Pumps are usually used to place concrete into above-grade ICF walls. Unlike traditional concrete forms, the foam blocks stay in place to become part of the wall

assembly, providing built-in insulation and an air-vapor barrier.

Windows, doors, floor, roof, and mechanical systems are easily installed. Then the interiors are completed just like a stick-built home. Today, there are more than a dozen different ICF systems available. Depending on the individual home design and specific application, different systems may be right for different projects. There is no pat answer to which system is better. ICFs are generally divided into two broad categories: panel and block. Interested builders should take the time to fully explore all their ICF options. There are a number of variations in both categories.

In the last 3 years, there has been an estimated fivefold increase in the use of ICFs. The increased popularity of the systems is due to numerous benefits to building with insulating concrete forms. But for home builders, the overall simplicity of the technology is what makes it such an attractive alternative to traditional framing. ICFs have many advantages for construction and allow for a short learning curve for installers. They are easy to learn and install by journeyman carpenters who are the traditional home builders. Once the crews have some practice, each ICF-built home requires less skilled labor and less total labor than a framed house does. And ICFs are very light, so crews stay fresh through the day and go home much less tired than with frame construction.

ICF construction reduces the energy needed to heat and cool a home in several ways. The ICF-built home has a continuous layer of insulation with no breaks. This provides superior insulation and a very low air infiltration rate. In addition, the very mass of the walls evens out temperature swings during the course of the day. Typically, air-conditioning costs can be cut by as much as 30% due to the insulation properties of the concrete walls. ICF construction also allows installation of smaller HVAC plants than comparable older models. Over time, a home owner will save a great deal of money heating and cooling an ICF-constructed home. Initial real-world tests of representative ICF wall systems suggest that to achieve equivalent performance in a traditional stemwall foundation home, you would need to build with R-40 to R-50 insulation to match the ICF walls shown in Figure 13-13.

**Figure 13-13
ICF Foundation Walls**

Another benefit of this simplified construction method is minimal waste disposal. Any scrap that is left over from the forming process can be returned to the form dealer for recycling. Any dimensional lumber used for bracing can be used later for framing interior walls, floors, and roofs. But many builders are finding they have virtually no scrap to discard. Builders report a 90% decrease in waste by using ICF systems versus traditional forming. Not only does that save time and effort, it saves money in both trucking and landfill fees because at the dump site, they charge by the ton for waste. Builders also are not on the job site cleaning up every day, like with stick-built jobs. Using ICF modular systems, any small piles of scrap left at the end of the day are thrown away in the job site dumpster.

All the benefits of ICF modular construction can be delivered regardless of the climate. Insulating concrete forms provide a thermos bottle effect, so the concrete can set up and cure in any weather. Pouring conventional concrete foundations and walls in under 32°F weather usually means greater expense and effort (see the adopted code-required cold weather foundation construction procedures). ICF walls can be poured in all seasons, at no additional expense, when other conventional foundations are stopped in their tracks. And if Mother Nature gets restless with seismic activity after the home is built, concrete provides still more insurance. Builders and home buyers are discovering the reasons why concrete is the building material of choice in commercial construction: greater durability in times of natural disaster and inherent resistance to both fire and animals and insects. Homes built with ICFs are extremely secure in hurricanes, earthquakes, and other natural disasters where conventionally framed homes are destroyed. ICF homes are fire-resistant as well, which gives home owners peace of mind - not to mention lower insurance rates. This is a big marketing advantage for savvy builders who know that safety and security are big consumer issues today.

For the builder, controlling the speed of construction with ICF systems becomes an important advantage, especially considering skyrocketing land costs, increasing federal regulation, and rising building material costs. ICFs allow crews to build a home more quickly than ever before. For builders using the systems for the first time, there is typically a learning curve of just a few houses.

Labor costs come down quite a bit because the concrete forming system is so easy. It can be done with very inexpensive labor. There needs to be one very knowledgeable person on the job, but then everyone else only needs to be able to handle these very simple blocks. Because additional insulation is not needed, both time and materials are reduced. And less labor is required to put the finishing touches on a job. Many builders around the country find the ability to use nonskilled laborers a particular advantage because they are having trouble finding enough skilled labor. Good framers are scarce because every time the construction industry has a downturn, more skilled laborers leave for other industries. It is hard to replenish that pool with new people when the upswing comes, as it always does. Insulating concrete form construction cuts way down on the need for skilled labor and still produces a high-quality home.

Chapter 14

Concrete

SECTION 1901.0 CONCRETE [1900.0; 1901.0]

Concrete without calcium additives or other hardeners added to the mix reaches its optimum design strength in 28 days. Forms can generally be stripped in 24 hr, but structural loads cannot be placed upon green (uncured) concrete. Every region has a different ambient environment and therefore the buildable curing time for concrete varies for different locales. Check with your local building department for curing times in your area. The national building codes recommend that newly placed concrete be maintained above 50^0F and kept moist to prevent shrinkage cracking for the first 7 days.
 Accelerated curing can be accomplished by chemical additives in the mix such as calcium; by constant steam applied at atmospheric pressure; by high-pressure steam; or by heat and moisture applied in continual application. Code requirements for accelerated curing are that the curing process must provide a compressive strength of the concrete at the load stage considered at least equal to the required design strength at that load stage.
 Slab or under-floor inspections are made after all in-slab or under-floor building service equipment, conduit, piping accessories, and other ancillary equipment items are in place, but before any concrete is placed or floor sheathing installed, including the subfloor, which would conceal pipes or equipment.
 Although the national building codes do not contain any new terms or definitions under this section, I have provided the following concrete-related terms and definitions for the benefit of the Reader:

Concrete - A mixture of portland cement or any other hydraulic cement, fine aggregate, coarse aggregate, and water, with or without admixtures.

Admixture - A material other than water, aggregate, or hydraulic cement used as an ingredient of concrete before or during its mixing to modify its properties.

Aggregate - Granular material, such as sand, gravel, crushed stone, and iron blast-furnace slag, and when used with a cementing medium forms a hydraulic cement concrete or mortar.

Deformed Reinforcement - Deformed reinforcing steel bars (rebar), bar and rod mats, deformed wire, welded smooth wire fabric, and welded deformed wire fabric used as reinforcement in the concrete.

Embedment Length - The length of embedded reinforcement required to be provided beyond a critical section (stress point).

Plain Concrete - Concrete that does not conform to the definition of reinforced concrete (contains no rebar).

Precast Concrete - Plain or reinforced concrete element cast in other than its final position in the structure and used as a modular component in the construction.

Prestressed Concrete - Reinforced concrete in which internal stresses have been introduced to reduce potential tensile stresses in concrete resulting from loads.

Reinforced Concrete - Concrete containing reinforcement (wire mesh and/or rebar), prestressed or nonprestressed, and designed so that the two materials act together in resisting forces.

There are eight code classifications of cement types:

Type I. For use in general concrete construction when the special properties specified for Types II, III, IV, and V are not required.

Type IA. Same use as Type I where air entrainment is desired.

Type II. For use in general concrete construction exposed to moderate sulfate action, or where moderated heat of hydration is desired.

Type IIA. Same use as Type II where air entrainment is desired.

Type III. For use when high early strength is desired.

Type IIIA. Same use as Type III where air entrainment is desired.

Type IV. For use when a low heat of hydration is required.

Type V. For use when high sulfate resistance is required.

Ratio proportions for concrete are given in three numbers, such as 2-5-7. The ratio is sack cement to cubic feet of sand to cubic feet of stone. In this example, 2 would indicate two sacks of cement, 5 indicates 5 cu ft of sand, and 7 indicates 7 cu ft of stone (aggregate). Nominal maximum sizing of aggregate is one-third the depth of the slab or three-fourths the minimum clear spacing between the rebar (reinforcement steel), or one-fifth the narrowest dimension between the sides of the formwork holding the concrete. All aggregate and mixing water must be clean and free of organic matter such as dirt or wood and pollutants such as oil or grease.

When pouring deep caissons or over embankments of differing elevations, care must be taken to prevent concrete from free-falling more than 4 ft. If concrete is allowed to do this, the aggregate will separate from the mix and the integrity ratio of the mix is destroyed. Deep drop concrete pours must be done with chutes or pump hoses to prevent separation of the concrete mix which will unacceptably weaken the design strength of the concrete.

1901.1 Scope

1901.1.1. Provisions of this section shall govern the materials, design, and construction of concrete used in buildings.

1901.1.2. Structural members of reinforced concrete, including prestressed concrete, shall be designed and constructed in accordance with the provisions of this section and ACI 318.

[ACI is the American Concrete Institute, publishers of the industry standards for concrete in conjunction with the ASTM specifications. A number of new provisions in the ACI code regarding concrete materials should be known to the designer responsible for structural specifications. Chloride in concrete from any source has been found to be detrimental to prestressed reinforcement (stress corrosion) and to aluminum embedments (galvanic corrosion). For prestressed concrete and concrete containing aluminum embedments of any type, the ACI code limits the accidental chloride ion content of mixing water (ACI section 3.4.2) and admixtures (ACI Section 3.6.3).

Nonpotable water is permitted only after specified comparative cube tests show that it will produce at least 90% of the design strength achieved with potable water (ACI section 3.4.3).

The code does not permit the use of calcium chloride as an admixture in concrete which will be exposed to severe or very severe sulfate-containing solutions (ACI section 4.5.3.1). Limits on chloride ion concentrations in hardened concrete contributed from the ingredients are given in ACI section 4.5.4. The limits on chloride ion may be more restrictive than is stated in that subsection when coated reinforcing bars are used.]

1901.1.3. Structural members of plain concrete shall be designed and constructed in accordance with the provisions of this section and ACI 318.1. Concrete that is either unreinforced or contains less reinforcement than the minimum amount specified for reinforced concrete shall be classified as plain concrete.

1903.0 Materials
1903.1 General
Materials used to produce concrete and admixtures for concrete shall comply with the requirements of this section and ACI 318.

1903.2 Cements
Cement shall conform to ASTM C 150 or to such other cements listed in ACI 318.

1903.3 Aggregates
1903.3.1. Concrete aggregates shall conform to ASTM C 33 or to ASTM C 330.

1903.3.2. Aggregates failing to meet the standards listed in Section 1903.3.1, but which have been shown by special test or actual service to produce concrete of adequate strength and durability, may be used where authorized by the building official.

1903.3.3. Nominal maximum size of coarse aggregate shall not be larger than:
- One-fifth the narrowest dimension between sides of forms
- One-third the depth of slabs
- Three-quarters the minimum clear spacing between individual reinforcing bars or wires, bundles of bars, or prestressing tendons or ducts

These limitations shall not apply if, in the judgment of the engineer, workability and methods of consolidation are such that concrete can be placed without honeycomb or voids.

1903.4 Water
1903.4.1. Water used in mixing concrete shall be clean and free from injurious amounts of oils, acids, alkalis, salts, organic

materials, or other substances that may be deleterious to concrete or reinforcement.

1903.4.2. Mixing water for prestressed concrete or for concrete that will contain aluminum embedments, including that portion of mixing water contributed in the form of free moisture on aggregates, shall not contain deleterious amounts of chloride ion.

1903.4.3. Nonpotable water shall not be used in concrete unless specific requirements of ACI 318 allowing the use of nonpotable water are satisfied.

1903.5 Metal Reinforcement

1903.5.1. Reinforcement shall be deformed reinforcement, except that plain reinforcement shall be permitted for spirals or tendons. Reinforcement consisting of structural steel, steel pipe, or steel tubing shall be permitted as specified in ACI 318.

1903.5.2. Reinforcing bars to be welded shall be indicated on the drawings, and the welding procedure to be used shall be specified. ASTM reinforcing bar specifications, except for ASTM A 706, shall be supplemented to require a report of material properties necessary to conform to welding procedures specified in AWS D1.4.

1903.5.3. Reinforcement shall conform to the applicable ASTM standards listed in ACI 318.

1903.6 Admixtures

1903.6.1. Admixtures to be used in concrete shall be subject to prior approval by the design engineer.

1903.6.2. An admixture shall be shown capable of maintaining essentially the same composition and performance throughout the work as the product used in establishing concrete proportions in accordance with Section 1905.2.

1903.6.3. Calcium chloride or admixtures containing chloride from other than impurities from admixture ingredients shall not be used in prestressed concrete, in concrete containing embedded aluminum, or in concrete cast against stay-in-place galvanized metal forms.

1903.6.4. Air-entraining admixtures, water-reducing admixtures, retarding admixtures, and water-reducing and accelerating admixtures shall conform to the applicable ASTM standards listed in ACI 318.

1903.6.5. Fly ash or other pozzolans used as admixtures shall conform to ASTM C 618. The building official shall require certification of all fly ash materials used in concrete as conforming to the ASTM C 618 specification.

1903.6.6. Ground granulated blast furnace slag used as an admixture shall conform to ASTM C 989.

1903.7 Storage of Materials

1903.7.1. Cementitious materials and aggregate shall be stored in such manner as to prevent deterioration or intrusion of foreign matter.

1903.7.2. Any material that has deteriorated or has been contaminated shall not be used for concrete.

1903.8 Tests of Materials

1903.8.1. The building official shall have the right to order testing of any materials used in concrete construction to determine if materials are of specified quality.

1903.8.2. Tests of materials and of concrete shall be made in accordance with ASTM standards listed in ACI 318. Laboratories conducting tests on concrete and concrete aggregates for use in construction shall comply with ASTM C 1077 except section 7.4.

1903.8.3. A complete record of tests of materials and of concrete shall be available for inspection during progress of work and for 2 years after completion of the project and shall be preserved by the inspecting engineer or architect for that purpose.

1904.0 DURABILITY REQUIREMENTS [1905.0; 1907.0]

1904.1 Cementitious Materials

1904.1.1. For the purposes of this section, a cementitious material is one specified in Section 1903 which has cementing value when used in concrete either by itself, such as portland cement or blended hydraulic cements, or in combination with fly ash, other raw or calcined natural pozzolans, and/or ground granulated blast furnace slag.

1904.1.2 Calculation of Water-Cementitious Materials Ratio

1904.1.2.1. To determine compliance with the maximum water-cementitious materials ratio requirement of Tables 1904B and 1904C, the weight of cement shall include the weights of any of the following if contained in the concrete mixture:
- Cement meeting ASTM C 150 or ASTM C 595

- Fly ash or pozzolan meeting ASTM C 618
- Ground granulated blast furnace slag meeting ASTM C 989

1904.2 Freezing and Thawing Exposures

1904.2.1 Air-entraining

1904.2.1.1. Normal-weight and lightweight concrete exposed to freezing and thawing or deicer chemicals shall be air-entrained with air content as delivered of $\pm 1.5\%$. For specified compressive strength f'_c greater than 5,000 psi (34.5 MPa), the air content indicated in Table 1904A may be reduced 1%.

[When finely divided materials of fly ash or natural pozzolans are used as mineral admixtures in air-entrained portland cement concrete, the building official will require air content tests to be made in accordance with ASTM C 231 to assure compliance with air content requirements of Table 1904A.]

1904.2.2 Low Water Permeability

Concrete that is intended to have low permeability to water or concrete that will be subject to freezing and thawing in a moist condition or will be exposed to deicing salts, brackish water, seawater, or spray from these sources shall conform to requirements of Table 1904B.

> **EXCEPTION:** Normal-weight aggregate concrete used in buildings or their appurtenances of Group R Occupancies three stories or less in height, and subject to weathering (i.e., freezing and thawing) or deicer chemicals, shall comply with the requirements of Table 1904C.

In addition, concrete that will be exposed to deicer chemicals shall conform to the limitations of Section 1904.2.3.

1904.2.3 Limitations on Use of Certain Cementitious Materials

For concrete exposed to deicing chemicals, the maximum weight of fly ash, other pozzolan, or ground granulated blast furnace slag that is included in the water-cementitious materials ratio shall not exceed the percentages of the total weight of cementitious materials specified in Sections 1904.2.3.1 through 1904.2.3.3.

1904.2.3.1 Concrete Containing Fly Ash or Pozzolan

The combined weight of fly ash and other pozzolan conforming to ASTM C 618 shall not exceed 25% of the total weight of cementitious materials. Fly ash or other pozzolan used to manufacture Type IP or IPM blended hydraulic cement conforming to ASTM C 595 shall be included with fly ash or other pozzolan added as an admixture.

1904.2.3.2 Concrete Containing Ground Granulated Blast Furnace Slag

The weight of ground granulated blast furnace slag conforming to ASTM C 989 shall not exceed 50% of the total weight of cementitious materials. Slag used to manufacture Type IS or ISM blended hydraulic cement conforming to ASTM C 595 shall be included with slag added as an admixture.

1904.2.3.3 Concrete Containing Fly Ash or Other Pozzolan and Slag

If fly ash or other pozzolan and slag are used in concrete, portland cement conforming to ASTM C 150 shall constitute not less than 50% of the total weight of cementitious materials. Fly ash or other pozzolan shall constitute no more than 25% of the total weight of cementitious materials. See Section 1904.2.3.1.

1904.3 Exposure to Sulfate-Containing Solutions

Concrete to be exposed to sulfate-containing solutions shall conform to requirements in SBC Table 1904D or be made with a cement that provides sulfate resistance and used in concrete with maximum water-cementitious materials ratio. Calcium chloride as an admixture shall not be used in concrete to be exposed to severe or very severe sulfate-containing solutions, as defined in Table 1904D.

1904.4 Water Soluble Chloride Ion Content

For corrosion protection, maximum water soluble chloride ion concentrations in hardened concrete at ages from 28 to 42 days contributed from the ingredients including water, aggregates, cementitious materials, and admixtures shall not exceed the limits of Table 1904E. Tests performed to determine water soluble chloride ion content shall conform to AASHTO T 260.

1904.5 Corrosion Protection for Reinforced Concrete

When reinforced concrete will be exposed to deicing salts, brackish water, seawater, or spray from these sources, requirements of Table 1904B for water-cementitious materials ratio or concrete strength and minimum concrete cover requirements of Section 1908.6 shall be satisfied. [Refer to ACI 318 for unbonded prestressing tendons.]

Table 1904A
Total Air Content for Frost-Resistant Concrete

	Air content, %	
Maximum Nominal Aggregate Size[1], in	Severe Exposure[3]	Moderate Exposure[3]
3/8	7-1/2	6
1/2	7	5-1/2
3/4	6	5
1	6	4-1/2
1-1/2	5-1/2	4-1/2
2[2]	5	4
3[2]	4-1/2	3-1/2

1 in = 25.4 mm

Notes:
1. See ASTM C 33 for tolerances on oversized aggregate for various nominal maximum size designations.
2. These air contents apply to total mix, as for the preceding aggregate sizes. When testing these concretes, however, aggregate larger than 1-1/2 in is removed by handpicking or sieving and the air content is determined on the minus 1-1/2 in-fraction of mix. (Tolerance on air of total mix is computed from the value determined on the minus 1-1/2-in fraction.)
3. The severe and moderate exposures referenced in this table are not based upon the weathering regions shown in Figure 6-3. For purposes of this table, severe and moderate exposures shall be defined as follows:
    a. Severe exposure occurs in a cold climate when concrete may be in almost continuous contact with moisture prior to freezing or where deicing salts are used. Examples are pavements, bridge decks, sidewalks, parking garages, and water tanks.
    b. Moderate exposure occurs in a cold climate when concrete will be only occasionally exposed to moisture prior to

freezing and where no deicing salts are used. Examples are certain exterior walls, beams, girders, and slabs not in direct contact with soil.

Table 1904B
Requirements for Special Exposure Conditions

Exposure Condition	Maximum Water-Cementitious Materials Ratio, by Weight, for Normal Weight Aggregate Concrete	Minimum f'_c for Normal-Weight and Lightweight Aggregate Concrete, psi
Concrete intended to have low permeability when exposed to water	0.50	4000
Concrete exposed to freezing and thawing in a moist condition or to deicing chemicals	0.45	4500
For corrosive protection for reinforced concrete exposed to chlorides from deicing chemicals, salts, brackish water, seawater, or spray from these sources	0.40	5000

1 psi = 6.8948 kPa

Table 1904C
Minimum Specified Compressive Strength of Concrete (f'_c)[1] Subject to Weathering and/or Deicer Chemicals, psi

Type and/or Location of Concrete Element	Weathering Probability[2]		
	Negligible	Moderate	Severe
Basement walls and foundations not exposed to the weather	2500	2500	2500[3]
Basement slabs and interior slabs and interior slabs-on-grade, except for garage floor slabs	2500	2500	2500[3]
Basement walls, foundation walls, exterior walls, and other vertical concrete surfaces exposed to the weather	2500	3000[4]	3000[4]
Porches, carport slabs, and steps exposed to the weather, and garage floor slabs	2500	3000[4]	3000[4]

1 psi = 6.8948 kPa

<u>Notes</u>:
1. At 28 days, psi.
2. See Section 1904 for weathering probability.
3. Concrete in these locations which may be subject to freezing and thawing during construction shall be air-entrained concrete in accordance with Table 1904A.
4. Concrete shall be air-entrained in accordance with Table 1904A.

1905.0 CONCRETE QUALITY [1905.1; 1901.0]

1905.1.1. Concrete shall be proportioned to provide an average compressive strength as prescribed in Section 1905.3.2 as well as satisfy the durability requirements in Section 1904. Concrete shall be produced to minimize the frequency of strengths below f_c as prescribed in Section 1905.6.2.3. The specified compressive strength f_c for concrete designed and constructed in accordance with this chapter shall not be less than 2500 psi (17.2 MPa).

1905.1.2. Requirements for f_c shall be based on tests of cylinders made and tested as prescribed in Section 1905.6.2.

1905.1.3. Unless otherwise specified, f_c shall be based on 28-day tests. If other than 28 days, the test age of f_c shall be as indicated in design drawings or specifications.

1905.2 Selection of Concrete Proportions

1905.2.1. Proportions of material for concrete shall be established to provide:
1. Workability and consistency to permit concrete to be worked readily into forms and around reinforcement under conditions of placement to be employed, without segregation or excessive bleeding.
2. Resistance to special exposures as required by Section 1904.0.
3. Conformance with strength test requirements of Section 1905.6.

1905.2.2. Where different materials are to used for different portions of proposed work, each combination shall be evaluated.

1905.2.3. Concrete proportions, including water-cementitious materials ratio, shall be established on the basis of field experience and/or trial mixtures with materials to be employed as required by Section 1905.3, except as permitted in Section 1905.4 or required by Section 1904.

1905.3 Proportioning on the Basis of Field Experience and/or Trial Mixtures

1905.3.1 Standard Deviation

1905.3.1.1. Where a concrete production facility has test records, a standard deviation shall be established. Test records from which a standard deviation is calculated:

1. Shall represent materials, quality-control procedures, and conditions similar to those expected. Changes in materials and proportions within the test records shall not have been more restricted than those proposed for the work.
2. Shall represent concrete produced to meet a specified strength or strengths f'_c within 1000 psi (6,900 kPa) of that specified for proposed work.
3. Shall consist of at least 30 consecutive tests or two groups of consecutive tests totaling at least 30 tests as defined in Section 1905.6.1.4, except as provided in Section 1905.3.1.2.

1905.3.1.2. Where a concrete production facility does not have test records meeting requirements of Section 1905.3.1.1 but does have a record based on 15 to 29 consecutive tests, a standard deviation may be established as the product of the calculated standard deviation and modification factor of Table 1905.3A. To be acceptable, the test record must meet requirements 1 and 2 of Section 1905.3.1.1 and represent only a single record of consecutive tests that span a period of not less than 45 calendar days.

1905.3.2 Required Average Strength

1905.3.2.1. Required average compressive strength f'_{cr} used as the basis for selection of concrete proportions shall be the larger of Eq. (1) or (2) using a standard deviation calculated in accordance with Sections 1905.3.1.1 or 1905.3.1.2.

$$f'_{cr} = f'_c + 1.34s \quad (1)$$

or $\quad f'_{cr} = f'_c + 2.33s - 500 \quad (2)$

where s is the standard deviation in pounds per square inch.

1905.3.2.2. When a concrete producing facility does not have field strength test records for calculation of the standard deviation meeting requirements of Sections 1905.3.1.1 and 1905.3.1.2., the required average strength f'_{cr} shall be determined from Table 1905.3B and documentation of average strength shall be in accordance with requirements of Section 1905.3.3.

Table 1905.3A
Modification Factor for Standard Deviation
When Less Than 30 Tests Are Available

Number of Tests[1]	Modification Factor[2] for Standard Deviation
Less than 15	Use Table 1905.3B
15	1.16
20	1.08
25	1.03
30 or more	1.00

Notes:
1. Interpolate for intermediate numbers of tests.
2. The modification factor is the modified standard deviation to be used to determine required strength f_{cr} from Section 1905.3.2.1.

Table 1905.3B
Required Average Compressive Strength
When Data Is Not Available to Establish
a Standard Deviation

Compressive Strength, f'_c (psi)	Required Average Compressive Strength f'_{cr} (psi)
Less than 3000	f'_c + 1000
3000 to 5000	f'_c + 1200
over 5000	f'_c + 1400

1 psi = 6.8948 kPa

1905.3.3 Document of Average Strength

This is documentation that proposed concrete proportions will produce an average compressive strength equal to or greater than the required average compressive strength test records or trial mixtures.

1905.3.3.1. When test records are used to demonstrate that proposed concrete proportions will produce the required average strength f'_{cr} (Section 1905.3.2), such records shall represent materials and conditions similar to those expected. Changes in materials, conditions and proportions within the test records shall not have been more restricted than those for the proposed work. For the purpose of documenting average strength potential, test records consisting of less than 30, but not less than 10, consecutive tests shall be permitted, provided test records encompass a period of time not less than 45 days.

Required concrete proportions shall be permitted to be established by interpolation between the strengths and proportions of two or more test records each of which meets the other requirements of Section 1905.3.

1905.3.3.2. When an acceptable record of field test results is not available, concrete proportions may be established based on trial mixtures meeting the following restrictions:

1. The combination of materials shall be that for proposed work.
2. Trial mixtures having proportions and consistencies required for proposed work shall be made using at least three different water-cementitious materials ratios or cementitious materials contents that will produce a range of strengths encompassing the required average strength f'_{cr}.
3. Trial mixtures shall be designed to produce a slump within ± 0.75 in (19 mm) of maximum permitted, and for air-entrained concrete, within $\pm 0.5\%$ of maximum allowable air content.
4. For each water-cementitious materials ratio or cementitious materials content, at least three test cylinders for each test age shall be made and cured in accordance with ASTM C 192. Cylinders shall be tested at 28 days or at the test age designed for determination of f'_c.

5. From results of cylinder tests a curve shall be plotted showing the relationship between water-cementitious materials ratio or cementitious materials content and compressive strength at the designed test age.
6. The maximum water-cementitious materials ratio or minimum cementitious materials content for the curve to produce the average strength required by Section 1905.3.2, should be used, unless a lower water-cementitious materials ratio or higher strength is required by Section 1904.

1905.4 Proportioning by Water-Cementitious Materials Ratio
1905.4.1. If data required by Section 1905.3 are not available, concrete proportions shall be based on water-cementitious materials ratio limits shown in Table 1905.4, if approved by the building official.
1905.4.2. Table 1905.4 shall only apply for concrete to be made with cements meeting strength requirements for Types I, IA, II, IIA, III, IIIA, or V of ASTM C 150, or Types IS, IS-A, IS(MS), IS-A(MS), I(SM), I(SM)-A, IP, IP-A, I(PM)-A, IP(MS), IP-A(MS), or P of ASTM C 595 and shall not apply for concrete containing lightweight aggregates or admixtures other than those for entraining air.
1905.4.3. Concrete proportioned by the water-cementitious materials ratio limits prescribed in Table 1905.4 shall also conform to durability requirements of Section 1904 and to compressive strength test criteria of Section 1905.6.

Table 1905.4
Maximum Permissible Water-Cementitious Materials Ratios for Concrete When Strength Data from Field Experience or Trial Mixtures Are Not Available

Specified Compressive Strength, f'_c, psi[1]	Absolute Water-Cementitious Materials Ratio by Weight	
	Non-air-entrained Concrete	Air-entrained Concrete

2500	0.67	0.54
3000	0.58	0.46
3500	0.51	0.40
4000	0.44	0.35
4500	0.38	See Note 2
5000	See Note 2	See Note 2

1 psi = 6.8948 kPa

Notes:
1. Full 28-day strength. With most materials, water-cementitious materials ratios shown will provide average strength greater than indicated in Section 1905.3.2 as being required.
2. For strengths above 4500 psi (non-air-entrained concrete) and 4000 psi (air-entrained concrete), concrete proportions shall be established by methods of Section 1905.3.

1905.5 Average Strength Reduction
As data become available during construction, the amount by which value f_{cr} must exceed the specified value of f_c may be reduced, provided:
1. Thirty or more test results are available and the average of test results exceeds that required by Section 1905.3.2.1 using a standard deviation calculated in accordance with Section 1905.3.1.1.
2. Fifteen to twenty-nine test results are available and the average of test results exceeds that required by Section 1905.3.2.1 using a standard deviation calculated in accordance with Section 1905.3.1.2.
3. Durability requirements of Section 1904 are met.

1905.6 Evaluation and Acceptance of Concrete
1905.6.1 Frequency of Testing
1905.6.1.1. Samples for strength tests of each class of concrete placed each day shall be taken not less than once a day, not less than once for each 150 cu yd (115 m³) of concrete, and not less

than once for each 5000 sq ft (465 m²) of surface area for slabs or walls.

1905.6.1.2. On a given project, if the total volume of concrete is such that the frequency of testing required by Section 1905.6.1.1 would provide less than five strength tests for a given class of concrete, tests shall be made from at least five randomly selected batches or from each batch if fewer than five batches are used.

1905.6.1.3. When the total quantity of a given class of concrete is less than 50 cu yd (38 m³), strength tests are not required when evidence of satisfactory strength is submitted to and approved by the building official.

1905.6.1.4. A strength test shall be the average of the strengths of two cylinders made from the same sample of concrete and tested at 28 days or at the test age designed for determination of f'_c.

1905.6.2 Laboratory-Cured Specimens

1905.6.3.1. The building official may require strength tests of cylinders cured under field conditions to check for the adequacy of curing and protection of concrete in the structure.

1905.6.3.2. Field-cured cylinders shall be cured under field conditions in accordance with ASTM C 31.

1905.6.3.3. Field-cured test cylinders shall be molded at the same time and from the same samples as laboratory-cured test cylinders.

1905.6.3.4. Procedures for protecting and curing cylinders shall be improved when the strength of field-cured cylinders at the test age designated for determination of f'_c is less than 85% of that of companion laboratory-cured cylinders. The 85% may be waived if the field-cured strength exceeds f'_c by more than 500 psi (3450 kPa).

1905.6.4 Investigation of Low-Strength Test Results

1905.6.4.1. If any strength test (Section 1905.6.1.4) of laboratory-cured cylinders falls below the specified value of f'_c by more than 500 psi (3450 kPa) or if tests of field-cured cylinders indicate deficiencies in protection and curing, steps shall be taken to assure that the load-carrying capacity of the structure is not jeopardized.

1905.6.4.2. If the likelihood of low-strength concrete is confirmed and computations indicate that load-carrying capacity may have been significantly reduced, tests of cores drilled from the area in question may be required in accordance with ASTM C 42. In such case, three cores shall be taken for each strength test more than 500 psi (3450 kPa) below the specified value of f'_c.

1905.6.4.3. If concrete in the structure will be dry under service conditions, cores shall be air dried [temperature 60 to 80°F (15.6 to 26.7°C), relative humidity less than 60%] for 7 days before testing and shall be tested dry. If concrete in the structure will be more than superficially wet under service conditions, cores shall be immersed in water for at least 40 hr and be tested wet.

1905.6.4.4. Concrete in an area represented by core tests shall be considered structurally adequate if the average of three cores is equal to at least 85% of f'_c and if no single core is less than 75% of f'_c. Additional testing of cores extracted from locations represented by erratic core strength results shall be permitted.

1905.6.4.5. If the criteria of Section 1905.6.4.4 are not met, and if structural adequacy remains in doubt, the engineer or the building official may order load tests as outlined in Chapter 20 of ACI 318 for the questionable portion of the structure or take other appropriate action.

1906.0 MIXING AND PLACING OF CONCRETE [1906.0; 1906.1]

1906.1 Preparation of Equipment and Place of Deposit

Preparation before concrete placement shall include the following:
1. All equipment for mixing and transporting concrete shall be clean.
2. All debris and ice shall be removed from spaces to be occupied by concrete.
3. Forms shall be properly coated.
4. Masonry filler units that will be in contact with concrete shall be well-drenched.
5. Reinforcement shall be thoroughly clean of ice or other deleterious coating.
6. Water shall be removed from place of deposit before concrete is placed unless a tremie is used or unless otherwise permitted by the building official.
7. All laitance and other unsound material shall be removed before additional concrete is placed against hardened concrete.

1906.2 Mixing
1906.2.1. All concrete shall be mixed until there is a uniform distribution of materials and shall be discharged completely before the mixer is recharged.
1906.2.2. Ready-mixed concrete shall be a uniform distribution of materials and shall be discharged completely before the mixer is recharged.
1906.2.3. Job-mixed concrete shall be in accordance with ACI 318.
1906.3 Conveying
1906.3.1. Concrete shall be conveyed from the mixer to the place of final deposit by methods that will prevent separation or loss of materials.
1906.3.2. Conveying equipment shall be capable of providing a supply of concrete at the site of placement without separation of ingredients and without interruptions sufficient enough to permit loss of plasticity between successive increments.
1906.4 Depositing
1906.4.1. Concrete shall be deposited as nearly as practical in its final position to avoid segregation due to rehandling or flowing.
1906.4.2. Concreting shall be carried on at such a rate that concrete is at all times plastic and flows readily into spaces between reinforcement.
1906.4.3. Concrete that has partially hardened or been contaminated by foreign materials shall not be deposited in the structure.
1906.4.4. Retempered concrete or concrete that has been remixed after initial set shall not be used unless approved by the building official.
1906.4.5. After concreting is started, it shall be carried on as a continuous operation until placing of a panel or section, as defined by its boundaries or predetermined joints, is completed except as permitted or prohibited by Section 1907.4.
1906.5 Curing
1906.5.1. Concrete (other than high-early-strength) shall be maintained above $50^{\circ}F$ ($10^{\circ}C$) and in a moist condition for at least the first 7 days after placement, except when cured in accordance with Section 1906.5.3.

1906.5.2. High-early-strength concrete shall be maintained above 50°F (10°C) and in a moist condition for at least the first 3 days after placement, except when cured in accordance with Section 1906.5.3.

1906.5.3. Accelerated curing shall conform to the following:
1. Curing by high-pressure steam, steam at atmospheric pressure, heat and moisture, or other accepted processes shall be permitted to accelerate strength gain and reduce time of curing.
2. Accelerated curing shall provide a compressive strength of the concrete at the load stage considered at least equal to required design strength at that load stage.
3. The curing process shall be such as to produce concrete with a durability at least equivalent to the curing method of Section 1906.5.1 or 1906.5.2.
4. Supplementary strength tests in accordance with Section 1905.6.3 may be required to assure that curing is satisfactory.

1906.6 Cold Weather Requirements

1906.6.1. Adequate equipment shall be provided for heating concrete materials and protecting concrete during freezing or near-freezing weather.

1906.6.2. All concrete materials and all reinforcement, forms, fillers, and ground with which concrete is to come in contact shall be free from frost.

1906.6.3. Frozen materials or materials containing ice shall not be used.

1906.7 Hot Weather Requirements

During hot weather, proper attention shall be given to ingredients, production methods, handling, placement, protection, and curing to prevent excessive concrete temperatures or water evaporation that may impair required strength or serviceability of the member or structure.

1907.0 FORMWORK, EMBEDDED PIPES, AND CONSTRUCTION JOINTS [1906.0; 1909.0]

1907.1 Design of Formwork

1907.1.1. Forms shall result in a final structure that conforms to shapes, lines, and dimensions of the members as required by the design drawings and specifications.

1907.1.2. Forms shall be substantial and sufficiently tight to prevent leakage of mortar.

1907.1.4. Forms and their supports shall be designed so as not to damage the previously placed structure.

1907.1.5. Design of formwork shall include consideration of the following factors:
1. Rate and method of placing concrete
2. Construction loads, including vertical, horizontal, and impact loads
3. Special form requirements for construction of shells, folded plates, domes, architectural concrete, or similar types of elements

1907.1.6. Forms for prestressed concrete members shall be designed and constructed to permit movement of the member without damage during application of prestressing force.

1907.2.1. No construction loads shall be supported on, nor any shoring removed from, any part of the structure under construction except when that portion of the structure in combination with the remaining forming and shoring system has sufficient strength to support safely its weight and loads placed thereon.

1907.2.2. Sufficient strength shall be demonstrated by structural analysis considering proposed loads, strength of forming and shoring system, and concrete strength data. Concrete strength data shall be based on tests of field-cured cylinders or, when approved by the building official, on other procedures to evaluate concrete strength. Structural analysis and concrete strength test data shall be furnished to the building official when so required.

1907.2.3. No construction loads exceeding the combination of superimposed dead load plus specified live load shall be supported on any unshored portion of the structure under construction, unless analysis indicates adequate strength to support such additional loads.

1907.2.4. Forms shall be removed in such manner as not to impair safety and serviceability of the structure. All concrete to be exposed by form removal shall have sufficient strength not to be damaged thereby.

1907.2.5. Form supports for prestressed concrete members shall not be removed until sufficient prestressing has been applied to enable prestressed members to carry their dead load and anticipated construction loads.

1907.3 Conduits and Pipes Embedded in Concrete

1907.3.1. Conduits, pipes, and sleeves of any material not harmful to concrete and within limitation of Section 1907.3 shall be permitted to be embedded in concrete with approval of the engineer, provided they are not considered to replace structurally the displaced concrete, except as provided in Section 1907.3.6.

1907.3.3. Conduits, pipes, and sleeves passing through a slab, wall, or beam shall not impair significantly the strength of the construction.

1907.3.4. Conduits and pipes, with their fittings, embedded within a column shall not displace more than 4% of the area of cross section on which strength is calculated or which is required for fire protection.

1907.3.5. Except when plans for conduits and pipes are approved by the engineer, conduits and pipes embedded within a slab, wall, or beam (other than those merely passing through) shall satisfy the following:

1. They shall not be larger in outside dimension than one-third the overall thickness of slab, wall, or beam in which they are embedded.
2. They shall not be spaced closer than diameters or widths on center.
3. They shall not impair significantly the strength of the total construction.

1907.3.6. Conduits, pipes and sleeves shall be permitted to be considered as replacing structurally in compression the displaced concrete provided:

1. They are not exposed to rusting or other deterioration.
2. They are of uncoated or galvanized iron or steel not thinner than standard Schedule 40 steel pipe.
3. They have a nominal inside diameter not over 2 in (51mm) and are spaced not less than three diameters on centers.

1907.3.7. In addition to other requirements of Section 1907.3, pipes that will contain liquid, gas, or vapor may be embedded in structural concrete under the following conditions:

1. Pipes and fittings shall be designed to resist effects of the material, pressure, and temperature to which they will be subjected.
2. No liquid, gas, or vapor, except water not exceeding 90^0F (32^0C) nor 50 psi (345 kPa) pressure, shall be placed in the pipes until the concrete has attained its design strength.
3. In solid slabs, piping, unless it is for radiant heating or snow melting, shall be placed between top and bottom reinforcement.
4. Concrete cover for pipes, conduit and fittings shall be not less than 1-1/2 in (38 mm) for concrete exposed to earth or weather nor less than 3/4 in (19 mm) for concrete not exposed to weather or in contact with ground.
5. Reinforcement with an area of not less than 0.002 times the area of concrete section shall be provided normal to piping.
6. Piping and conduit shall be so fabricated and installed that cutting, bending, or displacement of reinforcement from its proper location will not be required.

[Foundation inspections are made after excavations for piers and footings are finished, any required containment forms are in place, and all rebar (reinforcement steel) is installed. By the time this inspection is called for, all the materials for the foundation must be on the job site for inspection by the inspector, with the exception of ready-mixed concrete that will be brought in from an approved batch plant elsewhere. If the foundation is to be constructed of approved treated wood, additional inspections may be required.]

1907.4 Construction Joints

1907.4.1. The surface of concrete construction joints shall be cleaned and laitance removed.

1907.4.2. Immediately before new concrete is placed, all construction joints shall be wetted and standing water removed.

1907.4.3. Construction joints shall be so made and located as not to impair the strength of the structure. Provision shall be made for transfer of shear and other forces through construction joints.

1907.4.4. Construction joints in floors shall be located within the middle third of spans, beams, and girders. Joints in girders shall be offset a minimum distance of two times the width of intersecting beams.

1907.4.5. Beams, girders, or slabs supported by columns or walls shall not be cast or erected until concrete in the vertical support members is no longer plastic.

1907.4.6. Beams, girders, haunches, drop panels and capitals shall be placed monolithically as part of a slab system, unless otherwise shown in design drawings or specifications.

1908.0 DETAILS OF REINFORCEMENT [1907.1; 1910.0]

1908.1. Details of reinforcement shall comply with the requirements of this section and ACI 318.

1908.2. Bending Reinforcement

1908.2.1. All reinforcement shall be bent cold, unless otherwise approved by the engineer.

1908.2.2. Reinforcement partially embedded in concrete shall not be field bent except as shown on the design drawings or approved by the engineer.

1908.3 Surface Conditions of Reinforcement

1908.3.1. At the time concrete is placed, metal reinforcement shall be free from mud, oil, or other nonmetallic coatings that decrease the bond. Epoxy coating of bars in accordance with the standards listed in ACI 318 is permitted.

1908.3.2. Metal reinforcement, except prestressing tendons, with rust, mill scale, or a combination of both shall be considered satisfactory, provided the minimum dimensions (including height of deformations) and weight of a hand-wire-brushed test specimen are not less than the applicable specification requirements in the ASTM standards referenced in ACI 318.

1908.3.3. Prestressing tendons shall be clean and free of oil, dirt, scale, pitting, and excessive rust. A light oxide is permissible.

1908.4 Placing Reinforcement

1908.4.1. Reinforcement, prestressing tendons, and ducts shall be accurately placed, and adequately supported before concrete is placed, and shall be secured against displacement within tolerances permitted in Section 1908.4.2.

> **EXCEPTION**: When approved by the engineer, embedded items (such as dowels or inserts) of precast concrete members that either protrude from concrete or remain exposed for inspection may be embedded while the concrete is in a plastic state provided:

1. Embedded items shall not be required to be hooked or tied to reinforcement within plastic concrete.
2. Embedded items shall be maintained in correct position while concrete remains plastic.
3. Embedded items shall be properly anchored to develop required factored loads.

[Typical code compliance rebar placements are shown in Figures 14-1 and 14-2.]

Table 1908.6
Cast-in-Place Concrete Reinforcement Protection

Exposure	Minimum Concrete Cover, in (mm)
Concrete cast against and permanently exposed to earth	3 (76)
Concrete exposed to earth or weather:	
#6 through #18 bars	2 (51)
#5 bar, W31 or D31 wire, and smaller	1-1/2 (38)
Concrete not exposed to weather or in contact with ground:	
Slabs, walls, joists:	
#14 and #18 bars	1-1/2 (38)
#11 bar and smaller	3/4 (19)
Beams, columns:	
Primary reinforcement, ties, stirrups, spirals	1-1/2 (38)
Shells, folded plate members:	
#6 bar and larger	3/4 (19)
#5 bar, W31 or D31 wire, and smaller	1/2 (12.7)

Concrete 399

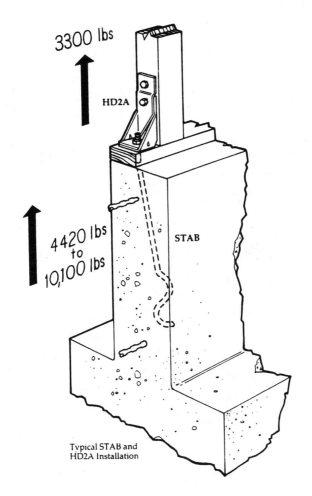

**Figure 14-1
Vertical Rebar Placement**

400 Chapter Fourteen

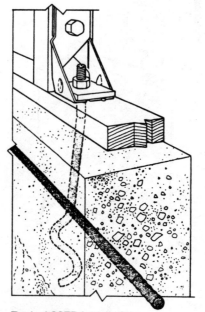

Typical SSTB Installation
with mudsill and
HD5A Holdown

**Figure 14-2
Horizontal Rebar Placement**

1908.6.2. The minimum cover for reinforcement in precast concrete manufactured under plant control conditions and for prestressed concrete shall be in accordance with ACI 318.

1908.6.3. In corrosive environments or other severe exposure conditions, the amount of concrete protection shall be suitably increased, and denseness and nonporosity of protecting concrete shall be considered, or other protection shall be provided.

1908.6.4. Exposed reinforcement, inserts, and plates intended for bonding with future extensions shall be protected from corrosion.

1908.6.5. When code requires a thickness of cover for fire protection greater than the minimum concrete cover specified in Section 1908.6 or ACI 318, such greater thicknesses shall be used.

1909.0 SLAB ON GROUND [1908.0; 1908.0]

1909.1 Minimum Thickness

The minimum thickness of concrete floor slabs supported directly on the ground shall be not less than 3-1/2 in (89 mm) unless designed by an architect or engineer.

1909.2 Vapor Retarder

A vapor retarder consisting of 6-mil (0.152 mm) minimum polyethylene with joints lapped 6 in (152 mm) and sealed, or other approved materials having a maximum perm rating of 0.5 (2.873 E-1 I kg/(Pa-s-M^2)) shall be installed underneath the slab.

EXCEPTIONS: The vapor retarder may be omitted:
1. From detached structures accessory to one- and two-family dwellings such as garages, utility buildings, or other unheated facilities
2. From buildings of other uses when migration of moisture through the slab from below will not be detrimental to the intended use of the building
3. From driveways, walks, patios, and other flat work not likely to be enclosed and heated at a later date
4. Where approved by the building official, based upon local site conditions

1910.0 GFRC EXTERIOR WALL PANELS [1900.0; 1900.0]

The minimum thickness of glass-fiber–reinforced concrete (GFRC) exterior wall panels shall be 3/8 in (9.5 mm).

EXCEPTIONS:
1. Sandwich wall panels
2. Glass-fiber–reinforced concrete wall forms which are left in place

1911.0 PARAPET WALLS [709.4; 1910.0]
Provisions for parapet walls are contained in Section 1507.

1912.0 SEISMIC PROVISIONS [1624.0 - 1632.1; 1904.0]
1912.1 General
The design and construction of reinforced concrete components that resist seismic forces shall conform to the requirements of this section and ACI 318, except as modified by Section 1912.1.1.

1912.1.1. Modifications to ACI 318. These sections of ACI 318 shall be modified as indicated in items 1 through 13.

1. Modify section 8.1.2 to read: "Except where load combinations of Standard Building Code section 1609 including seismic forces are used, design of nonprestressed reinforced concrete members using Appendix, Alternate Design Method, is permitted."
2. Replace ACI 318 section 9.2.3 with section 1600 of adopted code.
3. Add the following definitions to section 21.1 of ACI 318:

> **CONFINED REGION** - That portion of a reinforced concrete component in which the concrete is confined by closely spaced special transverse reinforcement restraining the concrete in directions perpendicular to the applied rooms.
> **JOINT** - That portion of a column bounded by the highest and lowest surfaces of the other members framing into it.
> **SPECIAL TRANSVERSE REINFORCEMENT** -
> Reinforcement composed of spirals, closed stirrups, or hoops and supplementary crossties provided to restrain the concrete and qualify the portion of the component, where used, as a confined region.

4. Replace ACI 318 sections 21.2.1.3 and 21.2.1.4 with the requirements of sections 1912.3 through 1912.6 of adopted code.

5. Modify section 21.2.1.5 to read: "A reinforced concrete structural system not satisfying the requirements of this chapter, including those composed of precast elements, is allowed if it is demonstrated by experimental evidence and analysis that the proposed system will have strength and toughness equal to or exceeding that provided by a comparable monolithic reinforced concrete structure satisfying this chapter."
6. Add the following to the end of section 21.2.5.1: "Posttensioning tendons are allowed in flexural members of frames provided the average prestress (fpc) calculated for a total area equal to the member's shortest cross-sectional dimension multiplied by the perpendicular dimension, does not exceed 350 psi (2413 kPa)."
7. Add a new part to section 21.3.2.5 to read: "For all members in which prestressing tendons are used together with ASTM A 706 or ASTM A 615 (Grades 40 or 60) reinforcement to resist earthquake-induced forces, prestressing tendons shall not provide more than one-quarter of the strength for both positive moments and negative moments at the joint face. Anchorages for tendons must be demonstrated to perform satisfactorily for seismic loadings. Anchorage assemblies shall withstand, without failure, a minimum of 50 cycles of loading ranging between 40 and 85% of the minimum specified anchored at the exterior face of the joint or beyond."
8. Modify section 21.3.3.4 to read: "Where hoops are not required, stirrups with $135°$ (2.356 rad) or greater hooks with 6-bar diameter, but not less than 3-in (76.2 mm) extensions, shall be located throughout the length of the member and spaced not more than one-half the distance from the extreme compression fiber to the centroid of tension reinforcement, d.
9. Add a new part to section 21.4.4.7 to read: "At any section where the nominal strength, P_n, of the column is less than the sum of the shear, V_e, computed in accordance with section 21.7 for all the beams framing into the column above the level under consideration, special transverse reinforcement shall be provided. For beams framing into opposite sides of the column, the moment components are allowed to be assumed to be of opposite sign. For determination of the nominal strength, P_n, of the column, these moments are allowed to be assumed to

result from the deformation of the frame in any one principal axis. [See Figures 14-3 and 14-4.]

Structural tie connections to concrete block are shown beginning with the block rebar placement, as seen in Figure 14-2, and then in various steel connector strap configurations as shown in Figures 14-6 through 14-9.

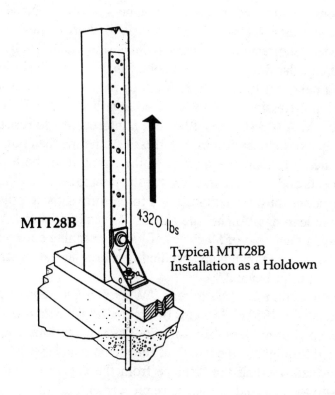

Figure 14-3
Framing Beam Tension Tie

Concrete 405

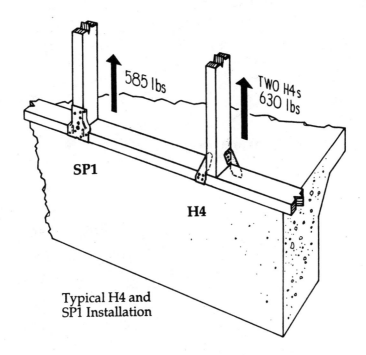

**Figure 14-4
Slab Tension Tie**

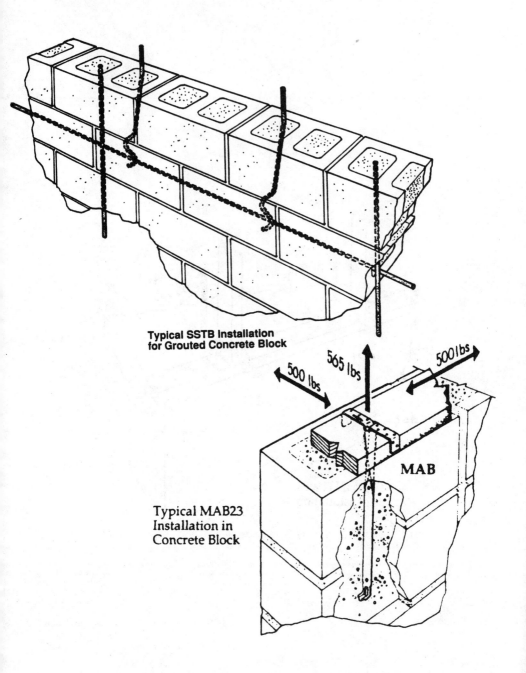

Figures 14-5 and 14-6
Hollow Block Rebar Placement / Plate to Block Seismic Tie

Concrete 407

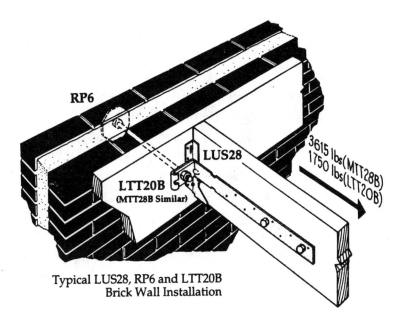

Figure 14-7
Plate to Brick Seismic Tie

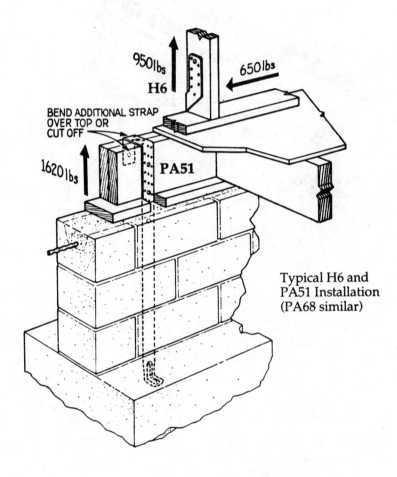

**Figure 14-8
Stemwall to Girder Seismic Tie**

Concrete 409

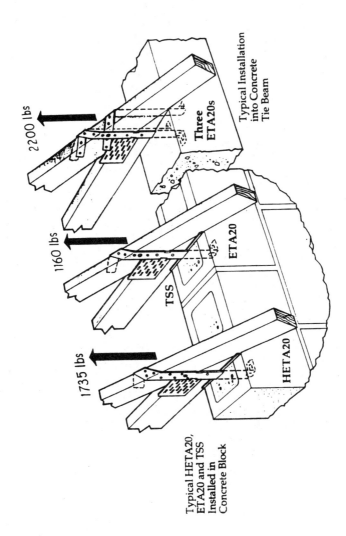

**Figures 14-9
Stemwall to Truss Seismic Tie**

Chapter 15
Electrical

SECTION 2700.0 GENERAL [2700.0; 2701.1]

2701.1 Provisions of this chapter shall govern the electrical systems employed in buildings and structures covered by code.
2701.2 Electrical systems shall comply with the provisions of NFiPA 70 and the National Electrical Code.

2702.0 DEFINITIONS [2702.0; 2702.0]
This chapter contains no unique definitions. For general definitions, see Chapter 2.

About the NEC
The National Electrical Code plays an important role in promoting safety and proper electrical installation in modern construction. The adherence of OSHA to the National Electrical Code is also clearly indicated. In particular, article 90 should be read and studied in its entirety as a necessary complement to this chapter. The *National Electrical Code* (NEC) as published by the National Fire Protection Association is a document of which the express purpose is "the practical safeguarding of persons and property from hazards arising from the use of electricity." This purpose is stated in section 90-1 of the NEC. Many organizations and people use the NEC and are involved in its preparation.

If we reflect on the variety of uses of electricity and where it is used, we can easily understand the need to have standards of control to ensure safety. Whether a builder is installing electrical work in an individual residence or in a highly complex multistoried building, safety is the primary concern. Proper installation procedures need to be standardized and controlled to ensure safety. In any given geographic location, there can be many different applications of electricity. It would be difficult, if not impossible, for local governments to write all the standards for electrical installations or even to know what all the standards should be. To facilitate the task of promoting safety in electrical work throughout the United States, the NEC has been developed to establish national standards. The *National Electrical Code* may be purchased from the

>National Fire Protection Association
>470 Atlantic Avenue
>Boston, Massachusetts 02210

The NEC is submitted for approval to the American National Standards Institute (ANSI), an organization that publishes a large number of approved standards used in the industry throughout the country and referred to in various other places within this book. In ANSI publications, the NEC is known as NPiPA 70-1981. The NPiPA authorizes a National Electrical Code Committee consisting of members and representatives of a wide variety of organizations concerned with electricity. This committee sets and reviews electrical installation standards. The names of the members of this committee may be found near the beginning of the National Electrical Code.

History
The NEC has been sponsored by the National Fire Protection Association since 1911. However, the original document actually preceded the participation of the NPiPA. According to the NEC, "the original Code document actually preceded the participation of the NPiPA." According to the NEC, "the original Code document was developed in 1897 as a result of the united efforts of various insurance, electrical, architectural, and allied interests."

Organization of the NEC
The NEC contains a very large quantity of information, much of which the electrician needs to know on a daily basis. An experienced electrician will, by studying the NEC and through daily application of this knowledge, know much of this information by memory. There will often be information, however, that the electrician will need to look up or research before beginning or continuing work. Safety is the most important concern of the NEC.

The main purpose of the NEC according to section 90-1 is "the practical safeguarding of persons and property from hazards arising from the use of electricity."

The NEC has been organized in a very conscious manner. As is true of all reference books, a very special concern is that of locating information in the text. It is essential that the reader know how and where to find the information desired. In addition, knowledge of where to find specific tables and how to interpret the information supplied is of utmost importance to the designer, builder, and inspector.

General Format
Some of the general organizational characteristics of the NEC are those commonly found in most books. In the first few pages, information is given about the book and its development. We are told who sponsors the NEC (National Fire Protection Association) and how it is produced, revised, and approved. Additional information is also given relating to the NEC itself and the fact that it has a legal status when adopted by lawmaking bodies.

Next comes the table of contents, followed by a list of the membership of the National Electrical Code Committee. After the membership section, we find approximately 600 pages of provisions, an index, and tentative interim amendments to the present edition of the NEC, as shown below:

    NEC Organization
        Publishing notes
        Table of contents
        Membership list (NEC committee)
        Code provisions (regulations)
        Index (alphabetical)
        Tentative interim amendments

Table of Contents and Index
The table of contents lists the major content areas in the order that they are found in the text. Each major content area is a chapter, and each chapter consists of smaller divisions. Note that each smaller division has a three-digit number: the first digit indicates which chapter the subdivision is located in.

Examine closely the major content areas. You will notice that they are in order from the general to the specific. The general (nearer the beginning) concerns itself with the more common aspects of electrical installations, such as common definitions which apply to the whole text and general requirements. Other common or general subjects found near the beginning are such areas as 210 Branch Circuits, 240 Overcurrent Protection, or 300 Wiring Methods.

On the other hand, the more specific areas are found farther toward the end of the text, such as 650 Organs or 660 X-Ray Equipment. In other words, these areas are more specialized (less general), since not all designers and builders install organs or x-ray equipment everyday. Advanced work will call for specialized electricians.

As you look at the table of contents and the three-digit numbers, it is important to realize that these subject areas are referred to as articles, not sections. Article 460, for example, is concerned with capacitors, article 402 with fixture wires, and article 480 with storage batteries.

To find out how a particular article is divided, you must look at the text. For example, look at article 240 Overcurrent Protection. According to the table of contents, this article begins on page 74. Every page number of the text is preceded by the number 70. The number 70 is the official NPiPA number of the NEC in the American National Standards Institute. To find Article 240, look on page 70-74. This article occupies about 10 pages in the text. Note also that the title of the article is always given at the top of each right-hand page.

You will note that each article is divided into smaller divisions. They are known as parts and sections. The number following the three-digit article number is the section number. Whenever we refer to chapters, articles, parts, or sections in this particular module, we are referring to the divisions as used in the NEC.

Even though section is the most common term used for referring to an NEC provision, it is important that the electrician know what a part is, since many articles are divided into parts. Remember that in this chapter, I shall reserve the use of the word part to indicate a subdivision in the NEC that is smaller than an article, yet larger than a section.

Note that article 210 is also divided into two large parts:
A. General provisions
B. Specific requirements

To view this type of division in more detail, look at article 410. This article is divided into parts lettered A through R. As usual, A is general. Then there are such parts as B Fixture locations, E Grounding, or F Wiring or Fixtures. At the end of the article, the final part is on "Special Provisions for Electric-Discharge Lighting Systems of More Than 1000 Volts." This part (part R) represents a high degree of specialization; it is appropriately put at the end of the article in order that all electricians do not have to page through all the sections to find the regulations for the more common situations.

For example, take the following designation: 364-2 Definition. The number 2 indicates a division (section) of article 364. In practice, one usually says that something is found in a particular section such as section 364-2.11. In other words, even though the article number is 364, one calls the paragraph section 364-2 to indicate where one finds the information.

Note that there are even smaller divisions in the text. For example, look at section 220-2, Computation of Branch Circuits, in article 220, Branch Circuit and Feeder Calculations. The whole section is ordered in a clear outline form. The following abbreviated outline is given:

220-2. Computation of Branch Circuits
    a. Continuous Loads
    b. Lighting Load for Listed Occupancies
    c. Other Loads - All Occupancies
        (1) ...
        (2) ...
        (3) ...
        (4) ...
    d. Loads for Additions to Existing Installations
        (1) ...
        (2) ...

In this example we would have to refer to one of the smallest divisions as section 220-2(c)(2) or section 220-2(d)(1) to tell someone exactly to what part of the NEC we were referring. Given the complexity of the electrical trade and the expertise of the writers and editors, this is probably one of the best systems of organization which is at the same time clear, inclusive, and specific. Continual exposure to the NEC is essential in order for the apprentice to feel confident in this system of information.

To help the electrician find the specific article, part, or section for a particular problem, an index is included at the end of the NEC. This index of key words is in alphabetical order and indicates the article number, section number, and sometimes the part letter. If you want to find the regulations concerning branch circuits in mobile homes, the index sends you to section 550-5. Look in the index of the NEC and find "Branch Circuits" under "Mobile Homes." The section can be easily found. Now that you understand how to find particular areas of content, you should examine in more depth the organization within particular areas. This will help you see the complexity and accuracy of detail in the NEC more comprehensively.

Specific Organizational Aspects
Chapter 1, which is called "General," follows an introduction to the text. Chapter 1 has only two major subdivisions (articles): "Definitions" and "Requirements for Electrical Installations." Article 100 begins like many articles and sections, with the word, scope. The scope provides an overview and tells the reader about the content of the article. The scope is a kind of introduction; it states any special information that might be of use to the reader of the text.

For example, the scope of article 100 gives some important information about the nature and limits of the article. You learn here that all definitions will not be included in this section and that, in many cases, definitions will be given throughout the text. You are also informed that equipment operating at more than 600 V nominal has a special status. There are two parts: Part A - General, and Part B - Over 600 V, Nominal. Note that there are often special regulations throughout the text pertaining to equipment operating at over 600 V nominal. It is the electrician's responsibility to be aware of such differences.

It is easy to find the definition of a word such as conductor, which understandably will be used often in the text. The same is true of capacity, approved, or building since these are such common terms. On the other hand, note that the term bussway is not found in this definition section, nor is wireway, type FC, or many other terms. However, if you look at the specific section involving these terms—bussways (section 364-2) and wireways (section 362-1)—you will find the appropriate definitions. Notice also that sometimes the term in article 100 will refer you to another place in the text, such as "Dust-Ignition-Proof: See Section 502-1." Such cross-references direct you to locations within the NEC.

To find tools and materials in the NEC's index, you must look for the correct generic names, and not for manufacturer's trade names or informal shop terms. By now, the reader may well be asking the question, "Why do I have to page all over the place to find one definition?" or "Why can't they put all the definitions in one place?" Several answers here are necessary and perhaps others exist as well.

First of all, a journeyman electrician will know the common definitions accurately. It would enlarge the book considerably to define the same terms repeatedly. On the other hand, suppose all the definitions were in one definition section. Since the electrician will commonly turn to a specific section for a job in order to learn all the provisions, it is helpful to have the important definitions right in the specific section. In addition, by having the uncommon, specialized definitions along with the related provisions, the electrician is more apt to read the specification's definition. This ensures accurate and safe installations based on fact. In construction design, this ping-ponging of cross-referencing sections is tiresome but necessary.

So at least three things may happen when you search for a word in say, article 100:

1. The definition may be given.
2. You may be referred elsewhere in article 100.
3. You may be referred elsewhere in the NEC (another section).

4. You may not find the word and will have to look in the NEC index.

It is important to read all provisions of the NEC carefully. The organizational structure of the NEC promotes conciseness and clarity, as well as ease in locating the desired information. Section 110-1, Mandatory Rules and Explanatory Material, states that "Mandatory rules of this Code are characterized by the use of the word 'shall' ". The NEC spells out its intentions in detail in order that it be legal (and in legal language) upon adoption by the local jurisdiction. This section also notes that explanatory material appears in fine-print notes.

An example of this can be found on pages 70-17, section 110-5. In this particular case, the reader is referred to some tables. Note the tables as you page through the National Electrical Code and make special note of Chapter 9 which consists of tables and examples.

Another special point of great importance is the use of italics throughout the NEC. The italics specify exceptions and are included within the sections. This information in italics is essential for achieving compliance with the regulations. Information in italics may contain the word shall just like any other regulation not in italics and must not be neglected.

An additional facet of organization needs to be mentioned to the reader concerning the sequence of numbers. It should be obvious that no chapter contains all 99 numbers that fall between the 100s (for example, 100, 200, 300, etc.). For example, chapter 2 has only nine articles and chapter 8 has only three. Less obvious is the fact that some numbers are lacking in the sequences in the section numbering. For example, 110-22 skips to 110-30 and 250-7 skips to 250-21. This is done on purpose. To provide for further regulations and subjects, additional slots (numbers) are made available for future expansion of the NEC. The benefit to the designer and builder is quite important. Even though additional materials may be added in the years to come, the numbering of what the journeyman electrician knows today does not necessarily have to change.

As the designer, builder, and inspector use the NEC, it is essential to pay close attention to the way in which the NEC is organized to facilitate finding material and to ensure the accuracy of the information obtained. Remember that the article number and the title are given at the top of each right-hand page. If, for example, you are looking at section 422-20, Disconnecting Means, you can tell immediately from the top of the right-hand page that you are in the Appliances article. Experience and practice will give the builder an appreciation for the organizational systems of the NEC.

Summary
It is clear that the NEC has been organized with a good deal of planning and practical usage in mind. The general aspects usually precede the specialized aspects with the benefit of most residential and commercial electricians in mind. Large content areas are called chapters. Divisions of chapters are called articles (each with a three-digit number, the first digit being the number of the chapter). Articles are often divided into parts in a capital-letter sequence: A, B, C, . . . The smaller divisions of articles are called sections, which are numbered and may be further divided into letter divisions (a, b, c, . . .) and then again into number divisions (1, 2, 3, . . .).

Exceptions are given in italics. Special explanatory materials are expressed in fine-print notes. Changes from the previous edition are indicated by a vertical line in the margin. The more common definitions are given in article 110, and other, more specialized definitions are given in the specific, related section such as the following job site basics.

Guarding of Live Parts
Section 110-17 of the NEC requires that live parts of electrical equipment operating at 50 V or more be guarded against accidental contact by approved enclosures, or by any of the following methods:
1. Location in a room, vault or similar enclosure accessible only to qualified persons
2. Substantial permanent partitions or screens so constructed as to minimize accidental contact with the live parts, and accessible only to qualified persons
3. Elevation of 8 ft (2.44 m) or more above the floor or working surface

This section provides for the protection of electrical equipment exposed to physical damage, and for the placing of conspicuous warning signs forbidding unqualified individuals to enter rooms or other guarded locations containing live parts. The grounding bar in a panel box that meets code compliance is shown in Figure 15-1.

Ground Fault Protection of Personnel

Section 210-1 of the NEC provides for ground fault protection for personnel. The specific and very important provisions are as follows for dwelling units:

1. All 125-V, single-phase, 15- and 20-ampere (A) receptacles installed in bathroom shall have ground-fault circuit interrupters (GFCI) protection for personnel.
2. All 125-V, single-phase, 15- and 20-A receptacles installed in garages shall have GFCI protection for personnel.
3. All 125-V, single-phase, 15- and 20-A receptacles installed outdoors where there is direct grade-level access to the dwelling unit and to the receptacles shall have GFCI protection for personnel.

For all construction sites, the provision for temporary wiring on a job site is mandated as follows:

All 125-V, single-phase, 15- and 20-A receptacle outlets which are not a part of the permanent wiring of the building or structure, and which are in use by employees, shall have GFCI protection for personnel. Ground-fault circuit interrupters are defined in Article 100 of the NEC. (Later in this chapter some OSHA requirements for GFCI will be presented.)

Ground Fault Protection of Equipment

People are, of course, the most important safety consideration, and the major thrust in safe practice should therefore be toward the preservation of life and limb. But equipment and property are also important, primarily because they represent some person's investment. Equipment damage due to faulty workmanship is costly. Indirectly also, improper installation procedures could cause injury (electric fires).

Electrical 421

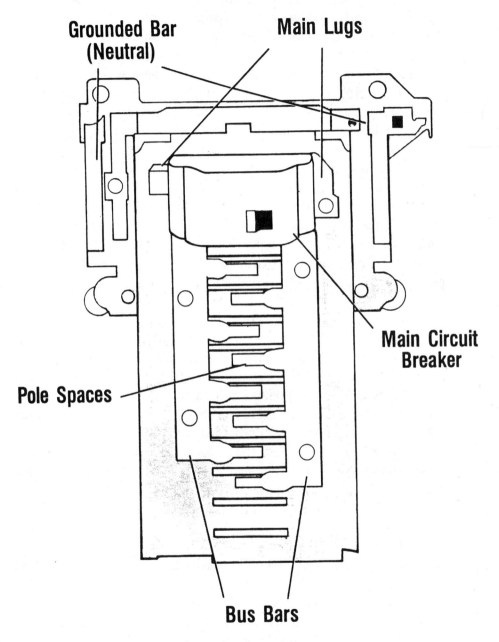

Figures 15-1
Code Compliant Panel Box

Ground-fault circuit interrupters are an important safeguard against property and equipment damage. Section 230-95 of the NEC addresses ground-fault protection of equipment. Specifically, this section states that ground-fault protection of equipment shall be provided for solidly grounded wye electric services of more than 150 V to ground but not exceeding 600 V phase-to-phase for each service disconnecting means rated 1000 A or more.

Article 280 of the NEC outlines installation and connection requirements for surge arrestors. A surge arrestor is defined in the NEC (section 280-2) as a protective device for limiting surge voltages by discharging or bypassing surge current, and it also prevents continued flow of follow current while remaining capable of repeating these functions. A good example of surge voltage would be lightning. When lightning strikes, unusually high voltages are introduced into normal circuits. Without arrestors, there is the risk of equipment damage or fire.

Grounding

Grounding of electric tools and equipment is one of the most important methods of controlling the hazards of low-voltage electricity. If the insulation in electrical equipment degenerates, or if a wire works loose and comes into contact with the frame or some other part that does not normally carry current, these parts can become energized. The electricity is now no longer controlled. It is ready to follow any path to the ground. The unprotected human body will complete the circuit and the consequences could be fatal. The NEC prescribes grounding for the following:
1. Motor frames
2. Elevators and cranes
3. Electric signs
4. The following residential appliances and equipment:
    a. Refrigerators, freezers, etc.
    b. Washers and dryers
    c. Hand-held, motor-operated tools
    d. Motor-operated appliances such as hedge clippers, lawn mowers, and snow blowers
5. Cord and plug-connected appliances used in damp or wet locations
6. Portable head lamps

These are just a few of the many NEC articles and sections that have a direct bearing on your personal safety. Builders and designers are strongly advised to read and become more familiar with all its provisions. The NEC also deals with some further stipulations for safe electrical practice as prescribed by OSHA. The references cited herein refer specifically to the construction industry.

OSHA was written into the law on December 29, 1970, and took effect 120 days after this date. It was designed with the intent to assure safe and healthy working conditions nationwide in the work environment for the American worker by authorizing enforcement of the standards developed under the Act; by assisting and encouraging the states in their efforts to assure safe and healthful working conditions; by providing for research, information, education, and training in the field of occupational safety and health; and for other purposes.

OSHA specifies the duties of both the employer and employee with respect to safety. Concerning the employer it states that the employer:

1. Shall furnish to each of his or her employees employment and a place of employment which are free from recognized hazards that are causing or likely to cause death or serious physical harm to his employees
2. Shall comply with occupational safety and health standards promulgated by OSHA

Hazardous (Classified) Locations
Articles 500 to 503 of the NEC deal with the requirements for electrical equipment and wiring for all voltages in locations where fire or explosion hazards may exist. These hazards include flammable liquids, combustible dust, or ignitable fibers.

Sections 501-2 and 502-2 recommend that transformers and capacitors containing a flammable liquid be installed only in approved vaults as specified in sections 450-41 and 450-48. Among the features of these prescribed vaults are walls, roof, and floor materials with a minimum fire-resistance of 3 hr. Section 502-2 also prescribes that no transformer or capacitor should be installed in locations where there is dust from magnesium, aluminum, aluminum-bronze powders, or other materials with similar hazardous characteristics.

Electrical Systems Test Procedure
Perform the following tests after all structural assembly, metal and trim installations, and electrical crossover connections are complete. The grounding continuity test is to be performed before connecting the home to the electrical service, and the polarity and operation tests are to be performed after the electrical installation is complete. Perform the following procedure checks for grounding continuity, polarity, and operation of the electrical system. Before the home is connected to 120/240 V service, proceed as follows: Connect one clip of a flashlight continuity tester to a convenient ground (metal skin, window frame on metal skinned units, floor duct rise, etc.) and touch the other clip to each fixture canopy. The continuity light should light if each fixture is properly grounded. Using the continuity tester, check every direct-connected appliance or fan. The tester must be hooked to a convenient ground and to the metal frame of the appliance. Using the continuity tester, check the continuity between the following:

- Between one riser of furnace duct and convenient ground
- Between metal roof and steel frame
- Between metal skin and steel frame
- Between metal gas piping and steel frame
- Between metal water piping and steel frame
- Between metal raceway below distribution panel and steel frame

Note: Continuity to ground is not required on metal inlet of plastic-piped water system.

When plumbing fixtures such as metallic sinks, tubs, faucets, and shower risers are connected only to plastic water piping and plastic drain piping, continuity to ground is not required. Any loss of grounding continuity will require investigation and correction. After the home is connected to the electrical service, proceed as follows: Plug an ac receptacle wiring tester into each receptacle in the home to check for reversed polarity, open grounds, and shorts. Any reverse polarity, open grounds, or shorts found will require investigation and repair. Install light bulbs and fluorescent tubes where not already installed. Make sure each light fixture is operable by turning the appropriate switch to the ON position. Shut off all light switches in the house and perform tests on smoke detectors. Repair or replace any defective items.

Chapter 16

Wood and Framing

SECTION 2300.0 GENERAL [2300.0; 2300.1]

Stud partition walls containing plumbing, heating, or other pipes must be so framed and the joists underneath so spaced as to give proper clearance for ductwork, conduit, and piping. Where a partition containing such runs parallel to the floor joists, the joists underneath such partitions must be doubled and spaced to permit the passage of such service components and must be bridged. Where plumbing, heating, or other pipes are placed in or partly in a partition, necessitating the cutting of sole plates or top plates, a metal tie not less than 1/8 in thick and 1-1/2 in wide must be fastened to the plate across and to each side of the opening with not less than four 16d nails. This is known as *preventive construction*. Preventive construction keeps Sheetrock nails and screws from penetrating wiring or piping inside the wall. Examples of these different area separation walls are shown in Figure 16-1.

 Frame inspections are made after the roof, all framing, fire-blocking and bracing are in place; all pipes, chimneys, and vents are complete; and the rough electrical, plumbing, and heating wires, pipes, and ducts are approved.

 Holes no greater than 40% of the stud width may be bored in any wood stud for running of wire, conduit, or pipes. Bored holes not greater than 60% of the width of the stud are permitted in nonbearing partition walls or in any wall where each bored stud is doubled, provided not more than two such successive doubled studs are so bored. You must provide fire-stops in openings around vents, pipes, ducts, chimneys, fireplaces, and similar openings which afford a passage for fire at ceilings and floor levels, with noncombustible materials.

 All piping in connection with a plumbing system must be so installed that a piping or connections will not be subject to undue strain or stresses, and provisions must be made for expansion and

428 Chapter Sixteen

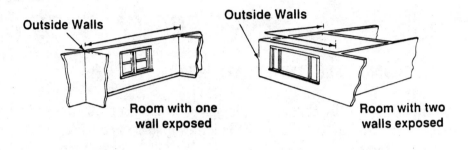

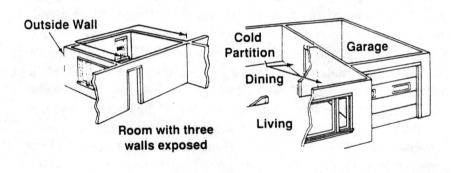

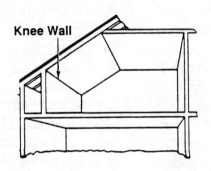

**Figure 16-1
Area Separation Walls**

structural settlement in concrete or masonry walls or footings. This means the studs must not pinch them as the new construction settles. More importantly though, no structural member can be seriously weakened or impaired by cutting, notching, or boring. Unless impractical due to structural conditions, all wood beams, girders, joists, studs, and similar construction must be bored with holes approximately the same diameter as the pipes passing through them.

2301.0 SCOPE [2301.0; 2301.0]
2301.1.1. The provisions of this chapter shall govern the material, design, quality, and construction of wood used in buildings or structures.
2301.1.2. The detailed structural requirements contained in this chapter are based on sound engineering principles such as those in the standards listed in this chapter and are intended for light frame construction in general use for structures having light loads [e.g., live loads of 40 psf (1.92 kPa) or less, locations in noncoastal areas] and closely spaced framing. Where additional structural requirements should be applied because of the nature of the structure, the standards in Section 2301.2.5 shall be accepted as good engineering practice.
2301.1.3. For heavily loaded or engineered timber construction, structural design based on the recommendations of the standards listed in Section 2301.2.5 shall be accepted as conformance with good engineering practices.
2301.1.4. Other sections of this chapter which are applicable shall apply to heavily loaded or engineered timber construction as well as light frame construction.
2301.2 Design
2301.2.1. The quality and design of wood members and their fastenings used for load supporting purposes shall conform to good engineering practice.
2301.2.2. All members shall be framed, anchored, tied and braced so as to develop the strength and rigidity necessary for the purposes for which they are used. [These structural design members are shown in Figure 16-2.]

2301.2.3. Preparation, fabrication, and installation of wood members and the glues, connectors, and mechanical devices for the fastening thereof shall conform to good engineering practices.

2301.2.4. For engineered wood structural panel diaphragm design, provisions of Section 2310 shall apply. For engineered particleboard structural diaphragm design, provisions of Section 2311 shall apply.

2301.2.5. The following standards shall be accepted as conforming to good engineering practice:

National Forest Products Association:
National Design Specification for Wood Construction
Wood Construction Data No.5,
  Heavy Timber Construction Details

American Institute of Timber Construction:
AITC 104: Typical Construction Details
AITC 110: Standard Appearance Grades for Structural Glued
  Laminated Timber
AITC 112: Standard for Tongue-and-Groove Heavy Timber Roof
  Decking
AITC 113: Standard for Dimensions of Structural Glued
  Laminated Timber
AITC 117: Structural Glued Laminated Timber of Softwood
  Species
AITC 119: Standard Specifications for Hardwood Glued
  Laminated Timber

American Plywood Association:
APA Design/Construction Guide-Residential & Commercial
  Cantilevered In-Line Joist System
Plywood Design Specifications
Plywood Design Specification Supplement 1 - Design &
  Fabrication of Plywood Curved Panels
Plywood Design Specification Supplement 2 - Design &
  Fabrication of Plywood-Lumber Beams
Plywood Design Specification Supplement 3 - Design &
  Fabrication of Plywood Stressed-Skin Panels
Plywood Design Specification Supplement 4 - Design &
  Fabrication of Plywood Sandwich Panels
Plywood Design Specification Supplement 5 - Design &

Wood and Framing

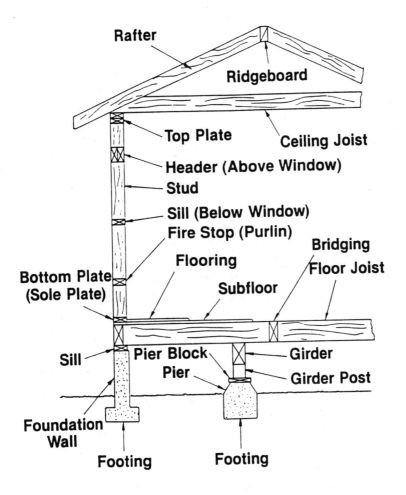

**Figure 16-2
Structural Members**

Fabrication of All-Plywood Beams
Truss Plate Institute, Inc.:
Design Specifications for Metal Plate Connected Wood Trusses
American Wood Preservers Institute:
Pole Building Design
National Particleboard Association:
Particleboard: The Ideal Floor Underlayment

2301.3 Determination of Required Sizes

2301.3.1. Wood structural members shall be of sufficient size to carry the dead and required live loads without exceeding the allowable working stresses as contained in the standards listed in Section 2301.2.5.

2301.3.2. Where applicable as determined by end use, allowable working stresses may be determined by machine stress rating as approved by the American Lumber Standards Committee as established by USDOC PS 20-70.

2301.3.3. Where minimum sizes of lumber members are shown herein, they shall be construed as meaning nominal sizes. Minimum dressed sizes corresponding to nominal sizes shall conform with the provisions of USDOC PS 20-70.

2301.3.4. For convenience, nominal sizes may be shown on the plans. If rough sizes or finished sizes greater or smaller than USDOC PS 20-70 dressed sizes are to be used, computations shall be predicated on such actual sizes, provided they are specified on the plans or in a statement appended thereto.

2301.4 Quality of Materials

2301.4.1. Lumber used for load-supporting purposes, including end-jointed or edge-glued lumber, shall be identified by the grade mark of a lumber grading or inspection agency which is accredited through a program which complies with USDOC PC 20-70 or the equivalent.

Grading practices and identification shall comply with rules published by an agency approved in accordance with procedures of USDOC PS 20-70 or equivalent procedures. In lieu of a grade mark on the material, a certificate of inspection as to species and grade issued by a lumber grading or inspection agency meeting the requirements of this section may be accepted for precut, remanufactured, or rough-sawn lumber, and for sizes larger than 3 in (75 mm) nominal thickness.

2301.4.2. Structural glue-laminated timber shall be manufactured and identified as required in ANSI/AITC A190.1.

2301.4.3. Wood structural panels when used structurally (including, among others, those used for siding, roof and wall sheathing, subflooring, diaphragms, and built-up members) shall conform to the requirements for its type in U.S. Product Standards PS 1 or PS 2. Wood structural panels shall include plywood, oriented strandboard (OSB), and composite panels.

Each panel or member shall be identified for grade and glue type by the trademarks of an approved testing and grading agency. Wood structural panel components shall be designed and fabricated in accordance with the applicable standards listed in Section 2301.2.5 and identified by the trademarks of an approved testing and inspection agency indicating conformance with the applicable standard. In addition, wood structural panels when permanently exposed in outdoor applications shall be of exterior type, except that wood structural panel roof sheathing exposed to the outdoors on the underside may be interior type bonded with exterior glue.

2301.4.4. Fiberboard for its various uses shall conform to ANSI/AHA A194.1. Fiberboard sheathing when used structurally shall be so identified by an approved agency as conforming to ANSI/AHA A194.1.

2301.4.5. Particleboard shall conform to ANSI A208.1. Particleboard shall be identified by the grade mark or certificate of inspection issued by an approved agency.

2301.4.5.1. Particleboard floor underlayment shall conform to Type 1-M-1 or Sanded Type 2-M-W of ANSI A208.1. Type 1-M-1 underlayment shall be not less than 1/4 in (6.4 mm) thick and shall be installed in accordance with the installation instructions of the National Particleboard Association. Sanded Type 2-M-W underlayment shall be not less than 1/4 in (6.4 mm) thick and shall be installed in accordance with the installation instructions of the Structural Board Association and the manufacturer.

2301.4.5.2. Particleboard subfloor or combination subfloor/underlayment shall conform to one of the grades in SBC Table 2307.6C.

2301.4.6. All lumber, sawn timber, plywood, piles, and poles supporting permanent structures required by Section 2304 to be pressure treated shall bear the quality mark of an approved inspection agency which maintains continued supervision, testing, and inspection over the quality of the product as described in the AWPA standards listed in SBC chapter 35. Quality-control inspection agencies for pressure-treated wood shall be certified as to competency and performance by an approved organization.

2301.4.7. Hardwood and decorative plywood shall be manufactured and identified as required in ANSI/HPMA HP.

2301.4.8. Wood flooring of the various types shall be manufactured and identified as required in the appropriate standard as listed:
- Laminated hardwood flooring - ANSI/HPMA LHF
- Flooring grading rules
- Oak, pecan, beech, birch, hard maple
- (Acer Saccharum)-NOFMA
- Mosaic-parquet hardwood slat flooring-ANSI/APA 1
- Hard maple, beech and birch-MPMA

2301.4.9. Hardboard siding used structurally shall be identified by an approved agency as conforming to ANSI/AHA A135.6. Hardboard underlayment shall meet the strength requirements of 7/32-in (5.6 mm) or 1/4-in (6.4 mm) service class hardboard planed or sanded on one side to a uniform thickness of not less than 0.200 in (5.1 mm). Prefinished hardboard paneling shall meet the requirements of ANSI/AHA A135.5. Other basic hardboard products shall meet the requirements of ANSI/AHA A135.4. Hardboard products shall be installed in full accordance with the manufacturer's recommendations.

2301.4.10. Metal-plate-connected parallel chord wood trusses shall be manufactured as required in the Truss Plate Institute Design Specification for Metal Plate Connected Parallel Chord Wood Trusses.

2301.4.11. Prefabricated wood I-joist design and structural capacities shall conform to ASTM D 5055.

2301.5 Minimum Lumber Grades

The minimum grade of lumber used for light frame construction shall be
- <u>For joists and rafters</u>: Those obtained in NFiPA Design Values for Joists and Rafters.

- **For load-bearing studs:** No.3 grade, standard grade, or stud grade. (Utility grade may be used to support roof and ceiling loads only.)
- For non-load-bearing studs: Utility grade.

2301.6 End-Jointed Lumber

End-jointed lumber may be used interchangeably with solid sawn lumber of the same grade and species. Such uses shall include, but are not limited to, light framing, studs, joists, planks, and decking.

2301.7 Moisture Content

All lumber and wood structural panel members, including pressure-treated and 2 in (51 mm) thick and less shall contain not more than 19% moisture at the time of permanent incorporation in a building or structure.

2301.8 Fire-Retardant-Treated Wood

2301.8.1. The allowable unit stresses for fire-retardant-treated wood, including fastener values, shall be developed from an approved method of investigation which considers the effects of anticipated temperature and humidity to which the fire-retardant-treated wood will be subjected, the type of treatment, and the redrying process.

2301.8.2. All fire-retardant-treated wood shall bear an identification mark showing the flamespread index thereof issued by an approved agency which audits the quality-assurance program of the treating facility each month that the facility is in production. Fire-retardant-treated wood shall bear the quality mark of an approved inspection agency which maintains continued supervision and inspection over the method of drying. The drying shall be done according to AWPA C20 for lumber and AWPA C27 for plywood.

2301.8.3. Where fire-retardant-treated wood is exposed to weather, it shall be further identified to indicate that there is no increase in the listed flamespread index as defined above when subjected to ASTM D 2898.

2301.8.4. Where experience has demonstrated a specific need for use of material of low hygroscopicity, fire-retardant-treated wood to be subjected to high-humidity conditions shall be identified as Type A in accordance with AWPA C20 or AWPA C27 to indicate the treated wood has a moisture content of not over 28% when tested in accordance with ASTM D 3201 procedures at 92% relative humidity.

2301.8.5. Fire-retardant-treated wood shall be dried to a moisture content of 19% or less for lumber and 15% or less for wood structural panels before use. The identification mark shall show the method of drying after treatment.

2302.0 DEFINITIONS
This chapter contains one unique definition. For general definitions, see Chapter 2.

PRESSURE-TREATED DOUGLAS FIR (PTDF) - Special construction grade Douglas Fir larch pressure treated to resist insects and dryrot.

2303.0 CONSTRUCTION PRACTICES [2301.0; 2303.0]
2303.1 Preparation of Building Site and Removal of Debris
2303.1.1. All building sites shall be graded so as to provide drainage under all portions of the building not occupied by basements.

2303.1.2. The foundation and the area encompassed therein shall have all vegetation, stumps, roots, and foreign material removed, and the fill material shall be free of vegetation and foreign material. The fill shall be compacted to assure adequate support of the foundation.

2303.1.3. After all work is completed, loose wood and debris shall be completely removed from all spaces under the building. All wood forms and supports shall be completely removed. Wood shall not be stored in contact with the ground under any building.

2303.2 Foundations
2303.2.1. Foundations shall be designed and constructed in accordance with the provisions of SBC sections 1606.4 and 1804. Where spot piers are used, spacing shall not exceed 8 ft (2438 mm) o.c. unless engineering analysis indicates a greater spacing is acceptable.

2303.2.2. A one-story building, except a dwelling, which does not exceed 400 sq ft (37 m^2) in area may be constructed without masonry or reinforced concrete foundation, provided such building is placed on a sill of approved wood of natural decay resistance or of pressure-treated wood and provided the structure is properly anchored to resist overturning and sliding as required in SBC section 1606.4. Mud sills shall be not less than a 2 X 6 or 3 X 4 PTDF.

2303.3 Moisture Protection
Surfaces exposed to the weather shall have an approved barrier to protect the structural frame and the interior wall covering. The barrier shall be at least Type 15 felt or kraft waterproof building paper. Building paper and felt shall be free from holes and breaks other than those created by fasteners and construction systems due to attaching of the barrier and shall be applied over studs or sheathing of all exterior walls. Such felt or paper shall be applied horizontally with the upper layer lapped over the lower layer not less than 2 in (51 mm). Where vertical joints occur, felt or paper shall be lapped not less than 6 in (152 mm).

> EXCEPTIONS: The approved barrier is not required in any of the following circumstances:
> - When exterior covering is of approved weatherproof panels
> - In back-plastered construction
> - When there is no human occupancy
> - Over water-repellent panel sheathing including wood structural panels complying with SBC Tables 2308.1B and 2308.1D, fiberboard not less than 7/16 in (11.1 mm) thick, particleboard complying with ANSI A208 grades in SBC Table 2308.1C, and gypsum not less than 1/2 in (12.7 mm) thick
> - Under approved paper-backed metal or wire fabric lath
> - Behind lath and portland cement plaster applied to the underside of roof and eave projections

2304.0 PROTECTION AGAINST DECAY AND TERMITES [2317.0; 2311.0]

2304.1 Protection

2304.1.1. Where protection of wood members is required by this section, protection shall be provided by using naturally durable or pressure-treated wood (PTDF).

2304.1.1.1. The expression "naturally durable wood" refers to the heartwood of the following species with the exception that an occasional piece with corner sapwood may be included if 90% or more of the width of each side on which it occurs is heartwood. These types include

1. Decay resistant: redwood, cedars, black locust
2. Termite resistant: redwood, eastern red cedar

2304.1.1.2. The expression "pressure-treated wood" refers to wood meeting the retention, penetration, and other requirements applicable to the species, products, treatment, and conditions of use in the approved standards of the American Wood Preservers Association (AWPA).

2304.1.1.3. Wood subject to damage from both decay and termites shall be a naturally durable species resistant to termites or pressure treated.

2304.1.2. In territories where the hazard of termite damage is known to be very heavy the building official may require floor framing of naturally durable wood, pressure-treated wood, soil treatment, or other approved methods of termite protection. Where the floor is constructed of a concrete slab on ground, the building official may require soil treatment under the slab.

2304.1.3. In geographical areas where experience has demonstrated a specific need, approved naturally durable or pressure-treated wood shall be used for those portions of wood members which form the structural supports of buildings, balconies, porches, or similar permanent building appurtenances when such members are exposed to the weather without adequate protection from a roof, eave, overhang, or other covering to prevent moisture or water accumulation on the surface or at joints between members.

Depending on local experience, such members may include horizontal members such as girders, joists, and decking and vertical members such as posts, poles, and columns.

2304.2 Wood in Ground Contact or Exposed to the Weather

2304.2.1. Wood in contact with ground or below ground level which supports permanent structures shall be approved pressure-treated wood suitable for ground contact use.

EXCEPTIONS:

1. Naturally durable wood used in contact with the ground for support of structures other than buildings and walking surfaces

Wood and Framing

2. Untreated wood used for supports which are entirely below groundwater level or continuously submerged in fresh water

2304.2.2. All posts, poles, and columns supporting permanent structures and embedded in concrete which is in contact with ground shall be approved pressure-treated wood suitable for ground contact use.

> **EXCEPTION:** Naturally durable wood used for posts, poles, and columns embedded in concrete for structures other than building and walking surfaces or in structures where wood is above ground level and not exposed to weather.

2304.2.3. Posts or columns supporting permanent structures which are closer than 8 in (203 mm) to exposed ground in enclosed crawl spaces or unexcavated areas located within the periphery of the building shall be of approved naturally durable or pressure-treated wood.

2304.2.4. Wood posts or columns exposed to the weather or in basements or cellars and which support permanent structures shall be supported by concrete piers or metal pedestals projecting at least 1 in (25.4 mm) above concrete or masonry floors or decks and 6 in (152 mm) above exposed earth and separated therefrom by an approved impervious barrier except when approved naturally durable or pressure-treated wood is used.

2304.2.5. Clearance between wood siding and earth on the exterior of a building shall be not less than 6 in (152 mm) except where siding, sheathing, and wall framing are of approved pressure-treated wood or approved naturally durable wood.

2304.2.6. Those portions of glue-laminated timbers which form the structural supports of a building or other structure and are exposed to weather and not properly protected by a roof, eave, or similar covering shall be pressure-treated with preservative or be manufactured from naturally durable wood.

2304.3 Crawl Space Construction

2304.3.1. Crawl spaces under buildings without basements shall be ventilated in accordance with Section 1804.6.3.1.

2304.3.2. All wood framing and sheathing less than 8 in (203 mm) from exposed earth in exterior walls that rest on treated wood, concrete, or masonry foundations shall be of approved naturally durable or pressure-treated wood.

2304.3.3. When the bottoms of wood structural floor elements, including joists, girders, and subfloor are less than 8 in (203 mm) above the horizontal projection of the outside ground level and extend toward the outside ground beyond the plane represented by the interior face of the foundation wall studs, such elements shall be of approved naturally durable or pressure-treated wood.

2304.3.4. When wood joists or the bottom of wood structural floors without joists are closer than 18 in (457 mm) or wood girders are closer than 12 in (305 mm) to exposed ground located within the periphery of the building over crawl space or unexcavated areas, they shall be of approved naturally durable wood or pressure-treated wood.

2304.4 Slabs

2304.4.1. Sleepers, sills, and sole plates on a concrete or masonry slab which is in direct contact with earth shall be approved naturally durable or pressure-treated wood (PTDF).

2304.4.2. Wood structural members supporting moisture permeable floors or roofs which are exposed to the weather, such as concrete or masonry slabs, shall be of approved naturally durable wood or pressure-treated wood unless separated from such floors or roofs by an approved impervious moisture barrier.

2304.5 Walls

2304.5.1. Ends of wood girders entering exterior masonry or concrete walls shall be provided with a 1/2-in (12.7 mm) air space on tops, sides, and ends unless approved naturally durable or pressure-treated wood is used.

2304.5.2. Wood furring strips or other wood framing members attached directly to the interior of exterior masonry or concrete walls below grade shall be of approved naturally durable or pressure-treated wood.

2304.5.3. Wood used in retaining or crib walls shall be approved pressure-treated wood.

> **EXCEPTIONS**:
> - Untreated wood may be used when the wall is not more than 2 ft (610 mm) high and is separated from the property line or a permanent building by a minimum distance equal to the height of the wall.

- Approved naturally durable wood may be used when the wall is not more than 2 ft (610 mm) high and is located on the property line.
- Approved naturally durable wood may be used when the wall is not more than 4 ft (1219 mm) high and is separated from the property line or a permanent building by a minimum distance equal to the height of the wall.

2305.0 FIRE PROTECTION [2303.2; 2305.1]
2305.1 Fireblocking
2305.1.1. Fireblocking shall be provided to cut off all vertical and horizontal concealed draft openings. Fireblocking shall be as indicated in this section and as provided in Section 705.3.

2305.1.2. Fireblocking, when of wood, shall effectively fill all spaces for the entire width or depth of the framing or structural member.

2305.1.3. Fireblocking, when of other materials as provided in Section 705.3, shall be securely and tightly fitted into place. Spaces between chimneys and wood framing shall be solidly filled with mortar or loose noncombustible matter on noncombustible supports.

2305.1.4. Fireblocking shall be installed in all wood frame construction in the following locations:

1. In concealed spaces of stud walls and partitions including furred spaces at ceiling and floor levels.
2. At all interconnections between concealed vertical and horizontal spaces such as occur at soffits, drop ceilings, cove ceilings, etc.
3. In concealed spaces between stair stringers at the top and bottom of the run.
4. At openings around vents, pipes, ducts, chimneys and fireplaces at ceiling and floor levels with approved materials in accordance with Section 705.46, except in the case of approved metal chimney installations as set forth in SBC Section 2804.4.
5. At all interconnections between concealed vertical stud wall or partition spaces and concealed spaces created by an assembly of floor joists, fireblocking shall be provided for the full depth of the joists at the ends and over the supports.

2305.1.5. Except as provided in Section 2305.1.4.4, fireblocking shall consist of 2-in nominal lumber; two thicknesses of 1-in nominal lumber with broken lap joints; one thickness of 23/32-in (18.3 mm) wood structural panel, with joints backed by 23/32-in (18.3 mm) wood structural panel, or one thickness of nominal 3/4-in 2-M-W particleboard with joints backed by 3/4-in particleboard or other approved materials.

2305.2 Draftstopping

2305.2.1. Draftstopping shall be provided in all wood frame construction in the locations listed in Sections 2305.2.2 through 2305.2.4.

2305.2.2 Floor-Ceiling Assemblies

1. <u>Single-family dwellings</u>. In floor-ceiling assemblies separating usable spaces into two or more approximate areas with no area greater than 500 sq ft (46.5 m^2), draftstopping shall be provided parallel to the main framing members.
2. <u>Multi-family (two or more) dwellings, motels, hotels</u>. In the floor/ceiling assemblies above and in line with the tenant separation, when tenant separation walls do not extend to the floor sheathing above.
3. <u>Other buildings</u>. In floor-ceiling assemblies so that horizontal areas do not exceed 1000 sq ft (92.9 m^2).

2305.2.3 Attics

1. <u>Single-family dwellings</u>. No draftstopping required.
2. <u>Multifamily (two or more) dwellings, motels, hotels</u>. In the attic, mansard, overhang, or other concealed roof space above and in line with the tenant separation when tenant separation walls do not extend to the roof sheathing above.

   EXCEPTIONS:
   - Where corridor walls provide a tenant separation, draftstopping shall only be required above one of the corridor walls.
   - Where flat roofs with solid joist construction are used, draftstopping over tenant separation walls is not required.
   - Where approved sprinklers are provided, draftstopping shall be required for areas over 9000 sq ft (836 m).

3. Other buildings. In attic spaces so that horizontal areas do not exceed 3000 sq ft (279 m).

EXCEPTIONS:
- Where flat roofs with solid joist construction are used, draftstopping over tenant separation walls is not required.
- Where approved sprinklers are provided, the area may be tripled.

2305.2.4. Continuous exterior cornices of wood, or of wood frames, shall be draftstopped at intervals not exceeding 20 ft (6096 mm).

EXCEPTION: Draftstopping of cornices is not required in single-family dwellings.

2305.2.5. Ventilation of concealed roof spaces shall be maintained in accordance with Section 2309.7

2305.2.6. Draftstopping materials shall be not less than 1/2 in (12.7 mm) gypsum board, 15/32 in (11.9 mm) wood structural panel, 1/2 in (12.7mm) particleboard, or other approved materials adequately supported.

2305.2.7. The integrity of all draftstops shall be maintained.

2305.3 Fire-Resistance Ratings

2305.3.1. When fire-resistance ratings are specified by code, they shall be provided in conformance with the requirements of Section 701.2.

2305.3.2. Where fire-resistance ratings are involved, wood spacers, backup cleats, or other devices shall not be used unless specifically approved for such use.

2305.4 Fire Cuts

Where joists, beams, or girders enter and terminate in a masonry wall, they shall be provided with a fire cut of 3 in (76 mm) or provided with wall plate boxes of self-releasing type or approved hangers. If located in a required fire-resistant wall, the joists, beams, or girders shall be separated from the opposite side of the wall by at least 4 in (102 mm) of solid masonry.

2306.0 FASTENINGS [2311.1; 2312.0]

2306.1 Nailing and Stapling Requirements

The number and size of nails or staples connecting wood members shall not be less than those specified in SBC Table 2306.1. Where nails of a type other then those shown in the table are used, the number and spacing shall be in accordance with the manufacturer's instructions.

2306.2 Other Fastenings
Where framing anchors, clips, staples, glues, or other methods of fastening are used, they shall be labeled, listed, and installed in accordance with their listing.

2306.3 In Treated Wood
Fastenings for pressure-treated and fire-retardant-treated wood shall be of hot dipped zinc-coated galvanized steel, stainless steel, silicon bronze, or copper. Fastenings for wood foundations shall be as required in NFiPA Technical Report No. 7.

> **EXCEPTION**: Fastenings in contact with pressure-treated wood products that are not exposed to rainfall or ground moisture.

2307.0 FLOOR FRAMING [2306.0; 2312.0]
Most modern home builders agree that the quality of lumber is getting worse. Many builders say that framing lumber quality is much worse than it was 10 years ago. And no one says it is getting any better. What are the most common defects in framing lumber deliveries? Wood that is warped, bent, and bowed tops the list by far. Many builders say knots and waned edges are also big problems. They also blame splits and checks, improper kiln drying, and inconsistent dimensioning for poor construction grade building lumber. The most common builder strategy mentioned at job sites for dealing with lousy lumber delivery drops is "Send it back and demand credit for returned lumber." Here are some of the other things competitive builders suggest doing with poor framing lumber:

- Order more material than you need, pick through it, and return unusable stock to suppliers. (Some builders eat the extra time this takes; some withhold payment to cover the cost of culling through the load.)
- Require framers to set aside poor-quality joists to use as headers.
- Specify better-grade lumber to increase percentage of usable stock in each delivery.
- Get suppliers to shop around for better mills.

- Use more engineered wood products like I-beam joists, LVL headers, glue-laminated beams, and finger-jointed studs.
- Look for alternative modular construction like ICF concrete block, steel framing, and autoclaved cellular concrete.

Why is all this happening? Most builders blame the federal government for the decline in lumber quality, saying limited harvests have hurt the industry. But many builders also point the finger at the wood products companies. They suspect the wood products companies are pushing wood engineered products (therefore neglecting dimension lumber products) and have not planted enough trees to replenish the resource. The top five defects, given by <u>Builder</u> magazine, of site-build frame lumber are:
- Warped, bent, and bowed lumber
- Wood with too many knots; loose knots
- Board with splits and checks
- Lumber not properly dried
- Dimensions not consistent

The number one defense against poor-quality site-build lumber used by today's builder is to pick it out, send it back, and do not pay for it. Some builders also think the industry exports the best logs to overseas markets, saving the worst for us at home. And some say the problems stem from lumber companies that process smaller, younger trees. Those of us who have been licensed contractors for more than a decade also question how well (if at all) the industry has maintained its grading and drying standards. All these factors are leading to new developments in the national building codes for floor framing such as web-joists and engineered buttress systems.

2307.1 Sill on Foundation

Sills on continuous foundation walls shall be not less than 2 in nominal thick and shall be anchored thereto by 1/2-in (12.7 mm) bolts spaced not more than 4 ft (1219 mm) apart and which are embedded at least 6 in (152 mm) in concrete or 8 in (203 mm) in masonry units. Except where wood of natural decay resistance or pressure-treated wood is used, an approved moisture barrier shall be provided between the sill and the foundation. Piers supporting girders shall provide a true and even bearing surface.

EXCEPTION: All buildings or structures located in seismic map areas having a peak-velocity-related acceleration, A_v, equal to or less than 0.05 by Section 1607.1; buildings of Group R3 located in seismic map areas having a peak-velocity-related acceleration, A_v, less than 0.15; and agricultural storage buildings which are intended only for incidental human occupancy are permitted to have maximum intermediate bolt spacings of 6 ft (1828 mm).

2307.2 Beams and Girders

Beams and girders shall be designed in accordance with Section 2301.1 or 2301.2.5. Where two or more pieces of 2-in (51 mm) lumber are nailed together to provide girders, the wide faces shall be vertical and the end joints shall occur over supports, provided that for a girder continuous over three or more supports the end joints may be staggered in adjacent pieces of one-fourth the distance from intermediate supports. Where a girder is spliced over the support, an adequate tie shall be provided. The ends of beams or girders supported on masonry or concrete shall have not less than 3 in (76 mm) of bearing. See Figure 16-3.

2307.3 Floor Joists

2307.3.1. Maximum spans for floor joists shall be in accordance with the NFiPA Span Tables for Joists and Rafters or may be designed in accordance with Section 2301.2 or 2301.3.1.

2307.3.2. Spans for field-glued wood structural panel-lumber floor systems using approved adhesives shall be as set forth in APA Design/Construction Guide Residential & Commercial. Approved adhesives shall be those meeting the requirements of APA AFG-01.

2307.3.3. Except where supported on a 1 X 4 ribbon strip and nailed to the adjoining stud, the ends of each joist shall have not less than 1-1/2 in (38 mm) bearing on wood or metal nor less than 3 in (76 mm) bearing on masonry.

2307.3.4. Except in one- and two-family and multifamily dwellings, floor joists with a depth-to-thickness ratio exceeding 6 or a design live load in excess of 40 psf (1.92 kPa) shall be supported laterally by bridging or blocking installed at intervals not exceeding 8 ft (2438 mm).

2307.3.5. Joists shall be supported laterally at the ends by solid blocks or diagonal struts except where the ends of joists are nailed to a beam (wood or steel with an attached nailer), header, band joists, or to an adjoining stud. See Figure 16-4.

Wood and Framing 447

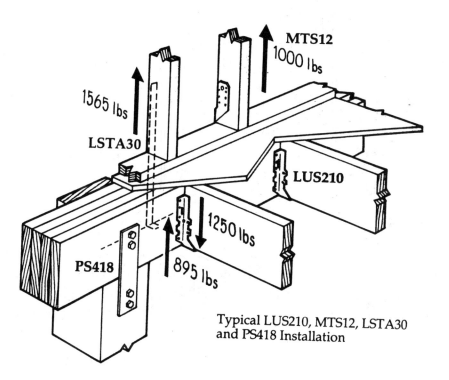

Figure 16-3
Beams and Girders Intersection

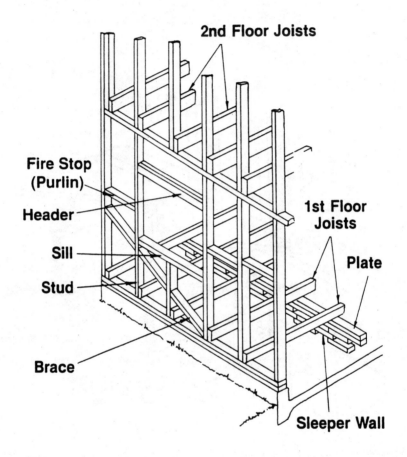

**Figure 16-4
Floor Joist System**

2307.3.6. Notches on the ends of joists shall not exceed one-fourth the depth. Holes bored for pipes or cable shall not be within 2 in (51 mm) of the top or bottom of the joist, and the diameter of any such hole shall not exceed one-third the depth of the joist. Notches for pipes in the top or bottom of joists shall not exceed one-sixth the depth and shall not be located in the middle one-third of the span.

2307.3.7. Joist framing from opposite sides of a beam, girder, or partition shall be lapped at least 3 in (76 mm) and fastened, or the opposing joists shall be tied together in an approved manner.

2307.3.8. Joists framing into the side of a wood girder shall be supported by framing anchors, on not less than 2 X 2 ledger strips or by other approved methods.

2307.4 Framing Around Openings

Trimmer and header joists shall be doubled when the span of the header exceeds 4 ft (1219 mm). The ends of header joists more than 6 ft (1829 mm) long shall be supported by framing anchors, joist hangers, or other approved methods unless bearing on a beam, partition, or wall. Tail joists over 12 ft (3658 mm) long shall be supported at the header by framing anchors or on not less than 2 X 2 ledger strips.

2307.5 Joists Supporting Partitions

Bearing partitions parallel to joists shall be supported on beams, girders, walls, or other bearing partitions. Bearing partitions perpendicular to joists shall not be offset from supporting girders, wall, or partitions more than the joist depth, unless such joists are of sufficient size to carry the additional load.

2307.6 Subfloors

2307.6.1. Except as provided in Section 2307.6.2, all floor joists shall be covered with subflooring of lumber as shown in SBC Table 2307.6A or wood structural panels applied in accordance with SBC Table 2307.6B and fastened in accordance with SBC Table 2306.1, or particleboard applied in accordance with SBC Table 2307.6C and fastened in accordance with SBC Table 2306.1.

2307.6.2. Subflooring may be omitted when joist spacing does not exceed 16 in (407 mm) and nominal 1-in tongue-and-grooved wood strip flooring is applied perpendicular to the joists.

2307.6.3. When resilient flooring is applied directly to wood structural panel subfloor, the subfloor shall be applied in accordance with SBC Table 2307.6D and fastened in accordance with Table 2306.1.

2307.7 Plank-and-Beam Framing

2307.7.1. Beams supporting plank floors shall not exceed the limitations set forth in Section 2307.2.

2307.7.2. The allowable span for 2-in planks shall be determined in accordance with Section 2301.2 or 2301.3.1.

2307.8 Floor Framing to Masonry Walls

Wood floor construction which rests on masonry walls shall be anchored thereto in accordance with Section 2110.3.

2307.9 Stair Framing

2307.9.1. Stair framing shall be supported adequately on floor framing or on walls or partitions.

2307.9.2. Except in public stairs where the number and size of stringers shall be determined by engineering analysis, two rough stringers shall be provided for each set of stairs, cut to receive finish treads and risers of uniform width and height.

2307.9.3. Unless stringers are supported on walls or partitions, the minimum effective depth at each notch shall be not less than 3-1/2 in (88.9 mm).

2308.0 VERTICAL FRAMING [2306.0; 2305.0]

2308.1 Exterior Wall Framing

2308.1.1. Studs in one- and two-story buildings shall be not less than 2 X 4 with the wide face perpendicular to the wall. In three-story buildings, studs in the first story shall be not less than 3 X 4 or 2 X 6. (See Figure 16-5.) Studs shall be spaced in accordance with SBC Table 2308.1A.

2308.1.2. Utility grade studs shall not be spaced more than 16 in (406 mm) o.c., not support more than a roof and ceiling, and not exceed 8 ft (2438 mm) in height for exterior loadbearing walls.

2308.1.3. Heights listed in Section 2308.1.1 are distances between points of horizontal lateral support placed perpendicular to the plane of the wall. Heights may be increased where justified by analysis.

2308.1.4. Where floor trusses, floor joists, or roof trusses are spaced more than 16 in (406 mm) o.c., and the bearing studs are spaced 24 in (610 mm) o.c., such joists or trusses shall bear within 5 in (127 mm) of the studs beneath.

Wood and Framing 451

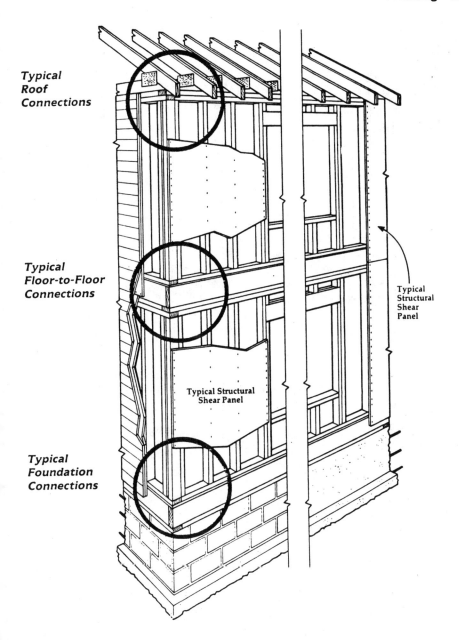

**Figure 16-5
Two-Story Vertical Framing**

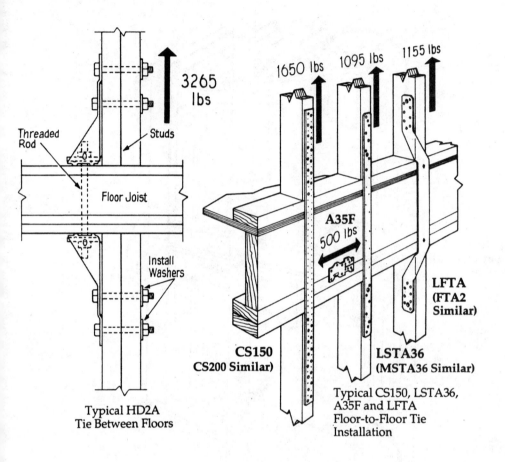

**Figure 16-6
Through-Floor Seismic Connectors**

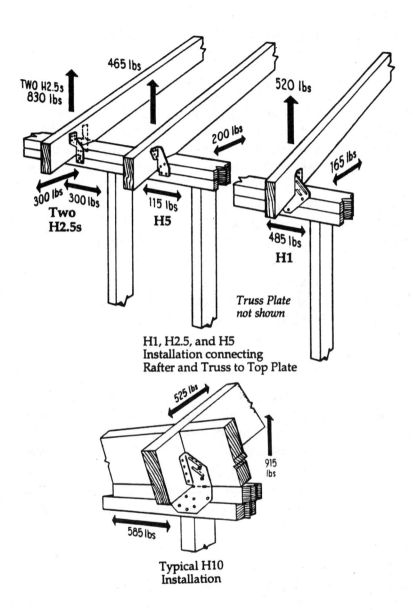

Figure 16-6a
Three-Way Truss Connectors

EXCEPTIONS:
- The top plates are two 2 X 6 or two 3 X 4 members.
- A third top plate is installed.
- Solid blocking equal in size to the studs is installed to reinforce the double top plate.

2308.1.5. Studs shall be capped with double top plates installed to provide overlapping at corners and at intersections with bearing partitions. End joints in double top plates shall be offset at least 24 in (610 mm). In lieu of double top plates, a continuous header may be used.

EXCEPTION: Buildings or structures located in seismic map areas having a peak-velocity-related acceleration, A_v less than 0.05, according to Section 1607.1; buildings of Group R3 located in seismic map areas having an A_v less than 0.15; and agricultural storage buildings which are intended only for incidental human occupancy are permitted to have a single 2 X 6 top plate installed in 2 X 6 bearing and exterior walls provided the plate is adequately tied at joints, corners, and intersecting walls by at least the equivalent of 3 X 6 X 0.036-in (76 X 152 X 0.91 mm) thick galvanized steel nailed to each wall or segment of wall by three 8d nails or equivalent, provided the rafters or joists are centered over the studs with a tolerance of no more than 1 in (25.4 mm). Top plates may be omitted over lintels which are adequately tied to adjacent wall sections as previously described or with 1 X 4 X 12-in (305 mm) wood members splice nailed to each wall section by three 8d nails or equivalent.

[Examples of these through-floor seismic connectors are shown in Figure 16-6.]

2308.1.6. Studs shall have full bearing on a plate or sill of not less than 2-in nominal thickness and having a width at least equal to the width of the studs.

2308.2 Bracing of Exterior Stud Walls
2308.2.1 Corner Bracing

2308.2.1.1. Not less than three studs shall be installed at every corner of an exterior wall, except that a third stud may be omitted through the use of a continuous wood spacer or backup cleat of 3/8-in (9.5 mm) thick wood structural panel or 2-M-W particleboard, 1-in (25 mm)-thick lumber, or other approved devices which will serve as an adequate backing for the attachment of facing materials.

EXCEPTION: Where fire-resistance or shear loads apply, wood spacers, backup cleats, or other devices shall not be used unless specifically approved.

2308.2.2 Wall Bracing

2308.2.2.1. Where structural analysis is not required, according to Sections 1607.1 and 2301.1.2, all exterior walls shall be in accordance with SBC Table 2308.2.2A. Structural sheathing shall be installed in accordance with the provisions of SBC Table 2308.2.2B when acting as wall bracing. To be considered effective as bracing, the sheathing shall be at least 48 in (1219 mm) in width covering three 16-in (406 mm) stud spaces or two 24-in (610 mm) stud spaces and be fastened to the wall studs according to SBC Table 2306.1.

All vertical joints of panel sheathing shall occur over studs, and all horizontal joints shall occur over blocking at least equal in size to the studs. All framing in connection with sheathing used for bracing shall be not less than 2-in (51 mm) nominal thickness.

2308.2.3.2 Framing

Four- by-eight-foot (1219 x 2438 mm) fiberboard sheathing shall be applied vertically to wood studs not less than 2-in (51 mm) nominal thick spaced 16 in (406 mm) o.c.

2308.2.3.3 Nailing

Nailing shall be in accordance with SBC Table 2308.2.3. Nails shall be spaced not less than 3/8 in (0.375 mm) from edges and ends of sheathing.

2308.3 Openings in Exterior Walls

2308.3.1. Headers shall be provided over each opening in exterior bearing walls. The spans in SBC Sections 2308.3A through 2308.3J may be used for one- and two-family residences. Headers for other buildings shall be designed in accordance with Section 2301.2 or 2301.3.1. Headers may be of two pieces of nominal 2-in (51 mm) framing lumber set on edge and nailed together or may be of solid lumber of equivalent size.

2308.3.2. A wall stud shall be at each side of the opening with the ends of the header supported as follows:

1. For openings 3 ft (914 mm) or less wide, each end of the header shall rest on a single header stud or may be supported by framing anchors attached to the wall stud.
2. For openings more than 3 ft (914 mm) but not more than 6 ft (1829 mm) wide, each end of the header shall rest on a single header stud.
3. For openings more than 6 ft (1829 mm) wide, each end of the header shall rest on two header studs.

2308.4 Post and Beam Framing

2308.4.1. Where post and beam framing is used in lieu of stud and joist construction, the posts shall be located to support the beams above and shall be designed in accordance with Section 2301.2 or 2301.3.1.

2308.4.2. Intermediate framing shall be attached to the posts and braced in the manner specified in Section 2306.2.

2308.5 Interior Bearing Partitions

2308.5.1. The provisions of Sections 2308.1.1, 2308.1.2, 2308.1.3 and 2308.1.4 shall apply to interior bearing partitions supporting more than a ceiling under an attic with no storage.

2308.5.2. Studs supporting a ceiling under an attic with no storage shall be installed with the wide face perpendicular to the partition and spaced not more than 24 in (610 mm) o.c.

2308.5.3. Headers shall be provided over each opening in interior bearing partitions as required in Section 2308.3.

2308.5.4. Studs shall be capped with double top plates installed to provide overlapping at corners and at intersections with exterior walls. End joints in double top plates shall be offset at least 24 in (610 mm). For platform frame construction, studs shall rest on a single bottom plate.

> **EXCEPTION**: A single top plate may be installed in accordance with Section 2308.1.5.

2308.6 Interior Nonbearing Partitions

2308.6.1. Framing for nonbearing partitions shall be of adequate size and spacing to support the finish applied thereto in accordance with the manufacturer's recommendations. In nonbearing walls and partitions, studs may be spaced not more than 28 in (711 mm) o.c. and may be set with the long dimension parallel to the wall.

2308.6.2. Openings in the nonbearing partitions may be framed with single studs and headers.

2308.7 Cutting, Notching and Bored Holes

2308.7.1. In exterior walls and bearing partitions, any wood stud may be cut or notched to a depth not exceeding 25% of its width. Cutting or notching of studs to a depth not greater than 40% of the width of the stud is permitted in nonbearing partitions supporting no loads other than the weight of the partition.

2308.7.2. A hole not greater in diameter than 40% of the stud width may be bored in any wood stud. Bored holes not greater than 60% of the width of the stud are permitted in nonbearing partitions or in any wall where each bored stud is doubled, provided not more than two successive double studs are bored.

2308.7.3. In no case shall the edge of the bored hole be nearer than 5/8 in (15.8 mm) to the edge of the stud. Bored holes shall not be located at the same section of the stud as a cut or notch.

2309.0 ROOF AND CEILING FRAMING [2307.0; 2305.0]
2309.1 Ceiling Joist and Rafter Framing

2309.1.1. Maximum spans for ceiling joists and rafters shall be in accordance with the NFiPA Span Tables for Joists and Rafters or may be designed in accordance with Section 2301.2 or 2301.3.1.

2309.1.2. Where rafters meet to form a ridge, they shall be placed directly opposite each other and nailed to a ridge board not less than 1-in (25.4 mm) thick and not less in depth than the cut end of the rafters.

2309.1.3. Ceiling joists and rafters shall be nailed to each other where possible, and the assembly shall be nailed to the top wall plate in an adequate manner to secure the roof framing to the walls.

2309.1.4. Ceiling joists shall be continuously or securely joined where they meet over interior partitions to provide a continuous tie across the building.

2309.1.5. Where ceiling joists are not parallel to rafters, subflooring or metal straps attached to the ends of the rafters shall be installed in a manner to provide a continuous tie across the building. Where ceiling joists are not provided at the top of the rafter support walls, the ridge formed by these rafters shall also be supported by a beam conforming to Section 2307.2.

2309.1.6. Valley and hip rafters shall be 2-in (51 mm) deeper than the jack rafters.

2309.1.7. Collar beams of 1 X 6 boards shall be installed in the upper third of the roof height to every third pair of rafters.

2309.1.8. Notches on the ends of joists shall not exceed one-fourth the depth. Holes bored for pipes or cable shall not be within 2 in (51 mm) of the top or bottom of the joist, and the diameter of any such hole shall not exceed one-third the depth of the joist. Notches for pipes in the top or bottom of joists shall not exceed one-sixth the depth and shall not be located in the middle one-third of the span.

2309.1.9. Where ends of rafters are not nailed to ridge boards, band joists, or similar framing members, the ends of the rafters shall be supported laterally by solid blocking or diagonal bracing.

2309.2 Trussed Rafters

2309.2.1. Trussed rafters shall be designed in accordance with accepted engineering practice. Members may be joined by nails, glue, bolts, timber connectors, or other approved framing devices.

2309.2.2. The design and manufacture of metal plate connected wood roof and floor trusses shall comply with Truss Plate Institute (TPI)-85 and addendum, and TPI PCT-80.

2309.2.3. The bracing of any metal plate connected wood trusses shall comply to their appropriate engineered design. In the absence of specific bracing requirements, trusses shall be braced in accordance with the TPI's Handling, Installing and Bracing Metal Plate Connected Wood Trusses, HIB-91.

2309.2.4. Truss members and components shall not be cut, notched, drilled, spliced, or otherwise altered in any way without written concurrence and approval of the design engineer. No additional loading of any member (e.g., HVAC equipment, water heater) shall be permitted without such additional load being incorporated in the engineering design.

2309.3 Roof Sheathing

2309.3.1. All rafters and roof joists shall be covered with one of the following sheathing materials:

- Lumber, solid sheathing of wood boards 5/8 in (15.9 mm) (net) minimum thick, or spaced sheathing of wood boards 3/4 in (19.1 mm) (net) minimum thick
- Wood structural panels applied in accordance with the provisions of SBC Table 2307.6B, SBC Table 2309.3A and nailed in accordance with SBC Table 2306.1

- Fiberboard insulating roof deck not less than 1-in (25 mm) nominal thickness
- Particleboard applied in accordance with the provisions of SBC Table 2309.3B and nailed in accordance with SBC Table 2306.1

2309.3.2. Joints in lumber sheathing shall occur over supports unless end-matched lumber or approved clips are used, in which case each piece shall bear on at least two rafters or joists.

2309.4 Plank-and-Beam Roofs
Beams shall be supported on posts, piers, or other beams and shall conform to Section 2307.2. Roof planks shall conform to Section 2307.7.

2309.5 Anchorage of Roof Framing to Masonry Walls
Wood roof construction which rests on masonry walls shall be anchored thereto in a manner equivalent to that specified in Section 2110.3.

2309.6 Access to Attic Space
Attic spaces shall be provided with an interior access opening not less than 22 X 36 in (559 X 914 mm). This access opening shall be accessible and provided with a lid or device that may be easily removed or operated. When mechanical equipment is to be installed in the attic, it shall be installed in accordance with section 304.4 of the Standard Mechanical Code [UMC; NMC]. Access is not required when the clear height of the attic space, measured at the roof peak, is less than 24 in (610 mm).

2309.7 Ventilation of Attic Space
2309.7.1. For gabled and hipped roofs, ventilation shall be provided to furnish cross ventilation of each separate attic space with weather protected vents. All vents shall be screened to protect the interior from intrusion of birds. The ratio of total net free ventilating area to the area of the ceiling shall be not less than 1:150. That ratio may be reduced to 1:300 provided:
- A vapor retarder having a permeance not exceeding one perm is installed on the warm side of the ceiling.
- At least 50% of the required ventilating area is provided by ventilators located in the upper portion of the space to be ventilated at least 3 ft (914 mm) above eave or cornice vents with the balance of the required ventilation provided by eave or cornice vents.

2309.7.2. For flat roofs, blocking and bridging shall be arranged so as not to interfere with the movement of air. Such roofs shall be ventilated along the overhanging eaves, with the net area of the opening being not less than 1:250 of the area of the ceiling below.

2309.7.3. All openings into the attic space of any habitable building shall be covered with screening, hardware cloth, or equivalent material to prevent the entry of birds, squirrels, rodents, etc. The openings therein shall not exceed 1/4 in (6.4 mm).

2310.0 WOOD STRUCTURAL PANEL DIAPHRAGMS [2306.1; 2305.0]

2310.1 General

2310.1.1. Wood structural panel diaphragms may be used to resist horizontal forces in horizontal and vertical distributing or resisting elements, provided the deflection in the plane of the diaphragm, as determined by calculations, tests, or analogies drawn therefrom, does not exceed the permissible deflection of attached distributing or resisting elements. [An example of one of these structural wall panels is shown in Figure 16-7.]

2310.1.2. Permissible deflection shall be that deflection up to which the diaphragm and any attached distributing or resisting element will maintain its structural integrity under assumed load conditions, i.e., will continue to support assumed loads without danger to occupants of the structure.

2310.1.3. Connections and anchorages capable of resisting the design forces shall be provided between the diaphragms and the resisting elements. Openings in diaphragms which materially affect their strength shall be fully detailed on the plans and shall have their edges adequately reinforced to transfer all shearing stresses.

2310.1.4. The size and shape of diaphragms shall be limited as set forth in SBC Table 2310.1. In buildings of wood construction where rotation is provided for, transverse shear resisting elements normal to the longitudinal element shall be provided at spacings not exceeding two times the width for wood structural panel diaphragms. In masonry or concrete buildings, wood structural panel diaphragms shall not be considered as transmitting lateral forces by rotation.

Wood and Framing 461

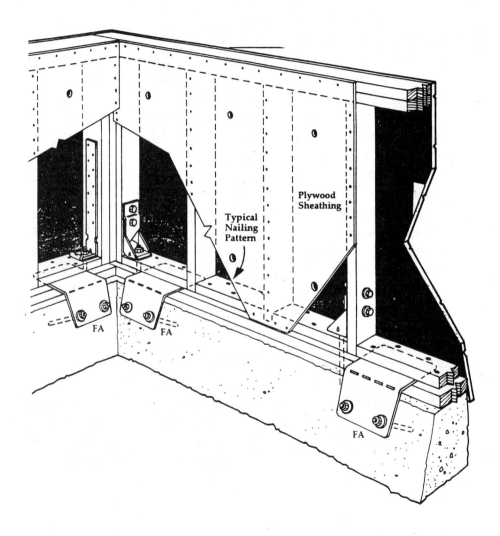

Figure 16-7
Structural Wall Panel

2310.2 Design

2310.2.1. Horizontal and vertical diaphragms sheathed with wood structural panel may be used to resist horizontal forces not exceeding those set forth in SBC Tables 2310.2A and SBC Tables 2310.2B or may be calculated by principles of mechanics without limitations by using values for nail strength and wood structural panel shear strength given elsewhere in code. Wood structural panel thickness for horizontal diaphragms shall be not less than set forth in SBC Tables 2307.6B and SBC Tables 2309.3A for corresponding joist spacing and loads, except that 1/4 in (6.4 mm) may be used where perpendicular loads permit.

2310.2.2. All boundary members shall be proportioned and spliced where necessary to transmit direct stresses. Framing members shall be at least 2 in nominal in width. In general, panel edges shall bear on the framing members and butt along their center lines. Nails shall be placed not less than 3/8 in (9.5 mm) from the panel edge, not more than 12 in (305 mm) apart along intermediate supports, and 6 in (152 mm) along panel edge bearings, and shall be firmly driven into the framing members. No unblocked panel less than 12 in (305 mm) wide shall be used.

2310.3 Wood Structural Panel Floor and Roof

2310.3.1. Nail size and spacing at diaphragm boundaries and at the edges of each sheet of wood structural panel shall be provided as shown in SBC Table 2310.2A and by the provisions of Sections 2310.3.2 through 2310.3.5. Nails of the same size shall be placed along all intermediate framing members at 12 in (305 mm) o.c.

2310.3.2. Shear capacities for fasteners in framing members of other wood species shall be calculated by multiplying the shear capacities for Structural I Panels by 0.82 for Group III species and 0.65 for Group IV species, as contained in the NFiPA NDS.

2310.3.3. The orientation of the structural framing and wood structural panels shall comply with case 1, 2, 3, 4, 5, or 6 of SBC Table 2310.2A. For blocked diaphragms, the maximum shear for cases 3, 4, 5, and 6 shall not exceed 1200 lb per linear foot (17.5 kN/m).

2310.3.4. Where either 2-in (51 mm) or 2-1/2-in (64 mm) fastener spacings are used with 2-in-wide framing members in accordance with SBC Table 2310.2A, the framing member adjoining panel edges shall be 3 inch (28.6 mm) nominal width and nails at the panel edges shall be staggered in two lines.

2310.3.5. Unblocked 1-1/8-in (28.6 mm) panels with tongue-and-grooved edges are permitted to use the blocked diaphragm shear capacities for 19/32-in (15.1 mm) minimum nominal panel thickness values where 1-in by 3/8-in crown by no.16 gauge staples are driven through the tongue-and-grooved edges 3/8 in (9.5 mm) from the panel edges so as to penetrate the tongue. Staples shall be spaced at one-half the boundary nail spacing for cases 3, 4, 5, and 6.

2310.4 Wood Structural Panel Shear Wall Construction

2310.4.1. Nail size and spacing at the edge of each panel shall be provided as shown in SBC Table 2310.2B and by the provisions of Section 2312. For 3/8-in (9.5 mm) and 7/16-in (0.438 mm) panels, installed on studs spaced 24 in (610 mm) o.c., nails shall be spaced at 6 in (152 mm) o.c. along intermediate framing members. For all other thicknesses and spacing of studs, intermediate framing members shall have nails of the same size spaced at 12 in (305 mm) o.c. All panel edges shall be backed with 2-in (51 mm) or wider framing members.

2310.4.2. In Category E buildings, the allowable shear capacity for wood structural panel shear walls, used to resist horizontal forces in buildings with masonry or reinforced concrete walls, shall be one-half of the allowable loads in SBC Table 2310.2B.

2310.4.3. Shear capacities for fasteners in framing members of other wood species shall be calculated by multiplying the shear capacities for Structural I panels by 0.82 for Group III species and 0.65 for Group IV species as contained in the NFiPA NDS. For galvanized casing nails, shear values shall be taken directly for SBC Table 2310.2B. These values shall be multiplied by 0.82 for Group III Lumber or 0.65 for Group IV lumber.

2310.4.4. Framing shall be 3 in nominal or wider, and the nails shall be staggered where nails are spaced 2 in (51 mm) o.c. or where 10d nails, having a penetration into framing of more than 1-5/8 in (41 mm), are used with a 3-in (76 mm) nail spacing.

2310.4.5. Nail spacings given in SBC Table 2310.2B are for common or galvanized box nails except that galvanized casing nails shall be permitted with plywood panel siding in grades included in PS 1.

2310.4.6. An increase in shear values shall be permitted for 3/8-in (9.5 mm) and 7/16-in (11.1 mm) panels with 8d nails to those shown for 15/32-in (11.9 mm) sheathing with the same nailing, provided the studs are spaced a maximum of 16 in (406 mm) o.c. or the panels are applied with long dimension across studs.

2311.0 PARTICLEBOARD DIAPHRAGM [2306.1; 2305.0]
2311.1 General
2311.1.1. Particleboard diaphragms may be used to resist horizontal forces in horizontal and vertical distributing or resisting elements, provided the deflection in the plane of the diaphragm, as determined by calculations, tests, or analogies drawn therefrom, does not exceed the permissible deflection of attached distributing or resisting elements.

2311.1.2. Permissible deflection shall be that deflection up to which the diaphragm and any attached distributing or resisting element will maintain its structural integrity under assumed load conditions, i.e., continue to support assumed loads without danger to occupants of the structure.

2311.1.3. Connections and anchorages capable of resisting the design forces shall be provided between the diaphragms and the resisting elements. Openings in diaphragms which materially affect their strength shall be fully detailed on the plans and shall have their edges adequately reinforced to transfer all shearing stresses.

2311.1.4. Size and shape of diaphragms shall be limited as set forth in SBC Table 2311.1. In buildings of wood construction where rotation is provided for, transverse shear resisting elements normal to the longitudinal element shall be provided at spacings not exceeding two times the width for particleboard diaphragms. In masonry or concrete buildings, particleboard diaphragms shall not be considered as transmitting lateral forces by rotation.

2311.2 Design
2311.2.1. Horizontal and vertical diaphragms sheathed with particleboard may be used to resist horizontal forces not exceeding those set forth in SBC Table 2311.2A for horizontal diaphragms and SBC Table 2311.2B for vertical diaphragms.

2311.2.2. Particleboard for horizontal diaphragms shall be as set forth in Table 2309.3B for roofs and SBC Table 2307.6C for floors.

2311.2.3. Particleboard for shear walls shall be as set forth in Table 2308.1C.

2311.2.4. Grades of particleboard and maximum spans for subfloor and underlayment shall be as set forth in SBC Table 2307.6C.

2311.2.5. All boundary members shall be proportioned and spliced where necessary to transmit direct stresses. Framing members shall be at least 2 in nominal in the dimension to which the particleboard is attached. In general, panel edges shall bear on the framing members and butt along their center lines. Nails shall be placed not less than 3/8 in (9.5 mm) from the panel edge, shall be spaced not more than 6 in (152 mm) o.c. along panel edge bearings, shall be 12 in (305 mm) apart along intermediate supports, and shall be firmly driven into the framing members. No unblocked panels less than 12 in (305 mm) wide shall be used.

2311.3 Particleboard Floor and Roof Diaphragm Construction

2311.3.1. The nail size and spacing at diaphragm boundaries and the edges of each sheet of particleboard shall be as shown in SBC Table 2311.2A and shall be designed in accordance with the provisions of this section. Nails of the same size shall be placed along all intermediate framing members at 12 in (305 mm) o.c.

2311.3.2. Shear capacities for fasteners in framing members of other wood species shall be calculated by multiplying the shear capacities by 0.82 for Group III species and 0.65 for Group IV species, contained in the NFiPA NDS.

2311.3.3. The orientation of the structural framing and particleboard panels shall comply with case 1, 2, 3, 4, 5, or 6 in SBC Table 2311.2A.

2311.3.4. When either 2-in or 2-1/2-in fastener spacings are used with 2-in-wide framing members in accordance with SBC Table 2311.2A, the framing member adjoining panel edges shall be 3 in nominal width and nails at panel edges shall be staggered in two lines.

2311.3.5. Framing at adjoining panel edges shall be 3 in nominal or wider, and nails shall be staggered where 10d nails having penetration into framing of more than 1-5/8 in (41 mm) are spaced 3 in (76 mm) or less o.c.

2311.4 Particleboard Shear Wall Construction
2311.4.1 Nailing

The required nail size and spacing in SBC Table 2311.2B apply to panel edges only. All panel edges shall be backed with 2-in nominal or wider framing. Sheets are permitted to be installed either horizontally or vertically. For 3/8-in (9.5 mm) particleboard sheets installed with the long dimension parallel to studs spaced 24 in (610 mm) o.c., nails shall be spaced at 6 in (152 mm) o.c. along intermediate framing members. For all other conditions, nails of the same size shall be spaced at 12 in (305 mm) o.c. along intermediate framing members.

2311.4.2 Other Wood Species
Shear capacities for fasteners in framing members of other wood species shall be calculated by multiplying the shear capacities by 0.82 for Group III species and 0.65 for Group IV species as contained in the NFiPA NDS.

2311.4.3 Framing
Framing shall be 3 in (76 mm) nominal or wider and the nails shall be staggered where nails are spaced 2 in (51 mm) o.c. or where 10d nails, having a penetration into framing of more than 1-5/8 in (41 mm), are used with a 3-in (76 mm) nail spacing.

2311.4.4 Shear Capacity Increase
The shear capacities for 3/8-in (9.5 mm) and 7/16-in (11.1 mm) particleboard applied directly to framing with 8d nails are permitted to be increased to the 1/2-in (12.7 mm) particleboard shear capacities of SBC Table 2311.2B when the framing studs are spaced a maximum of 16 in (406 mm) o.c. or the particleboard is applied with the long dimension perpendicular to the studs.

2311.4.5 Offset Joints
Where particleboard is applied to both faces of a wall and the nail spacing is less than 6 in (152 mm) o.c. on either side, panel joints shall be offset to be placed on different framing members or framing shall be 3 in nominal or thicker and nails on each side shall be staggered.

2312.0 SEISMIC PROVISIONS [2311.1; 2306.0]
2312.1 General
All buildings for which a seismic analysis is required, according to Section 1607.1, and which are constructed partially or wholly of wood or wood-based materials, shall be designed in accordance with the provisions of this section and SBC chapter 23.

2312.2 Definitions
The following words and terms shall apply to the provisions of this section and have the following meanings:

BLOCKED DIAPHRAGM - A diaphragm in which all sheathing edges not occurring on a framing member are supported on and connected to blocking.
DIAPHRAGM - A horizontal or nearly horizontal system designed to transmit lateral forces to the vertical elements of the seismic resisting system.
WOOD SHEAR PANEL - A wood floor, roof, or wall component sheathed to act as a shear wall or diaphragm.

2312.3 Strength of Members and Connections
The allowable load capacities of this section are to be used with allowable stress design load combinations.

2312.4 Engineered Timber Construction
When seismic analysis is required, according to Section 1607.1, the proportioning and design of wood systems, members, and connections shall be in accordance with this section, the NFiPA NDS, and APA Plywood Design Specification.

2312.4.1 Column Framing Requirements
All wood columns shall be provided with full end bearing. Columns shall be provided with adequate support to maintain stability. Where post and beam or girder construction is used, positive connections shall be provided to resist uplift and lateral displacement.

2312.4.2 Wood Shear Panels
Wood shear panels shall comply with Sections 2312.4.2.1 through 2312.4.2.5. Diaphragm construction shall comply with Section 2312.4.3. Shear wall construction shall comply with Section 2312.4.4. The construction of wood shear panels shall comply with Sections 2312.4.5 for diagonally sheathed lumber shear panels, 2312.4.6 for wood structural panel sheathed shear panels, 2312.4.7 for particleboard sheathed shear panels, or 2312.4.8 for other shear panel sheathing.

2312.4.2.1 General
All framing members used in shear panel construction shall be at least 2 in nominal thickness. Boundary members, chords, and

collector members shall be designed to transfer the axial forces. Boundary members shall be connected at all corners.

2312.4.2.2. Openings in shear panels shall be designed and detailed to transfer all shear loads.

2312.4.2.3. Positive connections and anchorages, capable of resisting the design forces, shall be provided between the shear panel and the attached components. [These design forces move in three directions resisting compression, tension, and shear forces. Examples of the positive connections that will resist in three directions required by this section are shown in Figure 16-8.]

2312.4.2.4. In Category E buildings, wood structural panels used for shear panels that are a part of the seismic-load-resisting system shall be applied directly to the framing members.

> **EXCEPTION**: Wood structural panels nailed over solid lumber decking or laminated decks.

2312.4.2.5. The diaphragms in buildings having torsional irregularity, where the lateral stiffness ratio of the structural members are greater than 4:1, or in buildings with one line of resistance in either orthogonal direction shall be sheathed with diagonal boards or wood structural panels. The length of the diaphragm normal to the soft side shall not exceed 25 ft (7.6 m) nor shall the diaphragm length-to-width ratio exceed 1 for one-story buildings or 0.67 for buildings over one story in height.

> **EXCEPTION**: Where calculations demonstrate that the diaphragm deflections can be tolerated, the length limit of 25 ft (7.6 m) does not apply and the length-to-width ratio of 1.5 is permitted for diaphragms sheathed with single diagonal boards and a ratio of 2 where sheathed with double diagonal boards or wood structural panels.

2312.4.3 Diaphragms

Wood diaphragms shall not be used to resist torsional forces induced by concrete or masonry construction in buildings of Category D or E which are more than two stories in height.

2312.4.3.1. Diaphragm sheathing shall not be used for providing ties and splices required by Sections 1607.3.6.1 and 1607.3.6.1.2.

> **EXCEPTION**: Diaphragm sheathing in Category A or B buildings, Category C buildings of Seismic Hazard Performance Groups I and II, and Category C buildings of

Wood and Framing 469

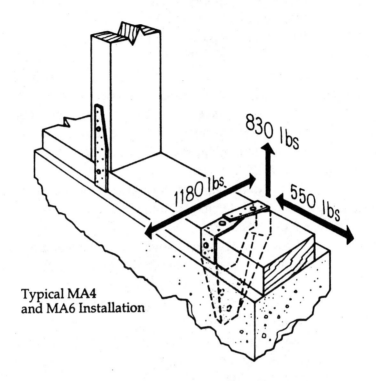

Figures 16-8
Three-Way Seismic Connectors

Seismic Hazard Performance Group III with an effective peak-velocity-related acceleration, A_v of less than 0.10.

2312.4.3.2. Unblocked diaphragms shall not be used as part of the seismic-load-resisting system in buildings of Category E.

2312.4.4. Shear wall construction shall comply with the requirements of this section.

2312.4.4.1. The design shear capacity is permitted to be doubled when identical materials are applied to both sides of the wall. Where the shear capacities of the materials are not equal, the allowable shear shall be considered to be equal to either the shear for the side with the higher capacity or twice the shear for the side with the lower capacity.

2312.4.4.2. Shear walls shall be sheathed with wood structural panels in Category B, C, D, and E buildings.

EXCEPTIONS:
- In Category B or C buildings, particleboard, gypsum sheathing, gypsum wallboard, fiberboard, and wire lath and cement plaster shear walls are permitted.
- In Category D buildings, particleboard, gypsum sheathing, gypsum wallboard, and wire lath and cement plaster shear walls in one-story buildings and in the top story of buildings two stories or more in height are permitted.

2312.4.5 Diagonally Sheathed Lumber Shear Panels

Diagonally sheathed lumber shear panels shall be nailed in accordance with SBC Table 2312.4.5.

2312.4.5.1. Single diagonally sheathed lumber shear panels shall be constructed of minimum 1-in thick nominal sheathing boards laid at an angle of approximately 45^0 (0.785 rad) to the supports. The shear capacity for single diagonally sheathed lumber shear panels of southern pine or Douglas fir-larch shall not exceed 200 lb per linear foot (2.92 kN/m) of width. The shear capacities shall be adjusted by reduction factors of 0.82 for Group III framing species and 0.65 for Group IV framing species, as contained in the NFiPA NDS.

2312.4.5.1.1. End joints in adjacent boards shall be separated by at least one stud or joist space and there shall be at least two boards between joints on the same support.

2312.4.5.1.2. Wood shear panels made up of 2-in nominal diagonal lumber sheathing fastened with 16d nails shall be designed with the same shear capacities as shear panels using 1-in boards fastened with 8d nails, provided there are no splices in adjacent boards on the same support and the supports are not less than 4 in nominal depth or 3 in nominal thickness.

2312.4.5.2. Double diagonally sheathed lumber shear panels shall be constructed of two layers of diagonal sheathing boards laid at 90^0 (1.57 rad) to each other on the same face of the supporting members. Each chord shall be considered as a beam loaded with uniform load per foot equal to 50% of the unit shear due to diaphragm action. The load shall be assumed as acting normal to the chord in the plane of the diaphragm in either direction.

The span of the chord or portion thereof shall be the distance between framing members of the diaphragm such as the joists, studs, and blocking that serve to transfer the assumed load to the sheathing. (See Figure 16-9.) The shear capacity of double diagonally sheathed diaphragms of southern pine or Douglas fir-larch shall not exceed 600 lb per linear foot (8.76 kN/m) of width. The shear capacity shall be adjusted by reduction factors of 0.82 for Group III framing species and 0.65 for group IV framing species, as contained in the NFiPA NDS.

2312.4.6 Wood Structural Panel Shear Panels
The design and shear capacity of wood structural panel shear panels shall be in accordance with SBC Tables 2310.2A for diaphragms and SBC Tables 2310.2B for shear walls or shall be calculated by using nail strengths in the NFiPA NDS and plywood design properties as given in the APA Plywood Design Specifications. Shear panels shall be constructed of wood structural panels, manufactured with exterior glue, not less than 4 X 8 ft (1.22 X 2.44 m), except at boundaries and changes in framing. Wood structural panels shall be designed to resist shear only, and chords, collector members, and boundary members shall be designed to transfer the axial forces. Boundary members shall be connected at all corners. Wood structural panels less than 12 in (305 mm) wide shall be blocked.

2312.4.7 Particleboard Shear Panels.
The design shear capacity of particleboard panels shall be in accordance with Sections 2311.3 for diaphragms and Section 2311.4 for shear walls. Shear panels shall be constructed with particleboard sheets not less than 4 X 8 ft (1.22 X 2.44 m), except

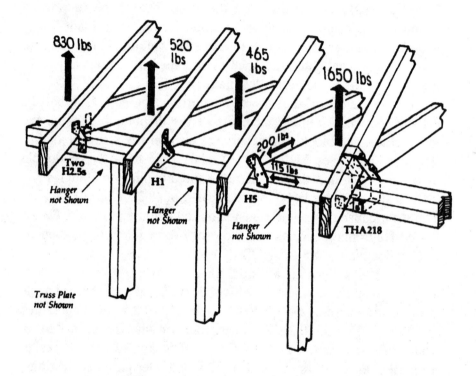

**Figures 16-9
Chord Truss to Top Plate Load Factors**

at boundaries and changes in framing. Particleboard panels shall be designed to resist shear only, and chords, collector members, and boundary members shall be connected at all corners. Particleboard panels less than 12 in (305 mm) wide shall be blocked.

2312.4.8 Shear Panels Sheathed with Other Materials. Wood stud wall sheathed with lath and plaster, gypsum sheathing board, gypsum wallboard constructed in accordance with Section 2506, or fiberboard sheathing constructed in accordance with Section 2308.2.2 shall be permitted to resist earthquake forces in wood framed buildings. Nails shall be spaced at least 3/8 in (9.5 mm) from the edges and ends of boards and panels. The maximum height-to-width ratio shall be 1.5. The shear capacities used in design shall not be cumulative with the shear capacities of other materials applied to the same wall.

Steel Framing

A 1995 survey by the National Association of Home Builders (NAHB) found that on national averages, builders are not shifting to steel framing for interior or exterior walls at the rate previously expected, but on the whole are incorporating more steel framing into standard construction. In 1993 and 1994, lumber prices were incredibly volatile and the price of framing lumber hit an all time high. The price spikes decimated builders' profit margins and many builders looked for ways to protect themselves. Many viewed steel-framed houses as a viable option.

Since 1994, the cost of lumber has declined and most builders no longer consider lumber prices a critical problem. The lumber market, however, might best be seen as a sleeping monster, ready to awaken and pounce at any time. Builders who do not plan for that eventuality could find themselves in a lot of trouble in future price escalations. In March 1993, following the first lumber price shock, NAHB surveyed builders to see if they were using wood substitutes. In March 1994, during the second price explosion, NAHB conducted a second survey. In October 1995, a third survey was done to see how much substitution actually had occurred.

In 1993, only 3% of the builders surveyed reported using steel for interior or exterior walls. Forty-five percent, however, said they were considering using steel. By 1994, a year after the first lumber price shock, 95% of builders said they used framing lumber for both interior and exterior walls. Only 1% reported using steel for the exteriors. Six percent reported using steel for interior walls. The remainder used concrete block or engineered wood for exterior wall supports. In October 1995, still only 1% of builders were using steel for exterior walls, and the proportion using steel for interior walls actually dropped to 4%. Most of those who use steel are in the south and west; use among builders building fewer than 100 units a year is negligible. The number of builders considering using steel has shrunk considerably since early 1994. In October of that year, only 16% were considering steel for exterior walls and only 18% were thinking of using steel for interior walls.

Among the alternatives to framing lumber, engineered wood products made the greatest inroads. Builders still describe these products, however, as "secondary" materials and continue to rely almost exclusively on framing lumber. Special factors (including feared and actual shrinkage in lumber supply from the federal lands in the northwest) created the extraordinary lumber price shocks of 1993 and 1994. And although the market has quieted down in later years, unanticipated shifts in supply or demand could drive the prices a long way quickly. To protect yourself against price shocks, lock in lumber pricing arrangements with dealers and shorten the length of time you guarantee prices to buyers. And do not forget about alternatives to framing lumber. Light-gauge steel may now be cost-effective for your operations; other alternatives, such as engineered lumber, will move in and out of the cost-effective range as prices fluctuate.

Modern truss and wall construction steel framing machines are mounted on raised stands, powered with an electric-hydraulic power system, and controlled by computer. The steel roll-forming machine will roll cee studs in two widths: 1-1/2 in X 3-1/2 in and 1-1/2 in X 5-1/2 in, with matching deep or shallow leg tracks. The machine will run steel coil in 16 through 24 gauge. Other machines will make hat channels, cee sections and angles. A variety of accessory tools and fastening systems are available as well. These wall and truss manufacturing machines also use a selection of computer controls for in-line punching and shearing and have the capability to input quantities and lengths. The hydraulic shear cuts to within a 1/16-in tolerance. Hole punches are automatic and allow selection of up to 10 holes at operator-selected positions on each component being formed.

The cee stud and track produced by the machine allows the manufacture of roof trusses, exterior wall framing or complete panelized wall sections, interior partitions, bearing partitions, and open floor truss style joists. The machine has a chain-and-gear drive; all forming dies are of high carbon steel and are chrome plated. The machine includes an oiler system to lubricate the forming dies, a heavy-duty welded steel frame, individual roll station bearing stands, and lifetime shielded ball bearings. Other functions allow for easy assembly of structurally engineered house framing studs, track, flat floor or pitched roof trusses, mansard frames, and assemblies for other common wall construction applications.

Cee stud and track component machines form 3-1/2 in X 1-1/2 in and 5-1/2 in X 1-1/2 in structural cee studs and matching deep or shallow leg track components. Structural steel in 16, 18, 20, 22, 24, and 26 gauge run through these machines at a rate of up to 40 ft/min. The machines are hydraulically driven with digital computers that selectively operate the in-line 1-23/64 in punches and the exit end hydraulic-powered shear. The computer-driven shear automatically cuts the formed cee stud or channel track components to desired lengths and quantities from batch inputs.

The standard machine is bench mounted, includes a sloped run-out table, steel coil stand and decoiler, oilers, computer, and two 1-23/64-in utility punches with consistent accuracy of at least 1/16 in. This machine forms three sizes: 1-1/2 in X 1-1/2 in, 1-1/2 in X 1-5/8 in and 1-1/2 in X 2-1/8 in. Four gauges of structural steel coil (18, 20, 22, and 24) are used. The machine has a computerized batch controller that operates the exit end hydraulic-powered shear for cutting formed components to desired lengths from batch inputs. Engineering load tables are provided for each stud component.

Structural steel framing is finished with the trusses fitting into saddles welded onto the top of C-channel metal studs. Two grade A bolts are then inserted through the saddle and truss. The trusses are placed at 4 ft centers. The panelized truss utilizes 1-1/2 in X 1-1/2 in tubular-square steel members. These members are welded together into wall and roof panels, trusses, and headers. The engineered components are fabricated within a precise tolerance and drilled at the factory. They are prepared with an anticorrosive wash, then electrostatically painted with a zinc chromate primer. The structural wall panels are designed to accommodate attachment of exterior materials on 16 in centers. C-channels are welded to the interior side of the panels so interior wallboard can be attached at 24 in centers.

Tubular steel bars and C-channels are manufactured from 20 gauge steel, while diagonal braces are manufactured from 18 gauge steel. The panels are manufactured 8 ft 4 in tall, 8 ft wide, and 5 in thick. A lip along the perimeter of the concrete foundation forms a recess into which the panels are erected and bolted. This type of system does not require bottom plates as the foundation key way recess inhibits water intrusion from the outside. Framing crews first attach the roof panel to the trusses. They start at one end of the structure and bolt the panels at the crest of the trusses, first one side and then the other, working their way toward the other end of the structure. The crew then bolts together two halves of a scissors truss that will provide vaulted ceiling rafters.

For homes located in a 70-mi/h or more wind exposure C-zone, special engineering must go into the design. Here the panelized construction framing system must be engineered and manufactured for a structure built to withstand 100-mi/h wind exposure D-zone criteria. For this rating the wall panels utilize 1-1/2 in X 1-1/2 in, 22 gauge tubular-square steel bars. C-channels are made from 22 gauge steel and the diagonal bars out of 20 gauge steel. C-channels allow for placement of interior wallboard at 24-in centers. The vertical tubular-steel members allow for placement of exterior shear wall sheathing at 16-in centers. A building shell constructed with this type of panelized construction can be a clear-span perimeter system, so all interior partitioning is non-load-bearing.

Steel systems allow for plenty of design flexibility. Standard trusses are now designed and manufactured to clear-span up to 48 ft and to accommodate a 4:12 roof pitch. The trusses fit into metal saddles welded in the top of the C-channel columns for placement at 4 ft centers. Two bolts attach the trusses and wall panels at each saddle. Scissors trusses for the vaulted ceilings are delivered in two parts and bolted together at the site. After the crew attaches the trusses to the wall panels, roof panels are bolted to the trusses. The framing crew starts at one end of the structure and attaches the panels at the crest of the trusses, first one side and then the other, working its way toward the other end of the structure. Then the framing crew works its way back across, attaching the panels to the lower end of the trusses.

Once assembled, panelized walls with steel framing systems absorb all live and dead loads and outside shear and seismic forces. The shell is so strong and durable that exterior walls are full load-bearing and interior partitions are non-load-bearing. The modular manufacturing plant will also supply off-the-shelf steel studs for the interior walls. The walls are typically finished with a direct-applied exterior finish system from the United States Gypsum (USG) Co. The USG system will provide the panels with a long-lasting, textured finish. The USG engineered the system to withstand high temperatures, humidity, water, wind, and repeated freeze-thaw cycles.

The steel system of structural wall panels, roof trusses and roof panels combines strength, durability, and precision. The system typically costs the same as wood and 40% less than concrete block. Fiber-reinforced cement panels used in the finishing system allow for quicker occupancy and reduced labor and labor-related costs. Fiber-reinforced cement panels are tough, flexible, will not rot or buckle, and are immune to water damage, snow, salt air, and termite attack. They are also the most fire-resistant siding money can buy. Cement board installation, joint treatment and base-coat application can be performed in 1 day. The exterior textured finish can be applied the next day to complete the system. Panelized construction of this type accommodates a variety of applications. Structural heights can reach three stories, and clear spans can reach 48 ft. Square footage can range from 400 to 8500 sq ft. The system can be employed for multifamily construction and large commercial buildings. The system can also be utilized in remodel or retrofit construction.

Chapter 17

Gypsum Board and Plaster

SECTION 2501.0 GENERAL [2500.0; 2501.1]

2501.1 Scope
2501.1.1. Provisions of this chapter shall govern the materials, design, construction, and quality of gypsum and plaster.
2501.1.2. Lathing, plastering, and gypsum construction shall be done in the manner and with the materials specified in this chapter and, when required for fire protection, shall also comply with the provisions of SBC chapter 7.
2501.2 Inspection
2501.2.1. No plaster shall be applied until the lathing has been inspected and approved by the building official.
2501.2.2. The building official may require that test holes be made in the wall for the purpose of determining the thickness and proportioning of the plaster, provided the permit holder has been notified 24 hr in advance of the time of making such tests.

2502.0 DEFINITIONS [2500.0; 2501.1]
This chapter contains no unique definitions. For general definitions, see Chapter 2.

2503.0 MATERIALS [2500.0; 2501.1]
Materials used in construction involving gypsum board and plaster shall conform to SBC Table 2503.

2504.0 APPLICATION [2500.0; 2501.1]
2504.1 Interior Lathing and Plastering
2504.1.1. Installation of interior gypsum lathing and furring shall comply with ASTM C 841.
2504.1.2. Interior gypsum plastering shall comply with ASTM C 842.

2504.1.3. Portland cement plaster shall comply with ASTM C 926.

2504.2 Exterior Lathing and Plastering

2504.2.1. Exterior use of portland cement plaster shall comply with the application requirements of ASTM C 926.

2504.2.2. Installation of exterior lathing and framing shall comply with ASTM C 1063.

2504.3 Pneumatically Placed Portland Cement Plaster

2504.3.1. Pneumatically placed portland cement plaster shall be a mixture of portland cement and aggregate conveyed by air through a pipe or flexible tube and deposited by air pressure in its final position.

2504.3.2. Rebound material may be screened and reused as aggregate in an amount not greater than 25% of the total sand in any batch.

2504.3.3. Pneumatically placed portland cement plaster shall consist of a mixture of one part cement to not more than five parts of aggregate. Plasticity agents may be used as specified elsewhere in this chapter. Except when applied to concrete or masonry, such plaster shall be applied in not less than two coats to a minimum total thickness of 7/8 in (22.2 mm).

2504.4 Application of Gypsum Wallboard

2504.4.1. Interior and exterior applications and finishing of gypsum board, other than gypsum veneer base and plaster, shall be done in accordance with Sections 701.4 and 2506.

2504.4.2. Gypsum veneer base and veneer plaster shall be applied and finished in compliance with Sections 701.4 and 2506 or ASTM C 843 and ASTM C 844.

2504.4.3. Joint treatment of gypsum wallboard shall not be applied until the wallboard application has been approved by the building official.

2504.5 Application of Steel Studs

2504.5.1. Non-load-bearing steel framing shall be installed in compliance with the provisions of ASTM C 754.

2504.5.2. Load-bearing (transverse and axial) steel studs and related accessories shall be installed in compliance with the provisions of ASTM C 1007.

2505.0 ALLOWABLE PARTITION HEIGHTS [2500.0; 2501.1]

Composite nonbearing partitions of gypsum wallboard and steel studs shall be limited in height in accordance with SBC Table 2505.

2506.0 VERTICAL GYPSUM BOARD DIAPHRAGMS [2500.0; 2501.1]

2506.1 General

2506.1.1. Gypsum wallboard, gypsum sheathing and gypsum veneer base may be used on wood studs for vertical diaphragms if applied in accordance with Section 2506. Shear resisting values shall not exceed those set forth in SBC Table 2506. When such construction is designed to resist seismic forces, panel size and arrangement provisions of Section 2312.4.8 shall be met in addition to this section.

2506.1.2. All studs, top and bottom plates, and blocking shall be nailed in accordance with Table 2506.

2506.1.3. The shear values tabulated shall not be cumulative with the shear value of other materials applied to the same wall. Cumulative allowable shear values for walls sheathed with more than one type of material shall be supported by engineering calculations or tests. The shear values may be doubled when identical materials applied as specified in Section 2506.3 are applied to both sides of the wall.

2506.2 Wall Framing

2506.2.1. Framing for vertical diaphragms shall comply with Section 2308.2 for bearing walls. Studs shall be spaced no farther apart than 16 in (406 mm) center to center. Marginal studs and plates shall be anchored to resist all design forces.

2506.2.2. The maximum allowable height-to-length ratio for the construction shall be 1-1/2:1.

2506.3 Application

2506.3.1. End joints of adjacent courses of gypsum board sheets shall not occur over the same stud.

2506.3.2. Where required in Table 2506, blocking having the same cross-sectional dimensions as the studs shall be provided at all joints that are perpendicular to the studs.

2506.3.3. The size and spacing of nails shall be as set forth in Table 2506. Nails shall be spaced not less than 3/8 inch (9.5 mm) from edges and ends of gypsum boards or sides of studs, blocking, and top and bottom plates.

2506.3.4. Gypsum sheathing 4 ft (1219 mm) wide may be applied parallel or perpendicular to studs. Pieces 2 ft (610 mm) wide shall be as set forth in Table 2506.

2506.3.5. Gypsum wallboard or veneer base may be applied parallel or perpendicular to studs. Maximum allowable shear values shall be as set forth in Table 2506.

2506.4 Masonry and Concrete Construction
Gypsum board shall not be used in vertical diaphragms to resist forces imposed by masonry or concrete construction.

Structural Wall Panels
In structural wall panelized construction, new materials such as engineered wood products and noncombustible structural fiber-reinforced cement board are code compliant are being used as shear wall sheathing and exterior siding. Structural wall panels are comprised of portland cement, mineralized cellulose fibers, and mineral additives. The cellulose fiber contained in cement board is derived from recycled materials. Noncombustible structural fiber-reinforced cement board has proven itself to be an excellent choice as an exterior wall sheathing behind many types of wall finishes. Because of its composition of silicone impregnated portland cement, it makes a highly cost effective product that will still perform to design specifications long after plywood or gypsum based sheathings have failed.

Structural fiber-reinforced cement board is code-compliant and highly impact resistant and is a perfect choice as an interior wall panel for schools, health care facilities, assembly, and high traffic areas where excessive abuse is a concern. The load-carrying capabilities of structural fiber-reinforced cement board allows its use as a floor component of a factory manufactured building or for it to be installed over light gauge metal framing at the job site. Structural fiber-reinforced cement board has proven to be a cost-effective alternative to plywood when a noncombustible, rot-proof, and vermin-proof product is required and a welcome substitute for costly metal deck and poured concrete flooring. Its high-quality ability to be easily machined and its unique clean cut make it well-suited for modern construction.

Noncombustible cement board is also used as a roof sheathing over metal roof decks, supported by light gauge metal trusses spaced 4 ft o.c. This has proven to be a working combination when a noncombustible roof assembly is required. The ability to staple-attach asphalt roof shingles to this cement board creates an aesthetically pleasing look that many architects desire when designing nursing homes and other health care facilities. The National Fire Protection Association also states that when a noncombustible roof assembly is incorporated into the overall building design, the building will probably qualify for a reduction of fire sprinkler requirements. Whether as a component of a factory manufactured building or installed over light gauge metal framing at the job site, this form of modern component product has proven itself to be very cost effective.

Portland cement structural wall panels have passed the requirements of ASTM E 136 and are rated for Zero Flame Spread and Zero Smoke Developed under ASTM E 84. They can be installed as a roof sheathing in lieu of plywood to provide a totally noncombustible roof assembly. Depending on local zoning codes, sprinkler requirements may be able to be reduced due to the noncombustible roof design. Additional savings could possibly be attained by lower insurance costs as well. Noncombustible structural fiber-reinforced cement board has been tested and approved under U.S. Testing Lab #107681, for the staple attachment of asphalt shingles. It meets Housing and Urban Development (HUD) requirements as a structural sheathing; can be used to construct floors, in some cases eliminating the need for metal deck and concrete; and is ideally suited for use as a subfloor over lightgauge metal framing. Tested as a component of a load-bearing floor-ceiling assembly, noncombustible structural fiber-reinforced cement board attains a 1-1/2-hr fire rating under ASTM E 119-88. Accordingly, it can be installed as exterior wall sheathing behind brick or exterior wall panels and as a base for exterior synthetic finishes. This product, being silicone-impregnated, will also resist surface water penetration. It is easily cut and installed using standard carpentry tools and fasteners. It can also be routered to produce reveals, and a synthetic finish can be applied to create a truly unique appearance.

Light gray in color, standard production wall panels are made with one smooth side and one rougher side. Calibrated noncombustible structural fiber-reinforced cement board sanded smooth on both sides is available by special factory order. Finishes are usually, but not necessarily, applied to the smooth side of boards, although its surface is not intended as a wear surface. Plycem board is manufactured and sold as a structural substrate sheathing material intended to be used as a component in finished roof, floor, and wall assemblies. Standard production is square-edged board. Some special edge shapes are possible on special orders. Standard production is 4 ft X 8 ft and 4 ft X 10 ft. The maximum width available is 4 ft and the maximum length is 10 ft. Standard stocked thicknesses are metric: 8 mm (5/16 in), 11 mm (7/16 in), 17 mm (5/8 in), and 22 mm (7/8 in). Additional thicknesses available on factory direct shipments are 6 mm (1/4 in), 20 mm (3/4 in), 25 mm (1 in), and 30 mm (1-3/16 in).

When transporting loose boards of structural fiber-reinforced cement board by truck, the boards must be laid flat, fully protected against edge damage, and covered with a waterproof covering. When hand-carrying single boards, they must be carried on edge with the 4-ft board dimension vertical. Boards should be stored flat on level supports not exceeding 32 in o.c. Never store on edges or upright. If stored outdoors, a protective covering must be secured over the pallets. If not in the original factory crate, do not stack more than 100 boards high. They come from the factory with a moisture content of 8% ($\pm$ 3%) and are in equilibrium when the temperature is 68°F (20°C) with relative humidity of 50 to 60%. The boards remain water vapor permeable and will adapt to the ambient humidity level. Plycem must be allowed to acclimatize for 24 to 48 hr before installation in the working environment of its intended use. Never install structural fiber-reinforced cement board while wet or damp. Although structural fiber-reinforced cement board is silicone-impregnated to repel surface water absorption, it remains vapor permeable to airborne humidity and other vapors. Extreme variances of humidity and temperature will manifest slight dimensional changes that should be considered in the design of the intended application. It is recommended to design joints that will allow for movement.

Although this stuff has the density of brick, it is easily cut and installed using standard carpentry tools. Use ordinary crosscut hand saws, jigsaws with coarse blades, or circular power saws with 18 tooth crosscut blades. Cut in a well-ventilated area, as it contains portland cement. It can easily be nailed or screwed to support framing members. A support backing must always occur behind the fastener location. Do not nail or screw collateral building materials to structural fiber-reinforced cement board without a secure backing surface behind the board (to prevent spalling of back side). It has a pH value of approximately 10 and consists primarily of portland cement, so screws and nails used to attach to supports must be treated for corrosion resistance (do not use standard phosphate-treated black "drywall screws"). Maintain maximum spacing along supports of 12 in o.c. with all fasteners following the adopted code fastener schedule. Center the line of fasteners to maintain a minimum distance from board edges of 3/4 in. Do not overdrive screw heads. Seat screw heads flush with the board surface. Use variable speed, 2500 rpm maximum, electric screwdrivers. Maintain a minimum 2-in distance from all board corners with any fastener. Do not locate fasteners at 45^0 from each other at board corners.

Alkaline-resistant latex paints are recommended as all products containing portland cement are highly alkaline; therefore, only alkali-resistant paints should be applied without a primer. Recommended types are polyacrylic, rubber latex, polyvinyl chloride, chlorinated rubber, and styrenebutadiene paints; plastic coatings of the neoprene and hypalon type; and water-based emulsion paints. The following paint types are not alkali-resistant: PVA emulsion paints, as well as oil-bound and size-bound distemper, oil paints, and alkyd paints. If these types of paints are to be used, then an alkali-resistant sealer must be applied as a primer prior to painting. If there is any question regarding the paint to be used, consult the paint manufacturer for compatibility with portland cement masonry surfaces. For successful painting of cement board, it is essential that the surface be free of loose matter, dust, oil, and grease. Wipe surfaces to be painted with a damp cloth. Remove oil and grease with a suitable solvent. Allow time for drying before painting.

Appendix
Specification Tables

Table 308.2A
Exempt Quantities of H1 Materials

Conditions	Explosives and Blasting Agents[1,5]		Liquid And Solid Oxidizers[2,3,4] (lbs) Class 4	Organic Peroxides[2,5] (lbs) UD	Unstable (Reactive) Materials (Detonatable)[2,5]					
					CLASS 3			CLASS 4		
	Solids (lbs)	Liquids (gals)			Solids & Liquids (lbs)		Gases (cu ft)	Solids & Liquids (lbs)		Gases (cu ft)
unprotected by sprinklers or cabinet	0	0	0	0	5		50	0		0
within cabinet in unsprinklered building	0	0	0	0	10		100	0		0
in sprinklered building, not in cabinet	1	0.10	1	1	10		100	1		10
in sprinklered building, within cabinet	2	0.20	2	2	20		200	2		20

UD = Unclassified Detonatable.

1 lb = 0.4536 kg
1 gal = 3.7854 L
1 cu ft = 0.02832 m³

Notes:
1. Storage of pyrotechnic special effect materials in motion picture, television, theatrical and group entertainment production when under permit amount as specified in Chapter 20 of the Standard Fire Prevention Code.
2. A conversion of 10 lbs/gal shall be used.
3. No exempt amounts are permitted in Group A, E, M, or R, or offices of Group B occupancies.
4. No exempt amounts are permitted in Group I occupancies or in classrooms or laboratories of Group B unless storage is within a hazardous material storage cabinet containing no other storage.
5. Except for laboratories in Group B occupancies, materials are not permitted in Group A, B, E, I, M and R occupancies.
6. Allowable quantities for retail display and storage are specified in Chapter 19 of the Standard Fire Prevention Code.

Table 308.2B
Exempt Quantities of H2 Materials

Conditions	Flammable and Combustible Liquids in Open Containers Or Containers Pressurized At More Than 15 Psig (Gal)[1]					Combustible Dusts Stored In Piles Or Open Containers (lbs)	Liquid and Solid Oxidizers (lbs)[3,4] Class 3	Organic Peroxides (lbs)[3,5] Class 1	Pyrophoric Materials[3,5]		Unstable Reactives[5] (Nondetonable) Class 3		Flammable Cryogenic Fluids (Gal)
	Class								Solid & Liquid (lbs)	Gases (Cu Ft)	Solid & Liquid (lbs)	Gases (Cu Ft)	
	IA	IB	IC	IA[2] IB[2] IC[2]	II	IIIA								
unprotected by sprinklers or cabinet	30	60	90	120	120	330	125	10	5	0	0	5	50	45
within cabinet in unsprinklered building	60	120	180	240	240	660	250	20	10	0	0	10	100	45
in sprinklered building, not in cabinet	60	120	180	240	240	660	250	20	10	4	50	10	100	90
in sprinklered building, within cabinet	120	240	360	480	480	1,320	500	40	20	8	100	20	200	90

1 lb = 0.4536 kg
1 gal = 3.7854 L
1 cu ft = 0.02832 m³

Notes:
1. For storage requirements see Chapter 9 of the Standard Fire Prevention Code.
2. Containing not more than the exempt amounts of Class IA, IB, IC, flammable liquids.
3. A conversion of 10 lbs/gal shall be used.
4. A maximum quantity of 200 lbs of solid or 20 gallons may be permitted in I, M, and R occupancies when necessary for maintenance purposes or operation of equipment.
5. Except for laboratories in Group B occupancies, material is not permitted in Group A, B, E, I, M and R occupancies.

Table 308.2C
Exempt Quantities of H3 Materials

Conditions	Flammable and Combustible Liquids in Closed Containers Pressurized at 15 Psig or Less (Gal)[1] Class						Flammable Solids (lbs)[4]	Liquid & Solid Oxidizers (lbs)[3] Class		Organic Peroxides[3] (lbs) Class				Unstable Materials[3]			Water Reactive[3] (lbs) Class			Flammable Or Oxidizing Cryogenic Fluids (Gals)
	IA	IB	IC	IA[2] IB[2] IC[2]	II	IIIA	IIIB		1	2	I	II	III	IV	Class 1 Solid & Liquid (lbs)	Gases (Cu Ft)	Class 2 Solid & Liquid (lbs)	Gases (Cu Ft)	1	2	3	
unprotected by sprinklers or cabinet	30	60	90	120	120	330	13,200	125	1,000	250	50	125	500		125	750	50	250	NL	50	5	45
within cabinet in unsprinklered building	60	120	180	240	240	660	26,400	250	2,000	500	100	250	1,000		250	1,500	100	500	NL	100	10	45
in sprinkered building, not in cabinet	60	120	180	240	240	660	NL	250	2,000	500	100	250	1,000		250	1,500	100	500	NL	100	10	90
in sprinkered building, within cabinet	120	240	360	480	480	1,320	NL	500	4,000	1,000	200	500	2,000		500	3,000	200	2,000	NL	200	20	90

1 lb = 0.4536 kg
1 gal = 3.7854 L
1 cu ft = 0.02832 m³

Notes:
1. For storage requirements see Chapter 9 of the Standard Fire Prevention Code.
2. Containing not more than the exempt amounts of Class IA, IB, or IC flammable liquids.
3. A conversion of 10 lbs/gal shall be used.
4. For baled combustible fibers the exempt quantities shall be 1,000 cu ft, 2,000 cu ft, 2,000 cu ft and 4,000 cu ft for the respective conditions.

Table 308.2D
Exempt Quantities of H4 Materials

Conditions	Highly Toxic Gases[1,2] (Cu Ft)	Toxic Compressed Gases[1,2,4] (Cu Ft)	Highly Toxic And Toxic Solids And Liquid[3] (lbs)		Corrosives, Irritants, Sensitizers, And Health Hazard Solids, Liquids, And Gases		
			Highly Toxic	Toxic	Solids (lbs)	Liquids (Gals)	Gases (Cu Ft)
unprotected by sprinklers or cabinets	0	650	1	500	5,000	500	650
within cabinet in unsprinklered building	20	1,300	2	1,000	10,000	1,000	1,300
in sprinklered building, not in cabinet	0	1,300	2	1,000	10,000	1,000	1,300
in sprinklered building, within cabinet	40	2,600	4	2,000	20,000	2,000	2,600

1 lb = 0.4536 kg
1 gal = 3.7854 L
1 cu ft = 0.02832 m³

Notes:
1. No exempt amounts are permitted in Group A, M, R and offices in Group B occupancies.
2. Except for cylinders not exceeding 20 cu ft stored within a gas storage cabinet or fume hood, no exempt amounts are permitted in Group E or I occupancies or in classrooms.
3. A conversion of 10 lbs/gal shall be used.
4. Compressed chlorine gas shall have an exempt amount of 810 cu ft.

Table 500
Allowable Heights and Building Areas

Lower case letters in table refer to Notes following table.
Height for types of construction is limited to the number of stories and height in feet shown.
Allowable building area (determined by definition of "Area, Building") is shown in thousands of square feet per floor.

TYPE CONSTRUCTION	I		II		III		IV 1-Hour		IV Unprot.		V 1-Hour		V Unprot.		VI 1-Hour		VI Unprot.	
Maximum Height In Feet:	NL		90'		65'		65'		55'		65'		55'		50'		40'	
OCCUPANCY	uns h	spr	uns h	spr	uns h	spr	uns h	spr	uns h	spr	uns h	spr	uns h	spr	uns h	spr	uns h	spr
A-1 ASSEMBLY LARGE (stage requiring proscenium opening protection) a,b																		
Max. No. of stories	NL	NL	NL	NL	0	0	0	0	0	0	0	0	0	0	0	0	0	0
Area: Multistory	UA	UA	UA	UA														
One Story only	UA	UA	UA	UA														
A-1 ASSEMBLY LARGE (no stage requiring proscenium opening protection) a, b																		
Max. No. of Stories	NL	NL	NL	NL	1	1	1	1	1	1	1	1	1	1	0	0	0	0
Area: Multistory	UA	UA	UA	UA														
One Story only	UA	UA	UA	UA	12.0	36.0	12.0	36.0	8.0	24.0	12.0	36.0	8.0	24.0				
A-2 ASSEMBLY SMALL (stage requiring proscenium opening protection) a,b																		
Max. No. of Stories	NL	NL	NL	NL	1	1	1	1	1	1	1	1	1	1	1	1	1	1
Area: Multistory	UA	UA	UA	UA														
One Story only	UA	UA	UA	UA	10.0	30.0	10.0	30.0	6.0	18.0	10.0	30.0	6.0	18.0	4.5	13.5	3.0	9.0
A-2 ASSEMBLY SMALL (no stage requiring proscenium opening protection) a, b																		
Max. No. of Stories	NL	NL	NL	NL	2	2	2	2	2	2	2	2	2	2	1	1	1	1
Area: Multistory	UA	UA	UA	UA	12.0	24.0	12.0	24.0	8.0	16.0	12.0	24.0	8.0	16.0				
One Story only	UA	UA	UA	UA	12.0	36.0	12.0	36.0	8.0	24.0	12.0	36.0	8.0	24.0	7.5	22.5	5.0	15.0
B BUSINESS a,b																		
Max. No. of Stories	NL	NL	NL	NL	5	5	5	5	2	2	5	5	2	5	2	2	2	2
Area: Multistory	UA	UA	UA	UA	25.5	51.0	25.5	51.0	17.0	34.0	21.0	42.0	14.0	28.0	13.5	27.0	9.0	18.0
One Story only	UA	UA	UA	UA	25.5	76.5	25.5	76.5	17.0	51.0	21.0	63.0	14.0	42.0	13.5	40.5	9.0	27.0

1 ft = 0.305 m
1 sq ft = 0.0929 m²

NL = No Limit
UA = Unlimited Area

Table 500
Allowable Heights and Building Areas (Continued)

Lower case letters in table refer to Notes following table.
Height for types of construction is limited to the number of stories and height in feet shown.
Allowable building area (determined by definition of "Area, Building") is shown in thousands of square feet per floor.

TYPE CONSTRUCTION	I		II 80'		III 65'		IV 1-Hour 65'		IV Unprot. 55'		V 1-Hour 65'		V Unprot. 55'		VI 1-Hour 50'		VI Unprot. 40'	
Maximum Height in Feet:	No Limit																	
OCCUPANCY	uns h	spr	uns h	spr	uns h	spr	uns h	spr	uns h	spr	uns h	spr	uns h	spr	uns h	spr	uns h	spr
E EDUCATIONAL a, b																		
Max. No. of stories	NL	NL	NL	NL	2	2	2	2	1	1	2	2	1	1	2	2	1	1
Area: Multistory	UA	UA	UA	UA	18.0	36.0	18.0	36.0			18.0	36.0			12.0	24.0		
One Story only	UA	UA	UA	UA	18.0	54.0	18.0	54.0	12.0	36.0	18.0	54.0	12.0	36.0	12.0	36.0	8.0	24.0
F FACTORY-INDUSTRIAL a,b,g																		
Max. No. of Stories	NL	NL	NL	NL	3	6	2	4	2	4	2	4	2	4	1	1	1	1
Area: Multistory	UA	30.0	UA	20.0	31.5	63.0	31.5	63.0	21.0	42.0	22.5	45.0	15.0	30.0	15.0	45.0	10.0	30.0
One Story only	UA	30.0	UA	20.0	31.5	94.5	31.5	94.5	21.0	63.0	22.5	67.5	15.0	45.0				
H-1 HAZARDOUS c																		
Max. No. of Stories	0	1	0	1	0	1	0	1	0	1	0	1	0	0	0	0	0	0
Area: Multistory		15.0		12.0		7.5		7.5		5.0		7.5						
One Story only																		
H-2 HAZARDOUS c																		
Max. No. of Stories	0	1	0	1	0	1	0	1	0	1	0	1	0	1	0	1	0	0
Area: Multistory		15.0		12.0		7.5		7.5		5.0		7.5		4.0		2.5		
One Story only																		
H-3 HAZARDOUS c																		
Max. No. of Stories	0	4	0	3	0	2	0	2	0	1	0	2	0	1	0	1	0	0
Area: Multistory		30.0		20.0		10.0		10.0		7.5		10.0		7.5		4.0		
One Story only		30.0		20.0		10.0		10.0				10.0						
H-4 HAZARDOUS c																		
Max. No. of Stories	0	NL	0	6	0	3	0	4	0	4	0	4	0	4	0	1	0	1
Area: Multistory		UA		UA		48.0		48.0		32.0		48.0		32.0		27.0		18.0
One Story only		UA		UA		72.0		72.0		48.0		72.0		48.0				

NL = No Limit
UA = Unlimited Area

1 ft = 0.305 m
1 sq ft = 0.0929 m²

Table 500
Allowable Heights and Building Areas (Continued)

Lower case letters in table refer to Notes following table.
Height for types of construction is limited to the number of stories and height in feet shown.
Allowable building area (determined by definition of "Area, Building") is shown in thousands of square feet per floor.

TYPE CONSTRUCTION	I		II		III		IV 1-Hour		IV Unprot.		V 1-Hour		V Unprot.		VI 1-Hour		VI Unprot.	
Maximum Height in Feet:	No Limit		80'		65'		65'		55'		65'		55'		50'		40'	
OCCUPANCY	uns h	spr j	uns h	spr j	uns h	spr j	uns h	spr j	uns h	spr j	uns h	spr j	uns h	spr j	uns h	spr j	uns h	spr j
I INSTITUTIONAL– RESTRAINED b																		
Max. No. of stories	NL	NL	NL	NL	0	2	2	3	0	2	0	3	0	2	0	3	0	2
Area: Multistory	UA	UA	UA	UA		24.0	15.0	30.0		20.0		21.0		14.0		15.0		10.0
One Story only	UA	UA	UA	UA		36.0	15.0	45.0		30.0		31.5		21.0		22.5		15.0
I INSTITUTIONAL– UNRESTRAINED b																		
Max. No. of Stories	0		0		0	2	0	3	0	1	0	1	0	0	0	1	0	0
Area: Multistory						24.0		30.0										
One Story only						36.0		45.0		30.0		31.5				22.5		
M MERCANTILE a,b																		
Max. No. of Stories	1	NL	1	NL	1	5	1	5	1	5	1	5	1	5	1	2	1	2
Area: Multistory	15.0	UA	15.0	UA	13.5	27.0	13.5	27.0	9.0	18.0	13.5	27.0	9.0	18.0	9.0	18.0	6.0	12.0
One Story only	15.0	UA	15.0	UA	13.5	40.5	13.5	40.5	9.0	27.0	13.5	40.5	9.0	27.0	9.0	27.0	6.0	18.0
R RESIDENTIAL a,b,d																		
Max. No. of Stories	NL	NL	NL	NL	3	3	5	5	2	5	5	5	2	5	3	3	2	2
Area: Multistory	UA	UA	UA	UA	18.0	36.0	18.0	36.0	12.0	24.0	18.0	36.0	12.0	24.0	10.5	21.0	7.0	14.0
One Story only	UA	UA	UA	UA	18.0	54.0	18.0	54.0	12.0	36.0	18.0	54.0	12.0	36.0	10.5	31.5	7.0	21.0
S STORAGE a,b,e,g																		
Max. No. of Stories	NL	NL	6	8	2	6	2	4	2	4	2	4	2	4	1	1	1	1
Area: Multistory	UA	UA	30.0	60.0	24.0	48.0	24.0	48.0	16.0	32.0	24.0	48.0	16.0	32.0	9.0	27.0	6.0	18.0
One Story only	UA	UA	30.0	90.0	24.0	72.0	24.0	72.0	16.0	48.0	24.0	72.0	16.0	48.0	9.0	27.0	6.0	18.0

NL = No Limit
UA = Unlimited Area

1 ft = 0.305 m
1 sq ft = 0.0929 m^2

Notes:

a. For height modifications and limitations by occupancy, see:
   1. Mezzanines 503.2.3
   2. Basements 503.2.4
   3. Assembly Basements 503.2.5
   4. Business 503.2.6
   5. Educational Basements 503.2.5
   6. Mercantile 503.2.6
   7. Residential 503.2.2, 503.2.6

b. For area modifications and limitations by occupancy see:
   1. Area increase for separation (All occupancies except H) 503.4.3, 503.4.4, 503.4.5, 503.4.6, 503.4.8
   2. Assembly 503.4.1, 503.4.8
   3. Business 503.4.2, 503.4.7
   4. Educational 503.4.1, 503.4.8
   5. Factory-Industrial 503.4.1, 503.4.8, 503.4.10
   6. Mercantile 503.3.2
   7. Storage 503.4.1, 503.4.8, 503.4.11

c. Modifications in height and area shall not be permitted in Group H occupancies.

d. See 903.7.5 and 903.7.6 for height limitations of unsprinklered R1 and R2 occupancies. Height and area increases in 503.2 are not permitted for NFiPA 13R sprinkler systems installed as an option in 903.7.6.

e. See 411.3.1 for allowable height and floor areas of Open Automobile Parking Structures.

f. Total area for unsprinklered Group M occupancies after increase permitted by 503.3 shall not exceed 15,000 sq ft.

g. Height in feet not applicable to Group S and Group F occupancies.

h. When all portions of buildings are sprinklered in accordance with the standards listed in 903.2, the height of buildings listed under this column may be increased one story. A general area increase provided for in 503.3.2 may be applied before using footnote h.

i. Automatic sprinkler protection required throughout all buildings where Use Condition 5 is used. See 409.2.3. and 1024.2.2.

j. When all portions of buildings are sprinklered in accordance with the standards listed in 903.2, the allowable heights and areas of buildings shall be as listed under this column.

Table 600
Fire Resistance Ratings
Required Fire Resistance in Hours

STRUCTURAL ELEMENT	Type I	Type II	Type III	Type IV 1-Hour Protected	Type IV Unprotected	Type V 1-Hour Protected	Type V Unprotected	Type VI 1-Hour Protected	Type VI Unprotected
PARTY AND FIRE WALLS (a)	4	4	4	4	4	4	4	4	4
INTERIOR BEARING WALLS									
Supporting columns, other bearing walls or more than one floor	(f)	3	2	1	NC	1 (h)	0 (h)	1	0
Supporting one floor only	3	2	1	1	NC	1	0	1	0
Supporting one roof only	3	2	1	1	NC	1	0	1	0
INTERIOR NONBEARING PARTITIONS	See 704.1, 704.2 and 705.2								
COLUMNS									
Supporting other columns or more than one floor	(f)	3	See 605 H(d)	1	NC	1	0	1	0
Supporting one floor only	3	2	H(d)	1	NC	1	0	1	0
Supporting one roof only	3	2	H(d)	1	NC	1	0	1	0
BEAMS, GIRDERS, TRUSSES & ARCHES									
Supporting columns or more than one floor	(f)	3	See 605 H(d)	1	NC	1	0	1	0
Supporting one floor only	3	2	H(d)	1	NC	1	0	1	0
Supporting one roof only	1 1/2(e,p)	1(e,f,p)	H(d)	1(e,p)	NC(e)	1	0	1	0
FLOORS & FLOOR/CEILING CONSTRUCTIONS	(f) 3	2	See 605 H(o)	(n) 1	(n,o) NC	(n) 1	(m,n,o) 0	1	(o) 0

(continued)

Table 600
Fire Resistance Ratings (continued)

STRUCTURAL ELEMENT	Type I	Type II	Type III	Type IV 1-Hour Protected	Type IV Unprotected	Type V 1-Hour Protected	Type V Unprotected	Type VI 1-Hour Protected	Type VI Unprotected
ROOFS & ROOF/CEILING CONSTRUCTIONS (g)	1 1/2(e,p)	1(e,f,p)	See 605 H(d)	1(e,p)	NC(e)	1	0	1	0
EXTERIOR BEARING WALLS and gable ends of roof (g, i, j)	(% indicates percent of protected and unprotected wall openings permitted. See 705.1.1 for protection requirements.)								
Horizontal separation (distance from common property line or assumed property line):									
0 ft to 3 ft (c)	4(0%)	3(0%)	3(0%)(b)	2(0%)	1(0%)	3(0%)(b)	3(0%)(b)	1(0%)	1(0%)
over 3 ft to 10 ft (c)	4(10%)	3(10%)	2(10%)(b)	1(10%)	1(10%)	2(10%)(b)	2(10%)(b)	1(20%)	0(20%)
over 10 ft to 20 ft (c)	4(20%)	3(20%)	2(20%)(b)	1(20%)	NC(20%)	2(20%)(b)	2(20%)(b)	1(40%)	0(40%)
over 20 ft to 30 ft	4(40%)	3(40%)	1(40%)	NC(40%)	NC(40%)	1(40%)	1(40%)	1(60%)	0(60%)
over 30 ft	4(NL)	3(NL)	1(NL)	NC(NL)	NC(NL)	1(NL)	1(NL)	1(NL)	0(NL)
EXTERIOR NONBEARING WALLS and gable ends of roof (g, i, j)	(% indicates percent of protected and unprotected wall openings permitted. See 705.1.1 for protection requirements.)								
Horizontal separation (distance from common property line or assumed property line):									
0 ft to 3 ft (c)	3(0%)	3(0%)	3(0%)(b)	2(0%)	1(0%)	3(0%)(b)	3(0%)(b)	1(0%)	1(0%)
over 3 ft to 10 ft (c)	2(10%)	2(10%)	2(10%)(b)	1(10%)	1(10%)	2(10%)(b)	2(10%)(b)	1(20%)	0(20%)
over 10 ft to 20 ft (c)	2(20%)	2(20%)	2(20%)(b)	1(20%)	NC(20%)	2(20%)(b)	2(20%)(b)	1(40%)	0(40%)
over 20 ft to 30 ft	1(40%)	1(40%)	1(40%)	NC(40%)	NC(40%)	1(40%)	1(40%)	0(60%)	0(60%)
over 30 ft (k)	NC (NL)	NC(NL)	NC(NL)	NC(NL)	NC(NL)	NC(NL)	NC(NL)	0(NL)	0(NL)

1 ft = 0.305 m

NC = Noncombustible
NL = No Limits
H = Heavy Timber Sizes

Notes:
a. See 704.5 for extension of party walls and fire walls.
b. See 704.5 for parapets.
c. See 705 for protection of wall openings.
d. Where horizontal separation of 20 ft or more is provided, wood columns, arches, beams, and roof deck conforming to heavy timber sizes may be used externally.
e. In buildings not over two stories approved fire retardant treated wood may be used.
f. In one story buildings, structural members of heavy timber sizes may be used as an alternate to unprotected structural roof members. Stadiums, field houses and arenas with heavy timber wood dome roofs are permitted. An approved automatic sprinkler system shall be installed in those areas where 20 ft clearance to the floor or balcony below is not provided.
g. See 1503 for penthouses and roof structures.
h. The use of combustible construction for interior bearing partitions shall be limited to the support of not more than two floors and a roof.
i. Exterior walls shall be fire tested in accordance with 601.3. The fire resistance requirements for exterior walls with 5 ft or less horizontal separation shall be based upon both interior and exterior fire exposure. The fire resistance requirements for exterior walls with more than 5 ft horizontal separation shall be based upon interior fire exposure only.
j. Where Appendix F is specifically included in the adopting ordinance, see F102.2.6 for fire resistance requirements for exterior walls of Type IV buildings in Fire District.
k. Walls or panels shall be of noncombustible material or fire retardant treated wood, except for Type VI construction.
l. For Group A - Large Assembly, Group A - Small Assembly, Group B, Group E, Group F, Group R occupancies and Automobile Parking Structures, occupancies of Type I construction, partitions, columns, trusses, girders, beams, and floors may be reduced by 1 hour if the building is equipped with an automatic sprinkler system throughout, but no component or assembly may be less than 1 hour.
m. Group A - Large Assembly (no stage requiring proscenium opening protection) and Group A - Small Assembly occupancies of Type V Unprotected construction shall have 1 hour fire resistant floors over any crawl space or basement.
n. For Group B and Group M occupancies of Type IV or Type V construction, when five or more stories in height a 2 hour fire resistant floor shall be required over the basement.
o. For unsprinklered Group E occupancies of Type III, Type IV Unprotected, Type V Unprotected or Type VI Unprotected, floors located immediately above useable space in basements shall have a fire resistant rating of not less than 1 hour.
p. In buildings of Group A, B, E, and R occupancies, fire resistance may be omitted where structural members support a roof only and are 20 ft or more clear above any floor or balcony.

Table 709.2.1.4A
Multiplying Factor for Finishes on Nonfire-Exposed Side of Wall

Type of Finish Applied to Wall	Type of Aggregate Used in Concrete or Concrete Masonry				
	Concrete: Siliceous or Carbonate Concrete Masonry: Siliceous or Calcareous Gravel	Concrete: Sand-Light-Weight Concrete Masonry: Limestone, Cinders or Unexpanded Slag	Concrete: Lightweight Concrete Masonry: Expanded Shale, Clay or Slate	Concrete Masonry: Pumice, or Expanded Slag
Portland Cement-Sand Plaster	1.00	0.75[1]	0.75[1]	0.50[1]
Gypsum-Sand Plaster or Gypsum Wallboard	1.25	1.00	1.00	1.00
Gypsum-Vermiculite or Perlite Plaster	1.75	1.50	1.25	1.25

Note:
1. For portland cement-sand plaster 5/8 inch (15.9 mm) or less in thickness and applied directly to the concrete masonry on the nonfire-exposed side of the wall, the multiplying factor shall be 1.00.

Table 709.2.1.4B
Time Assigned to Finish Materials on Fire-Exposed Side of Wall

Finish Description	Time, (min.)
Gypsum Wallboard	
3/8 in	10
1/2 in	15
5/8 in	30
2 layers of 3/8 in	25
1 layer 3/8 in, 1 layer 1/2 in	35
2 layers 1/2 in	40
Type X Gypsum Wallboard	
1/2 in	25
5/8 in	40
Portland Cement-sand plaster applied directly to concrete masonry	See Note.1
Portland cement-sand plaster on metal lath	
3/4 in	20
7/8 in	25
1 in	30
Gypsum sand plaster on 3/8 in gypsum lath	
1/2 in	35
5/8 in	40
3/4 in	50
Gypsum sand plaster on metal lath	
3/4 in	50
7/8 in	60
1 in	80

1 in = 25.4 mm

Note:
1. The actual thickness of Portland cement-sand plaster, provided it is 5/8 inch (15.9 mm) or less in thickness, may be included in determining the equivalent thickness of the masonry for use in Table 709.3.1.

Table 709.3.1
Minimum Equivalent Thickness[1] (in) of Bearing Or Nonbearing Concrete Masonry Walls[2,3,4]

Type of Aggregate	Fire Resistance Rating (hours)														
	0.50	0.75	1	1.25	1.50	1.75	2	2.25	2.50	2.75	3	3.25	3.50	3.75	4
Pumice or Expanded Slag	1.5	1.9	2.1	2.5	2.7	3.0	3.2	3.4	3.6	3.8	4.0	4.2	4.4	4.5	4.7
Expanded Shale, Clay or Slate	1.8	2.2	2.6	2.9	3.3	3.4	3.6	3.8	4.0	4.2	4.4	4.6	4.8	4.9	5.1
Limestone, Cinders, or Unexpanded Slag	1.9	2.3	2.7	3.1	3.4	3.7	4.0	4.3	4.5	4.8	5.0	5.2	5.5	5.7	5.9
Calcareous Gravel	2.0	2.4	2.8	3.2	3.6	3.9	4.2	4.5	4.8	5.0	5.3	5.5	5.8	6.0	6.2
Siliceous Gravel	2.1	2.6	3.0	3.5	3.9	4.2	4.5	4.8	5.1	5.4	5.7	6.0	6.2	6.5	6.7

1 in = 25.4 mm

Notes:
1. Equivalent thickness is the average thickness of the solid material in the unit. Determine the equivalent thickness in accordance with ASTM C 140.
2. Values between those shown in the table can be determined by direct interpolation.
3. Where combustible members are framed into the wall, the thickness of solid material between the end of each member and the opposite face of the wall, or between members set in from opposite sides, shall not be less than 93% of the thickness shown in the table.
4. Requirements of ASTM C 55, C 73 or C 90 shall apply.

Table 709.4.1A
Fire Resistance Periods of Bearing and Nonbearing Clay Brick Masonry Walls or Partitions[1,2]

Wall or Partition Assembly (minimum nominal thickness)	Members Framed into Wall or Partition	
	Combustible (minutes)	Noncombustible (minutes)
CLAY OR SHALE, SOLID		
4 in brick	—	75
6 in brick	—	153
8 in brick	120	240
12 in brick	240	—
CLAY OR SHALE, HOLLOW		
8 in brick,		
71% solid	120	180
60% solid, cells filled with loose-fill insulation	—	240
12 in brick		
64% solid	—	240
CLAY OR SHALE, ROLOK		
8 in Hollow Rolok	60	150
12 in Hollow Rolok	180	240
8 in Hollow Rolok Bak	—	240
CAVITY WALLS, CLAY OR SHALE		
8 in wall		
two 3-inch (actual) brick wythes separated by 2-inch air space; masonry joint reinforcement spaced 16 in o.c. vertically	—	180
9 in wall		
two nominal 4-inch wythes separated by 2-inch air space; 1/4-inch metal ties for each 3 sq ft of wall area	60[2]	240
CLAY OR SHALE BRICK, METAL FURRING CHANNELS		
5 in wall		
4-inch nominal brick (75% solid) backed with a hat-shaped metal furring channel 3/4-inch thick formed from 0.021-inch sheet metal attached to brick wall on 24-inch centers with approved fasteners; and 1/2-inch Type X gypsum board attached to the metal furring strips with 1-inch long Type S screws spaced 8 inches on centers	—	120
HOLLOW CLAY TILE, BRICK FACING		
8 in wall		
4-inch units (40% solid)[3] plus 4-inch solid brick	60	210
12 in wall		
8-inch units (40% solid)[3] plus 4-inch solid brick	120	240

1 inch = 25.4 mm.
Notes:
1. Units shall comply with the requirements of ASTM C 62, C 126, C 216 or C 652.
2. A 9-inch wall has a 120-minute rating if the hollow spaces near combustible members are filled with fire resistance materials for the full thickness of the wall and for at least 4 inches above and below and between the combustible members.
3. Units shall comply with the requirements of ASTM C 34.

Table 709.4.1B
Fire Resistance Periods of Bearing And Nonbearing Clay Tile Masonry Wall or Partitions[1]

Wall or Partition Assembly (minimum nominal thickness)	Members Framed into Wall or Partition	
	Combustible (minutes)	Noncombustible (minutes)
HOLLOW CLAY TILE		
8-inch unit		
2 cells in wall thickness, 40% solid	45	75
2 cells in wall thickness, 43% solid	45	90
2 cells in wall thickness, 46% solid	60	105
2 cells in wall thickness, 49% solid	75	120
3 or 4 cells in wall thickness, 40% solid	45	105
3 or 4 cells in wall thickness, 43% solid	45	120
3 or 4 cells in wall thickness, 48% solid	60	150
3 or 4 cells in wall thickness, 53% solid	75	180
12-inch unit		
3 cells in wall thickness, 40% solid	120	150
3 cells in wall thickness, 45% solid	150	180
3 cells in wall thickness, 49% solid	180	210
12-inch wall		
2 units with 3 or 4 cells in wall thickness, 40% solid	120	210
2 units with 3 or 4 cells in wall thickness, 45% solid	150	240
2 units with 3 or 4 cells in wall thickness, 53% solid	180	240
16-inch wall		
2 or 3 units with 4 or 5 cells in wall thickness, 40% solid	240	240
STRUCTURAL CLAY TILE		
4-inch unit		
1 cell in wall thickness, 40% solid[2, 3]	—	75
1 cell in wall thickness, 40% solid[3, 4]	—	75
6-inch unit		
1 cell in wall thickness, 30% solid[2, 3]	—	120
1 cell in wall thickness, 30% solid[3, 4]	—	120
2 cells in wall thickness, 45% solid[4]	—	60
HOLLOW STRUCTURAL CLAY TILE		
8-inch unit		
2 cells in wall thickness, 40% solid	45	75
2 cells in wall thickness, 49% solid	75	120
2 cells in wall thickness, 46% solid	60	105
3 or 4 cells in wall thickness, 53% solid	75	180
12-inch unit		
3 cells in wall thickness, 40% solid	120	150
3 cells in wall thickness, 45% solid	150	180
3 cells in wall thickness, 49% solid	180	210
12-inch wall		
2 units, with 3 cells in wall thickness, 40% solid	120	210
2 units with 3 or 4 cells in wall thickness, 45% solid	150	240
16-inch wall		
2 units with 4 cells in wall thickness, 43% solid	240	240
2 or 3 units with 4 or 5 cells in wall thickness, 40% solid	240	240

1 inch = 25.4 mm.

(continued)

Notes *(continued):*
1. Units shall comply with the requirements of ASTM C 34, C 56, C 212 or C 530.
2. Ratings are for dense hard-burned clay or shale.
3. Cells filled with tile, stone, slag, cinders or sand mixed with mortar.
4. Ratings are for medium-burned clay tile.

Table 709.4.1C
Fire Resistance Ratings for Bearing Steel Framed Brick Veneer Walls or Partitions

Wall or Partition Assembly	Plaster Side Exposed (hr)	Brick Faced Side Exposed (hr)
Outside facing of steel studs: 1/2-in wood fiberboard sheathing next to studs, 3/4-in air space formed with 3/4 x 1 5/8-in wood strips placed over the fiberboard and secured to the studs; metal or wire lath nailed to such strips, 3 3/4-in brick veneer held in place by filling 3/4-in air space between the brick and lath with mortar. Inside facing of studs: 3/4-in unsanded gypsum plaster on metal or wire lath attached to 5/16-in wood strips secured to edges of the studs.	1 1/2	4
Outside facing of steel studs: 1-in insulation board sheathing attached to studs, 1-in air space, and 3 3/4-in brick veneer attached to steel frame with metal ties every 5th course. Inside facing of studs: 7/8-in sanded gypsum plaster (1:2 mix) applied on metal or wire lath attached directly to the studs.	1 1/2	4
Same as above except use 7/8-in vermiculite—gypsum plaster or 1-in sanded gypsum plaster (1:2 mix) applied to metal or wire.	2	4
Outside facing of steel studs: 1/2-in gypsum sheathing board, attached to studs, and 3 3/4-in brick veneer attached to steel frame with metal ties every 5th course. Inside facing of studs: 1/2-in sanded gypsum plaster (1:2 mix) applied to 1/2-in perforated gypsum lath securely attached to studs and having strips of metal lath 3-in wide applied to all horizontal joints of gypsum lath.	2	4

1 in = 25.4 mm

Table 709.4.1D
Values of $R_n^{0.59}$
(For Use in Eq. 709.4.1.2, 709.4.1.3 or 709.4.3)

R, minutes	$R_n^{0.59}$
60	11.20
120	16.85
180	21.41
240	25.37

Table 709.4.1E
Coefficients for Plaster[1]

Thickness of plaster, inches	One Side	Two Side
1/2 (12.7 mm)	0.3	0.6
5/8 (15.9 mm)	0.37	0.75
3/4 (19.1 mm)	0.45	0.90

Note:
1. Values listed in table are for 1:3 sanded gypsum plaster.

Table 709.4.2
Minimum Equivalent Thickness[1] (In) of Bearing or Nonbearing Clay Masonry Walls[2,3]

Type of Material	Fire Resistance Rating			
	1 hr	2 hr	3 hr	4 hr
Hollow brick[4] of clay or shale, not filled	2.3	3.4	4.3	5.0
Hollow brick[4] of clay or shale, grouted or filled with perlite, vermiculite, or expanded shale aggregate	3.0	4.4	5.5	6.6

1 in = 25.4 mm

Notes:
1. Equivalent thickness is the average thickness of the solid material in the wall. It may be found by taking the total volume of a wall unit, subtracting the volume of core spaces and dividing this by the area of the exposed face of the unit.
2. Values between those shown in the table can be determined by direct interpolation.
3. Where combustible members are framed in the wall, the thickness of solid material between the end of each member and the opposite face of the wall, or between members set in from opposite sides, shall be not less than 93% of the thickness shown in the table.
4. Requirements of ASTM C 652 shall apply.

Table 709.5.1A
W/D Ratios for Steel Columns

Structural Shape	Contour Profile	Box Profile	Structural Shape	Contour Profile	Box Profile
W14x233	2.49	3.65	W10x112	1.78	2.57
x211	2.28	3.35	x100	1.61	2.33
x193	2.10	3.09	x 88	1.43	2.08
x176	1.93	2.85	x 77	1.26	1.85
x159	1.75	2.60	x 68	1.13	1.66
x145	1.61	2.39	x 60	1.00	1.48
x132	1.52	2.25	x 54	0.91	1.34
x120	1.39	2.06	x 49	0.83	1.23
x109	1.27	1.88	x 45	0.87	1.24
x 99	1.16	1.72	x 39	0.76	1.09
x 90	1.06	1.58	x 33	0.65	0.93
x 82	1.20	1.68			
x 74	1.09	1.53	W8x67	1.34	1.94
x 68	1.01	1.41	x58	1.18	1.71
x 61	0.91	1.28	x48	0.99	1.44
x 53	0.89	1.21	x40	0.83	1.23
x 48	0.81	1.10	x35	0.73	1.08
x 43	0.73	0.99	x31	0.65	0.97
			x28	0.67	0.96
W12x190	2.46	3.51	x24	0.58	0.83
x170	2.22	3.20	x21	0.57	0.77
x152	2.01	2.90	x18	0.49	0.67
x136	1.82	2.63			
x120	1.62	2.36	W6x25	0.69	1.00
x106	1.44	2.11	x20	0.56	0.82
x 96	1.32	1.93	x16	0.57	0.78
x 87	1.20	1.76	x15	0.42	0.63
x 79	1.10	1.61	x12	0.43	0.60
x 72	1.00	1.48	x 9	0.33	0.46
x 65	0.91	1.35			
x 58	0.91	1.31	W5x19	0.64	0.93
x 53	0.84	1.20	x16	0.54	0.80
x 50	0.89	1.23			
x 45	0.81	1.12	W4x13	0.54	0.79
x 40	0.72	1.00			

1 plf/in = 0.059 kg/m/mm

Table 709.5.1C
Minimum Cover (in) for Steel Columns Encased in Normal Weight Concrete[1]
(Figure 709.5.1F(c))

Structural Shape	Fire Resistance Rating (Hours)				
	1	1 1/2	2	3	4
W14x233			1	1 1/2	2
x176					2 1/2
x132		1			2 1/2
x 90	1			2	
x 61			1 1/2		3
x 48					3
x 43		1 1/2		2 1/2	
W12x152			1		2 1/2
x 96		1		2	
x 65	1				
x 50			1 1/2		3
x 40		1 1/2		2 1/2	
W10x 88	1			2	
x 49					3
x 45	1	1 1/2	1 1/2		
x 39				2 1/2	3 1/2
x 33			2		
W8x 67		1			3
x 58			1 1/2		
x 48	1			2 1/2	
x 31		1 1/2			3 1/2
x 21			2		
x 18				3	4
W6x 25		1 1/2	2		3 1/2
x 20				3	
x 16	1	2			4
x 15					
x 9	1 1/2		2 1/2	3 1/2	

1 in = 25.4 mm

Note:
1. The tabulated thicknesses are based upon the assumed properties of normal weight concrete given in Table 709.5.1B.

Table 709.5.1D
Minimum Cover (in) for Steel Columns Encased in Structural Lightweight Concrete[1]
(Figure 709.5.1F(c))

Structural Shape	Fire Resistance Rating (Hours)				
	1	1 1/2	2	3	4
W14x233	1	1	1	1	1 1/2
x193					
x 74	1	1	1	1 1/2	2
x 61					
x 43	1	1	1 1/2	2	2 1/2
W12x 65	1	1	1	1 1/2	2
x 53	1	1	1	1 1/2	2
x 40	1	1	1 1/2	2	2 1/2
W10x112	1	1	1	1 1/2	2
x 88	1	1	1	1 1/2	2
x 60	1	1	1	1 1/2	2
x 33	1	1	1 1/2	2	2 1/2
W8x 35	1	1	1 1/2	2	2 1/2
x 28	1	1	1 1/2	2	2 1/2
x 24	1	1	1 1/2	2	3
x 18	1	1 1/2	1 1/2	2 1/2	3

1 in = 25.4 mm

Note:
1. The tabulated thicknesses are based upon the assumed properties of structural lightweight concrete given in Table 709.5.1B.

Table 709.5.1E
Minimum Cover (in) for Steel Columns
In Normal Weight Precast Covers[1]
(Figure 709.5.1F(a))

Structural Shape	Fire Resistance Rating (Hours)				
	1	1 1/2	2	3	4
W14x233	1 1/2	1 1/2	1 1/2	2 1/2	3
x211					
x176					3 1/2
x145			2	3	
x109					
x 99					
x 61					4
x 43		2	2 1/2	3 1/2	4 1/2
W12x190	1 1/2	1 1/2	1 1/2	2 1/2	3 1/2
x152					
x120			2		
x 96				3	
x 87					4
x 58					
x 40		2	2 1/2	3 1/2	4 1/2
W10x112	1 1/2	1 1/2	2	3	3 1/2
x 88					
x 77					4
x 54		2	2 1/2	3 1/2	
x 33					4 1/2
W 8x 67	1 1/2	1 1/2	2	3	4
x 58					
x 48		2	2 1/2	3 1/2	
x 28					4 1/2
x 21					
x 18		2 1/2	3	4	
W 6x 25	1 1/2	2	2 1/2	3 1/2	4 1/2
x 20					
x 16			3		
x 12	2	2 1/2		4	
x 9					5

1 in = 25.4 mm

Note:
1. The tabulated thicknesses are based upon the assumed properties of normal weight concrete given in Table 709.5.1B.

Table 709.5.1F
Minimum Cover (in) for Steel Columns
In Structural Lightweight Precast Covers[1]
(Figure 709.5.1F(a))

Structural Shape	Fire Resistance Rating (Hours)				
	1	1 1/2	2	3	4
W14x233	1 1/2	1 1/2			2 1/2
x176				2	
x145			1 1/2		
x132					3
x109					
x 99				2 1/2	
x 68			2		
x 43				3	3 1/2
W12x190	1 1/2	1 1/2			2 1/2
x152				2	
x136			1 1/2		
x106					3
x 96				2 1/2	
x 87					
x 65			2		
x 40				3	3 1/2
W10x112	1 1/2	1 1/2		2	
x100			1 1/2		3
x 88					
x 77				2 1/2	
x 60			2		
x 39				3	3 1/2
x 33		2			
W8x 67	1 1/2		1 1/2	2 1/2	3
x 48		1 1/2			
x 35			2		3 1/2
x 28				3	
x 18		2	2 1/2		4
W 6x 25	1 1/2		2	3	3 1/2
x 15		2			
x 9			2 1/2	3 1/2	4

1 in = 25.4 mm

Note:
1. The tabulated thicknesses are based upon the assumed properties of structural lightweight concrete given in Table 709.5.1B.

Table 709.5.1G
Properties of Concrete Masonry Units

Density (r) lb/ft^3	Thermal Conductivity (K) Btu/hr/ft. °F
80	0.207
85	0.228
90	0.252
95	0.278
100	0.308
105	0.340
110	0.376
115	0.416
120	0.459
125	0.508
130	0.561
135	0.620
140	0.685
145	0.758
150	0.837

1 lb/ft^3 = 16.0185 kg/m^3
1 Btu/hr/ft · °F = 1.731 W/(m · K)

Table 709.5.2
Weight to Heated Perimeter Ratios (W/D)
For Typical Wide Flange Beam and Girder Shapes

Structural Shape	Contour Profile	Box Profile	Structural Shape	Contour Profile	Box Profile
W36x300	2.47	3.33	W24x162	1.85	2.57
x280	2.31	3.12	x146	1.68	2.34
x260	2.16	2.92	x131	1.52	2.12
x245	2.04	2.76	x117	1.36	1.91
x230	1.92	2.61	x104	1.22	1.71
x210	1.94	2.45	x 94	1.26	1.63
x194	1.80	2.28	x 84	1.13	1.47
x182	1.69	2.15	x 76	1.03	1.34
x170	1.59	2.01	x 68	0.92	1.21
x160	1.50	1.90	x 62	0.92	1.14
x150	1.41	1.79	x 55	0.82	1.02
x135	1.28	1.63			
			W21x147	1.83	2.60
W33x241	2.11	2.86	x132	1.66	2.35
x221	1.94	2.64	x122	1.54	2.19
x201	1.78	2.42	x111	1.41	2.01
x152	1.51	1.94	x101	1.29	1.84
x141	1.41	1.80	x 93	1.38	1.80
x130	1.31	1.67	x 83	1.24	1.62
x118	1.19	1.53	x 73	1.10	1.44
			x 68	1.03	1.35
W30x211	2.00	2.74	x 62	0.94	1.23
x191	1.82	2.50	x 57	0.93	1.17
x173	1.66	2.28	x 50	0.83	1.04
x132	1.45	1.85	x 44	0.73	0.92
x124	1.37	1.75			
x116	1.28	1.65	W18x119	1.69	2.42
x108	1.20	1.54	x106	1.52	2.18
x 99	1.10	1.42	x 97	1.39	2.01
			x 86	1.24	1.80
W27x178	1.85	2.55	x 76	1.11	1.60
x161	1.68	2.33	x 71	1.21	1.59
x146	1.53	2.12	x 65	1.11	1.47
x114	1.36	1.76	x 60	1.03	1.36
x102	1.23	1.59	x 55	0.95	1.26
x 94	1.13	1.47	x 50	0.87	1.15
x 84	1.02	1.33	x 46	0.86	1.09
			x 40	0.75	0.96
			x 35	0.66	0.85

(continued)

Table 709.5.2 (continued)
Weight to Heated Perimeter Ratios (W/D)
For Typical Wide Flange Beam and Girder Shapes

Structural Shape	Contour Profile	Box Profile	Structural Shape	Contour Profile	Box Profile
W16x100	1.56	2.25	W10x112	2.14	3.38
x 89	1.40	2.03	x100	1.93	3.07
W16x 77	1.22	1.78	x 88	1.72	2.75
x 67	1.07	1.56	x 77	1.52	2.45
x 57	1.07	1.43	x 68	1.35	2.20
x 50	0.94	1.26	x 60	1.20	1.97
x 45	0.85	1.15	x 54	1.09	1.79
x 40	0.76	1.03	x 49	0.99	1.64
x 36	0.69	0.93	x 45	1.03	1.59
x 31	0.65	0.83	x 39	0.94	1.40
x 26	0.55	0.70	x 33	0.77	1.20
			x 30	0.79	1.12
W14x132	1.83	3.00	x 26	0.69	0.98
x120	1.67	2.75	x 22	0.59	0.84
x109	1.53	2.52	x 19	0.59	0.78
x 99	1.39	2.31	x 17	0.54	0.70
x 90	1.27	2.11	x 15	0.48	0.63
x 82	1.41	2.12	x 12	0.38	0.51
x 74	1.28	1.93			
x 68	1.19	1.78	W8 x 67	1.61	2.55
x 61	1.07	1.61	x 58	1.41	2.26
x 53	1.03	1.48	x 48	1.18	1.91
x 48	0.94	1.35	x 40	1.00	1.63
x 43	0.85	1.22	x 35	0.88	1.44
x 38	0.79	1.09	x 31	0.79	1.29
x 34	0.71	0.98	x 28	0.80	1.24
x 30	0.63	0.87	x 24	0.69	1.07
x 26	0.61	0.79	x 21	0.66	0.96
x 22	0.52	0.68	x 18	0.57	0.84
			x 15	0.54	0.74
W12x 87	1.44	2.34	x 13	0.47	0.65
x 79	1.32	2.14	x 10	0.37	0.51
x 72	1.20	1.97			
x 65	1.09	1.79	W 6x 25	0.82	1.33
x 58	1.08	1.69	x 20	0.67	1.09
x 53	0.99	1.55	x 16	0.66	0.96
x 50	1.04	1.54	x 15	0.51	0.83
x 45	0.95	1.40	x 12	0.51	0.75
x 40	0.85	1.25	x 9	0.39	0.57
x 35	0.79	1.11			
x 30	0.69	0.96	W 5x 19	0.76	1.24
x 26	0.60	0.84	x 16	0.65	1.07
x 22	0.61	0.77			
x 19	0.53	0.67	W 4x 13	0.65	1.05
x 16	0.45	0.57			
x 14	0.40	0.50			

1 plf/in = 0.059 kg/m/mm

Table 709.6.2A
Time Assigned to Wallboard Membranes[1,2,3]

Description of Finish	Time, Min.
3/8-inch wood structural panel bonded with exterior glue	5
15/32-inch wood structural panel bonded with exterior glue	10
19/32-inch wood structural panel bonded with exterior glue	15
3/8-inch gypsum wallboard	10
1/2-inch gypsum wallboard	15
5/8-inch gypsum wallboard	30
1/2-inch type X gypsum wallboard	25
5/8-inch type X gypsum wallboard	40
Double 3/8-inch gypsum wallboard	25
1/2 + 3/8-inch gypsum wallboard	35
Double 1/2-inch gypsum wallboard	40

1 in = 25.4 mm

Notes:
1. These values apply only when membranes are installed on framing members which are spaced 16 inches o.c.
2. Gypsum wallboard installed over framing or furring shall be installed so that all edges are supported, except 5/8-inch Type X gypsum wallboard may be installed horizontally with the horizontal joints staggered 24 inches each side and unsupported but finished.
3. On wood framed floor/ceiling or roof/ceiling assemblies, gypsum board shall be installed with the long dimension perpendicular to framing members and shall have all joints finished.

Table 709.6.2B
Time Assigned for Contribution of Wood Frame[1,2,3]

Description	Time Assigned To Frame, Min.
Wood studs 16 inches (406 mm) o.c.	20
Wood floor and roof joists 16 inches (406 mm) o.c.	10

Notes:
1. This table does not apply to studs or joists spaced more than 16 inches (406 mm) o.c.
2. All studs shall be nominal 2x4 and all joists shall have a nominal thickness of at least 2 inches (51 mm).
3. Allowable spans for joists shall be determined in accordance with 2307.3.1 and 2309.3.1.

Table 709.6.2C
Membrane[1] on Exterior Face of Wood Stud Walls

Sheathing	Paper	Exterior Finish
5/8-in T & G lumber 5/16-in exterior glue plywood 1/2-in gypsum wallboard 5/8-in gypsum wallboard 1/2-in fiberboard	Sheathing paper	Lumber siding Wood shingles and shakes 1/4-in wood structural panels exterior type 1/4-in hardboard Metal siding Stucco on metal lath Masonry veneer
None		3/8-in exterior grade wood structural panels

1 in = 25.4 mm

Note:
1. Any combination of sheathing, paper and exterior finish listed may be used.

Table 709.6.2D
Flooring or Roofing Over Wood Framing[1]

Assembly	Structural Members	Subfloor or Roof Deck	Finish Flooring or Roofing
Floor	Wood	15/32-in wood structural panels or 11/16-in T & G softwood	Hardwood or softwood flooring on building paper Resilient flooring, parquet floor felted-synthetic-fiber floor coverings, carpeting, or ceramic tile on 3/8-in thick panel-type underlay Ceramic tile on 1 1/4-in mortar bed
Roof	Wood	15/32-in wood structural panels or 11/16 in T & G softwood	Finish roofing material with or without insulation

1 in = 25.4 mm

Note:
1. This table applies only to wood joist construction. It is not applicable to wood truss construction.

Table 1003.1
Minimum Occupant Load

Use	Area per Occupant[2, 3] (sq ft)
Assembly without fixed seats	
Concentrated (includes among others, auditoriums, churches, dance floors, lodge rooms, reviewing stands, stadiums)	7 net
Waiting Space	3 net
Unconcentrated (includes among others conference rooms, exhibit rooms, gymnasiums, lounges, skating rinks, stages, platforms)	15 net
Assembly with fixed seats	Note 1
Bowling alleys, allow 5 persons for each alley, including 15 ft of runway, and other spaces in accordance with the appropriate listing herein	7 net
Business areas	100 gross
Courtrooms other than fixed seating areas	40 net
Educational (including Educational Uses Above the 12th Grade)	
Classroom areas	20 net
Shops and other vocational areas	50 net
Industrial areas	100 gross
Institutional	
Sleeping areas	120 gross
Inpatient treatment and ancillary areas	240 gross
Outpatient area	100 gross
Resident housing areas	120 gross
Library	
Reading rooms	50 net
Stack area	100 gross
Malls	Section 413
Mercantile	
Basement and grade floor areas open to public	30 gross
Areas on other floors open to public	60 gross
Storage, stock, shipping area not open to public	300 gross
Parking garage	200 gross
Residential	200 gross
Restaurants (without fixed seats)	15 net
Restaurants (with fixed seats)	Note 1
Storage area, mechanical	300 gross

1 sq ft = 0.0929 m^2

Table 1004
Travel Distance, Dead-End Length, Exit and Means of Egress Width

Occupancy Classification	Maximum Travel Dist. To Exit (ft)		Maximum Dead End Corridor Length (ft)	Egress Width Per Person Served (in)		Minimum Corridor/ Aisle Width (in)	Minimum Clear Op'g Of Exit Doors (in)	Minimum Stair Width[10] (in)
	Unsprk.	Sprk.		Level[12]	Stairs			
Group A	200	250	20	0.2	0.37[14]	44[1,10]	32	44
Group B	200	250	20	0.2	0.37[14]	44[10]	32	44
Group E	200	250	20	0.2	0.37[14]	72[2]	32	44
Group F	200	250[7]	20	0.2	0.37[14]	44[10]	32	44
Group H	NP	100[13]	20	0.4	0.7	44[10]	32	44
Group I Restrained	Varies[11]	Varies[11]	20	0.2	0.37[14]	48	32	44
Group I Unrestrained	150	200	20	0.2	0.37[14]	44[3]	36[9]	44
Group M	200	250	20	0.2	0.37[14]	44[4,10]	32	44
Group R	200	250	20[8]	0.2	0.37[14]	45[5,10]	32	44
Group S	200[6]	250[6,7]	20	0.2	0.37[14]	44[10]	32	44

1 in = 25.4 mm
1 ft = 0.305 m

Notes:
1. See 1019.10.2.
2. For occupant loads less than 100 persons, 44 inches may be used.
3. 96 inches shall be provided in areas requiring the movement of beds.
4. See 413 for covered mall buildings.
5. 36 inches shall be permitted within dwelling units.
6. Maximum travel distance shall be increased to 300 ft if unsprinklered and 400 ft if sprinklered for Group S2 occupancies and open parking structures constructed per 411.
7. See 1004.1.4 for exceptions.
8. See 1026.1.1 for exceptions.
9. 44 inches required in areas requiring movement of beds.
10. 36 inches acceptable if stair or corridor serves occupant load of less than 50.
11. See 1024.2.6.
12. Applies to ramps, doors and corridors.
13. For HPM Facilities, as defined in 408, the maximum travel distance shall be 100 ft.
14. Use 0.3 for stairs having tread depths 11 inches or greater and riser heights between 4 inches minimum and 7 inches maximum.

Table 1004.1.4
Roof Vent Size and Spacing for Increased Travel Distance in Groups F and S

Occupancy	Hazard Classification[1]	Vent Height H[2]	Minimum Curtain Board Depth[3]	Maximum Area Formed by Curtain Boards (sq ft)	Vent Area to Floor Area Ratio	Maximum Spacing of Vent Centers	Maximum Distance From Wall or Curtain Boards	Maximum Distance Between Curtain Boards
F	—	—	0.2H (4 ft min.)	50,000	1:100	120 ft	60 ft	8H but ≤250 ft
S	I through IV	20 ft or less	6 ft	10,000	1:100	100 ft	60 ft	8H
S	I through IV	Over 20 ft to 40 ft	6 ft	8,000	1:75	100 ft	55 ft	8H but ≤250 ft
S	I through IV	20 ft or less	4 ft	3,000	1:75	100 ft	55 ft	8H
S	I through IV	Over 20 ft to 40 ft	4 ft	3,000	1:50	100 ft	50 ft	8H but ≤250 ft
S	V	20 ft or less	6 ft	6,000	1:50	100 ft	50 ft	8H
S	V	Over 20 ft to 30 ft	6 ft	6,000	1:40	90 ft	45 ft	8H
S	V	30 ft or more	4 ft	2,000	1:30	75 ft	40 ft	8H but ≤100 ft

1 ft = 0.305 m
1 sq ft = 0.0929 m^2

Table 1604.1
Minimum Uniformly Distributed Live Loads

Occupancy or Use	Live Load (psf)
Apartments (see Residential)	
Armories and drill rooms	150
Assembly halls and other places of assembly:	
Fixed seats	50
Movable seats	100
Balcony and decks (exterior) same as occupancy but not less than	60
On one and two family dwellings	40
Bowling alleys, poolrooms and similar recreational areas	75
Corridors:	
First floor	100
Other floors, same as occupancy served except as indicated	100
Dance halls and ballrooms	100
Dining rooms and restaurants	100
Dwellings (see Residential)	
Fire escapes	100
On multi- or single-family residential buildings only	40
Garages (passenger cars only)	50
For trucks and buses use AASHTO[1] lane loads	
Grandstands (see Reviewing Stands)	
Gymnasiums, main floors and balconies	100
Hospitals:	
Operating rooms, laboratories	60
Private rooms	40
Wards	40
Corridors, above first floor	80
Hotels (see Residential)	
Libraries:	
Reading rooms	60
Stack rooms (books and shelving at 65 pcf)	125
Corridors, above first floor	80
Manufacturing:	
Light	100
Heavy	150
Marquees	75
Office Buildings:	
Offices	50
Lobbies	100
Corridors, above first floor	80
File and computer rooms require heavier loads based upon anticipated occupancy	

(continued)

Table 1604.1 (continued)
Minimum Uniformly Distributed Live Loads

Occupancy or Use	Live Load (psf)
Penal institutions:	
Cell blocks	40
Corridors	100
Residential:	
Multifamily houses:	
Private apartments	40
Public rooms	100
Corridors	80
Dwellings:	
Sleeping rooms	30
Attics with storage	30
Attics without storage	10
All other rooms	40
Hotels:	
Guest rooms	40
Public rooms	100
Corridors serving public rooms	100
Corridors	80
Reviewing stands and bleachers[2]	100
Schools:	
Classrooms	40
Corridors	80
Sidewalks, vehicular driveways and yards, subject to trucking	200
Skating rinks	100
Stairs and exitways	100
Storage warehouse:	
Light	125
Heavy	250
Stores:	
Retail:	
First floor, rooms	75
Upper floors	75
Wholesale	100
Theaters:	
Aisles, corridors and lobbies	100
Orchestra floors	50
Balconies	50
Stages and platforms	125
Catwalks	40
Followspot, projection and control rooms	50
Yards and terraces, pedestrians	100

1 psf = 47.8803 Pa

Table 1606
Use Factors for Buildings and Other Structures

Nature of Occupancy	Use Factor
All buildings and structures except those listed below	1.0
Buildings and structures where the occupant load is 300 or more in any one room.	1.15
Buildings and structures designated as essential facilities, including, but not limited to: (1) Hospital and other medical facilities having surgery or emergency treatment areas (2) Fire or rescue and police stations (3) Primary communication facilities and disaster operation centers (4) Power stations and other utilities required in an emergency	1.15
Buildings and structures that represent a low hazard to human life in the event of failure, such as agricultural buildings, certain temporary facilities, and minor storage facilities	0.9

Table 1606.2A
Velocity Pressure (q)[1] (psf)

Mean Roof Height H[3] (ft)	Fastest Mile Wind Speed, (V)[2] in mph (From Figure 1606)				
	70	80	90	100	110
0-15	10.0	13.1	16.6	20.4	24.7
16	10.2	13.3	16.9	20.8	25.2
17	10.4	13.6	17.2	21.2	25.6
18	10.5	13.8	17.4	21.5	26.1
19	10.7	14.0	17.7	21.9	26.5
20	10.9	14.2	18.0	22.2	26.8
22	11.2	14.6	18.5	22.8	27.6
24	11.5	15.0	18.9	23.4	28.3
26	11.7	15.3	19.4	23.9	28.9
28	12.0	15.6	19.8	24.4	29.6
30	12.2	15.9	20.2	24.9	30.1
33	12.5	16.4	20.7	25.6	31.0
35	12.8	16.7	21.1	26.0	31.5
40	13.3	17.3	21.9	27.0	32.7
45	13.7	17.9	22.7	28.0	33.8
50	14.1	18.4	23.3	28.8	34.9
55	14.5	19.0	24.0	29.6	35.8
60	14.9	19.4	24.6	30.4	36.7

1 mph = 0.447 m/s
1 psf = 47.8803 Pa
1 ft = 0.305 m

Notes:
1. A single value for velocity pressure (q) is used for the entire building.

$$q = 0.00256 \, V^2 \left(\frac{H}{33}\right)^{2/7}$$

2. V = Fastest mile wind speed in miles per hour determined from Figure 1606.
3. H = Mean height of roof above ground or 15 ft whichever is greater. Eave height may be substituted for mean roof height if roof angle "a" is not more than 10°.

Specification Tables 523

Table 1606.2B
GC_p Coefficients For MWFRS
Providing Resistance In Transverse Direction[1]

	Roof Angle[5]	Notes	End Zone Coefficients				Interior Zone Coefficients			
			1E	2E	3E	4E	1	2	3	4
Enclosed Building	0 < a ≤ 10°	2	+.50	-1.4	-.80	-.70	+.25	-1.0	-.65	-.55
		3	+.90	-1.0	-.40	-.30	+.65	-.60	-.25	-.15
	10° < a ≤ 20°	2	+.70	-1.4	-1.0	-.95	+.40	-1.0	-.75	-.70
		3	+1.1	-1.0	-.60	-.55	+.80	-.60	-.35	-.30
	20° < a ≤ 30°	2	+.70	-1.0	-1.0	-.95	+.40	-.75	-.75	-.70
		3	+1.1	-.60	-.60	-.55	+.80	-.35	-.35	-.30
	30° < a ≤ 45°	2	+.60	+.10	-.80	-.75	+.45	+.05	-.70	-.65
		3	+1.0	+.50	-.40	-.35	+.85	+.45	-.30	-.25
		4	-.75	-1.4	-.80	-.75	-.70	-1.0	-.65	-.70
	a = 90°[5]	2	+.58	+.58	-.74	-.74	+.43	+.43	-.62	-.62
		3	+.98	+.98	-.34	-.34	+.83	+.83	-.22	-.22
		4	-.74	-1.4	-.80	-.74	-.71	-.98	-.62	-.71
Partially Enclosed	0 < a ≤ 10°	2	+.10	-1.8	-1.2	-1.1	-.15	-1.4	-1.1	-.95
		3	+1.0	-.90	-.30	-.20	+.75	-.50	-.15	-.05
	10° < a ≤ 20°	2	+.30	-1.8	-1.4	-1.4	0.0	-1.4	-1.2	-1.1
		3	+1.2	-.90	-.50	-.45	+.90	-.50	-.25	-.20
	20° < a ≤ 30°	2	+.30	-1.4	-1.4	-1.4	0.0	-1.2	-1.2	-1.1
		3	+1.2	-.50	-.50	-.45	+.90	-.25	-.25	-.20
	30° < a ≤ 45°	2	+.20	-.30	-1.2	-1.2	-.05	-.35	-1.1	-1.1
		3	+1.1	+.60	-.30	-.25	+.95	+.55	-.20	-.15
		4	-1.2	-1.8	-1.2	-1.2	-1.1	-1.4	-1.1	-1.1
	a = 90°[5]	2	+.28	+.28	-1.0	-1.0	+.03	+.03	-1.0	-1.0
		3	+1.1	+1.1	-.24	-.24	+.93	+.93	-.12	-.12
		4	-1.1	-1.8	-1.2	-1.1	-1.1	-1.4	-1.0	-1.1
Completely Open	0 < a ≤ 10°	6,11	+1.8	-.70	-.70	0	+1.8	-.70	-.70	0
		7,11	+1.8	-.30	-.80	0	+1.8	-.30	-.80	0
	10° < a ≤ 25°	6,11	+1.8	-.70	-.70	0	+1.8	-.70	-.70	0
		7,11	+1.8	+.70	-.70	0	+1.8	+.70	-.70	0
		7,11	+1.8	+.20	-.90	0	+1.8	+.20	-.90	0
	25° < a ≤ 45°	6,11	+1.8	-.70	-.70	0	+1.8	-.70	-.70	0
		7,11	+1.8	+2.0	+.30	0	+1.8	+2.0	+.30	0

See notes following Figure 1606.2.B2

Table 1606.2C
GC_p Coefficients For MWFRS
Providing Resistance In Longitudinal Direction
(All Roof Angles)[1]

Building Classification	Notes	End Zone Coefficients				Interior Zone Coefficients			
		2E	3E	5E	6E	2	3	5	6
Enclosed	8,9	-1.40	-0.80	+0.50	-0.70	-1.00	-0.65	+0.25	-0.55
	8,10	-1.00	-0.40	+0.90	-0.30	-0.60	-0.25	+0.65	-0.15
Partially Enclosed	8,9	-1.80	-1.20	+0.10	-1.10	-1.40	-1.05	-0.15	-0.95
	8,10	-0.90	-0.30	+1.00	-0.20	-0.50	-0.15	+0.75	-0.05
Open	6,11	-0.70	-0.70	+1.8	0.00	-0.70	-0.70	+1.8	0.00
	7,11	-0.30	-0.80	+1.8	0.00	-0.30	-0.80	+1.8	0.00

See notes following Figure 1606.2B2

Notes to Accompany Tables 1606.2B, 1606.2C
And Figures 1606.2B1, 1606.2B2

1. The building must be designed for all wind directions. Transverse and longitudinal directions denote directions perpendicular and parallel to ridge, respectively. For buildings having flat roofs, a ridge line normal to the wind direction shall be assumed at the mid-length dimension of the roof for the direction considered. Each corner must be considered in turn as the windward corner shown in the figures. For all roof slopes, Load Case A and Load Cases B2 and B3 are required as separate conditions to generate the wind actions, including torsion, to be resisted by the structural systems. If the roof slope is 30° or more, a third loading condition B1 is also required.
2. Load Case A with internal pressure; wind generally perpendicular to ridge.
3. Load Case A with internal suction; wind generally perpendicular to ridge.
4. Load Case B1 with internal pressure; wind generally parallel to ridge.
5. For roof angles, 45° < a < 90°, the coefficient GC_p may be obtained by linear interpolation.
6. Uplift equal on both roof surfaces.
7. Unbalanced loading on roof surfaces.
8. Load Case B2 with internal pressure; wind generally parallel to ridge.
9. Load Case B3 with internal pressure; wind generally parallel to ridge.
10. Load Case B3 with internal suction; wind generally parallel to ridge.
11. Coefficient to be applied to the windward side of the effective solid area of every vertical surface exposed to the wind.
12. For buildings whose widths are greater than 5 times their eave heights, coefficient 2, when negative, shall be applied to the windward slope of the roof starting at the windward eave for a distance of 2 1/2 times the eave height. Coefficient 3 shall be applied to the remainder of the roof.
13. Notations:
    a: Roof angle from horizontal, in degrees.
    B: Width of building, in feet.
    H: Reference height for assessing design pressures given by mean height of roof or 15 feet, whichever is greater. Eave height may be substituted for mean height when angle of the roof "a" is less than or equal to 10°.
    N: Number of transverse frames.
    S: Bay spacing, in feet.
    X: Zone width defined as 2Z, where Z is edge strip distance defined below. All areas not within end zone are considered interior zones.
    Z: Edge strip defined as the lesser of 10% of least horizontal dimension of building or 40% of height H but not less than 4% of least horizontal dimension of building and at least 3 ft.
    Ø: Ratio of solid area of frame to gross area outlined by frame on vertical plane.

1 degree = 0.01745 rad
1 ft = 0.305 m

Table 1606.2D
GC_p Coefficients for Roof Overhangs

Load Case	Windward Roof Overhang		Leeward Roof Overhang	
	Zone	GC_p	Zone	GC_p
A	2 & 2E	+0.2, -1.5	3 & 3E	Coefficients given in Tables 1606.2B and 1606.2C
B1, B2	2E & 3E	+0.2, -1.5	2 & 3	
B3	2 & 2E	+0.2, -1.5	3 & 3E	

Table 1607.3.3[1]
Structural Systems

BASIC STRUCTURAL SYSTEM	Response modification factor R	Deflection amplification factor C_d	Structural system limitations and building height (ft) limitations[2]			
			Seismic Performance Category			
Seismic resisting system			A & B	C	D[4]	E[5]
BEARING WALL SYSTEM						
Light framed walls w/shear panels	6 1/2	4	NL	NL	160	100
Reinforced concrete shear walls	4 1/2	4	NL	NL	160	100
Reinforced masonry shear walls	3 1/2	3	NL	NL	160	100
Concentrically braced frames	4	3 1/2	NL	NL	160	100
Unreinforced masonry shear walls	1 1/4	1 1/4	NL	Note 3	NP	NP
Plain Concrete Shear Walls	1 1/2	1 1/2	NL	Note 3	NP	NP
BUILDING FRAME SYSTEM						
Eccentrically braced frames, moment resisting connections at columns away from link beam	8	4	NL	NL	160	100
Eccentrically braced frames, non-moment resisting connections at columns away from link beam	7	4	NL	NL	160	100
Light framed walls with shear panels	7	4 1/4	NL	NL	160	100
Concentrically braced frames	5	4 1/2	NL	NL	160	100
Reinforced concrete shear walls	5 1/2	5	NL	NL	160	100
Reinforced masonry shear walls	4 1/2	4	NL	NL	160	100
Unreinforced masonry shear walls	1 1/2	1 1/2	NL	Note 3	NP	NP
Plain Concrete Shear Walls	2	2	NL	Note 3	NP	NP
MOMENT RESISTING FRAME SYSTEM						
Special moment frames of steel	8	5 1/2	NL	NL	NL	NL
Special moment frames of reinforced concrete	8	5 1/2	NL	NL	NL	NL
Intermediate moment frames of reinforced concrete	5	4 1/2	NL	NL	NP	NP
Ordinary moment frames of steel	4 1/2	4	NL	NL	160	100
Ordinary moment frames of reinforced concrete	3	2 1/2	NL	NP	NP	NP

(continued)

Table 1607.3.4.2
Vertical Structural Irregularities

Irregularity Type and Description	Reference Section	Seismic Performance Category Application
1. Stiffness Irregularity–Soft Story A soft story is one in which the lateral stiffness is less than 70% of that in the story above or less than 80% of the average stiffness of the three stories above.	1607.3.5.3	D and E
2. Weight (Mass) Irregularity Mass irregularity shall be considered to exist where the effective mass of any story is more than 150% of the effective mass of an adjacent story. A roof that is lighter than the floor below need not be considered.	1607.3.5.3	D and E
3. Vertical Geometric Irregularity Vertical geometric irregularity shall be considered to exist where the horizontal dimension of the lateral force-resisting system in any story is more than 130% of that in an adjacent story.	1607.3.5.3	D and E
4. In-Plane Discontinuity in Vertical Lateral Force-Resisting Elements An in-plane offset of the lateral force-resisting elements greater than the length of those elements.	1607.3.6.4.2	D and E
5. Discontinuity in Capacity–Weak Story A weak story is one in which the story lateral strength is less than 80% of that in the story above. The story strength is the total strength of all seismic resisting elements sharing the story shear for the direction under consideration.	1607.3.6.2.4	B, C, D and E

Table 1607.1.6
Seismic Hazard Exposure Group

Group Type	Nature of Occupancy
Group I	All buildings except those listed below
Group II Seismic Hazard Exposure Group II buildings are those which have a substantial public hazard due to occupancy or use, including buildings containing any one or more of the indicated uses.	1. Group A in which more than 300 people congregate in one room. 2. Group E with an occupant load greater than 250. 3. Group B used for college or adult education with an occupant load greater than 500. 4. Group I Unrestrained with an occupant load greater than 50, not having surgery or emergency treatment facilities. 5. Group I Restrained. 6. Power generating stations and other public utility facilities not included in Group III Seismic Hazard Exposure Group. 7. Any other occupancy with an occupant load greater than 5,000.
Group III Seismic Hazard Exposure Group III buildings are those having essential facilities which are required for postearthquake recovery, including buildings containing any one or more of the indicated uses.	1. Fire or rescue and police stations. 2. Group I Unrestrained having surgery or emergency treatment facilities. 3. Earthquake emergency preparedness centers. 4. Postearthquake recovery vehicle garages. 5. Power generating stations and other utilities required as emergency backup facilities. 6. Primary communication facilities. 7. Highly toxic materials as defined by 308.2.1 as an H4 occupancy where the quantity of the material exceeds the exempt amounts of Table 308.2D.

Table 1607.1.8
Seismic Performance Categories

Effective Peak Velocity-Related Acceleration, A_v	Seismic Hazard Exposure Group		
	I	II	III
$A_v < 0.05$	A	A	A
$0.05 \leq A_v < 0.10$	B	B	C
$0.10 \leq A_v < 0.15$	C	C	C
$0.15 \leq A_v < 0.20$	C	D	D
$0.20 \leq A_v$	D	D	E

Table 1607.3.3[1] (continued)
Structural Systems

BASIC STRUCTURAL SYSTEM	Response modification factor R	Deflection amplification factor C_d	Structural system limitations and building height (ft) limitations[2]			
Seismic resisting system			Seismic Performance Category			
			A & B	C	D[4]	E[5]
DUAL SYSTEM WITH A SPECIAL MOMENT FRAME CAPABLE OF RESISTING AT LEAST 25% OF THE PRESCRIBED SEISMIC FORCES						
Eccentrically braced frames, moment resisting connections at columns away from link beam	8	4	NL	NL	NL	NL
Eccentrically braced frames non-moment resisting connections at columns away from link beam	7	4	NL	NL	NL	NL
Concentrically braced frames	6	5	NL	NL	NL	NL
Reinforced concrete shear walls	8	6 1/2	NL	NL	NL	NL
Reinforced masonry shear walls	6 1/2	5 1/2	NL	NL	NL	NL
Wood sheathed shear walls	8	5	NL	NL	NL	NL
DUAL SYSTEM WITH AN INTERMEDIATE MOMENT FRAME OF REINFORCED CONCRETE OR AN ORDINARY MOMENT FRAME OF STEEL CAPABLE OF RESISTING AT LEAST 25% OF THE PRESCRIBED SEISMIC FORCES						
Concentrically braced frames	5	4 1/2	NL	NL	160	100
Reinforced concrete shear walls	6	5	NL	NL	160	100
Reinforced masonry shear walls	5	4 1/2	NL	NL	160	100
Wood sheathed shear walls	7	4 1/2	NL	NL	160	100
INVERTED PENDULUM STRUCTURES						
Special moment frames of structural steel	2 1/2	2 1/2	NL	NL	NL	NL
Special moment frames of reinforced concrete	2 1/2	2 1/2	NL	NL	NL	NL
Ordinary moment frames of structural steel	1 1/4	1 1/4	NL	NL	NP	NP

1 ft = 0.305 m

NL = not limited
NP = not permitted

Notes:
1. Response modification Factor R for use in 1607.4 and 1607.5. Deflection amplification factor C_d for use in 1607.4 and 1607.5.
2. The building height is not to exceed the general height limitation of Table 500 based on the type of construction.
3. The masonry shear walls shall have nominal reinforcement as required by Section A.3 of ACI 530/ASCE 5. See 2115.3.
4. See 1607.3.3.4.1 for description of building systems which are limited to buildings with a height of 240 ft or less.
5. See 1607.3.3.5 for description of building systems which are limited to buildings with a height of 160 ft or less.

Table 1607.3.4.1
Plan Structural Irregularities

Irregularity Type and Description	Reference Section	Seismic Performance Category Application
1. Torsional Irregularity–to be considered when diaphragms are rigid in relation to the vertical structural elements which resist the lateral seismic forces. Torsional irregularity shall be considered to exist when the maximum story drift, computed including accidental torsion, at one end of the structure transverse to an axis is more than 1.2 times the average of the story drifts at the two ends of the structure.	1607.3.6.4.2 1607.4.3.1	D and E C, D and E
2. Re-entrant Corners Plan configuration of a structure and its lateral force-resisting system contain re-entrant corners, where both projections of the structure beyond a re-entrant corner are greater than 15% of the plan dimension of the structure in the given direction.	1607.3.6.4.2	D and E
3. Diaphragm Discontinuity Diaphragms with abrupt discontinuities or variations in stiffness, including those having cutoff or open areas greater than 50% of the gross enclosed area diaphragm, or changes in effective diaphragm stiffness of more than 50% from one story to the next.	1607.3.6.4.2	D and E
4. Out-of-Plane Vertical Element Offsets Discontinuities in a lateral force resistance path, such as out-of-plane offsets of the vertical elements which resist the lateral seismic forces.	1607.3.6.4.2	D and E
5. Nonparallel Systems The vertical lateral force-resisting elements are not parallel to, or are not symmetric about, the major orthogonal axes of the lateral force-resisting system.	1607.3.6.3.1	C, D and E

Table 1607.3.5.3
Analysis Procedures for Seismic Performance Categories D and E

Building Description	Referenced Section and Procedures
1. Buildings designated as regular which do not exceed 240 ft in height.	1607.4
2. Buildings that have only vertical irregularities of Type 1, 2 or 3 in Table 1607.3.4.2 and have a height exceeding 5 stories or 65 ft, and all buildings exceeding 240 ft in height.	1607.5
3. All other buildings designated as having plan or vertical irregularities in accordance with Table 1607.3.4.1 and Table 1607.3.4.2.	1607.4 or 1607.5
4. Buildings in Seismic Hazard Exposure Groups II and III in areas with A_a greater than 0.40 within 6.2 miles (10 km) of faults having the capability of generating magnitude 7 or greater earthquakes.	A site-specific response spectrum shall be used but the design base shear shall not be less than that determined from 1607.4.
5. Buildings in areas with A_v of 0.2 or greater with a building period of 0.7 seconds or greater, located on type S_4 soils.	A site-specific response spectrum shall be used but the design base shear shall be not less than that determined from 1607.4. The modal seismic design coefficient (C_{sm}) shall not be limited in accordance with 1607.5.5.

1 ft = 0.305 m

Index

Index

Index

Acceleration, seismic design, 316
Access floor system, 34
Accessibility, egress areas, 178–179
Accessible, 34, 53
Accessory structure, 34
Acoustical ceilings, 153
Additions, 34
Adhesives for roof materials, 275
Administration of codes, 1–31
    appeals, 27–31
    applicability of code, 3–5
    Building Departments established, 5–8
    buildings covered, 4
    certificates, 25–27
    electrical, 4
    enforcement of codes, 1
    federal and state authority vs., 5
    gas, 4
    inspections, 3, 20–25
    inspector power and duty, 8–10
    local vs. national codes, 2
    mechanical systems, 4
    penalties, 31
    permits, 3, 10–20
    plumbing, 4
    quality control, 3
    remedial nature of national codes, 3
    scope of national building codes, 1–2
    standards, 5
    technical codes, 2
    tests required, 27
    unsafe buildings, 30
    variances, 29–30
    violations, 31
    waiver permission, 1–2
Aerial supports, 80
Affidavits for permit process, 16, 18
Aggregates (see Concrete)
Agricultural buildings, 34
Aisles (see Assembly occupancy)
Alarms, fire/smoke alarms, 156, 164–167
Alley, 34
Alterations, 34, 55
American Concrete Institute (ACI), 34, 375

American Institute of Timber Construction, 430
American Plywood Association, 430, 432
American Society for Testing and Materials (ASTM), 36
American Wood Preservers Institute, 432
Annular space, 103, 132
Apartments, 35, 75, 117–118
Appeals process, 27–31
Applicable governing body, 35
Approved, 33, 35
Appurtenant structures, 35
Architects, 14, 35, 53
Architectural trim, 35
    area calculations/limitations, 35–36
    fire resistance and area, 86–87, 103
    floors, 42
    limitations to area, 77–85, 98, 492–495
    modifications to area, 83–85
Asbestos shingles, 252
Asphalt shingles, 269–272
Assembly occupancy, 36
    aisle capacity/positioning, 214–225
    balconies, galleries, 211
    bleachers and grandstands, 36, 50, 82, 217–218, 225
    doors, 212
    egress requirements, 210–225
    fire safety separations, 111–112
    foyers and lobbies, 210
    grandstands (see Bleachers and grandstands)
    guardrails, 214, 225
    handrails, 222
    projection rooms, 212
    seating capacity/positioning, 214–225
    smoke-protected assembly seating, 218–225
    stages, 211
    stairways, 213–214
    tents, 212
ASTM Design for Fire Resistance of Precast Prestressed Concrete, 142

534 Index

ASTM Standards in Building Codes, 5
Atriums, 36
Auditoriums (*see* Assembly occupancy)
Authorized work, 12
Automatic systems, 36
Awnings, 36

Backfilling, 342
Balconies, 36, 206, 255
Base/base shear, seismic design, 316, 330–331, 336–338
Basements, 36, 79, 81–82
Bay windows, 255
Beams, 94–95, 446–447
    link beams, 317
    W/D ratios, 512–513
Bearing walls, 57, 316, 328
Blasting materials, 488
Bleachers and grandstands, 36, 50, 82, 217–218, 225
Boarding houses, 75
Boilers, 37, 47, 48
    egress requirements, 181
    fire safety separation, 112
Brick, 105
Building Departments established, 5–8
Building line, 37
Buildings, terms and definitions, 37
Bulk handling, 37
Burglar bars, 187
Business occupancy, 37, 65–67, 226

Carpeting, 150–151
Cast stone, 37
Ceilings, 153, 457–460
    (*See also* Roofs)
Cement board, 287–290
Certificates, 25–27
    completion certificate, 26
    floor loads, 27
    occupancy certificates, 25–26, 27
    utilities connection, 26
Chimneys, 80, 283–284
Churches (*see* Assembly occupancy)
Closed systems, 56
Code of Federal Regulations (CFR), 38
Code officials, 38
Columns, 93–94
    fire resistance, 102
    minimum concrete covering, 507–510
    W/D ratios table, 506

Combustible materials, 38, 159–160, 489, 490
    (*See also* Fire safety; Flammable)
Completion certificate, 26
Concealed, 33
Concentrically braced frame (CBF), 316
Concrete, 373–409
    admixtures, 374, 377–378
    aggregates, 374, 376
    air-entraining, 379
    cold weather pours, 393
    compressive strength, 383
    corrosion protection, 380
    curing process, 373, 392–393
    deformed reinforcement, 374
    durability requirements, 378–383
    embedded pipes, 393, 395–396
    embedment length, 374
    fire resistance, 137–138, 140–142
    formwork, 393–395
    foundation walls, 347
    freeze-thaw strength, 379
    frost resistant concrete, 381–382
    glass-fiber reinforced concrete (GFRC) wall panels, 401–402
    hot weather pours, 393
    inspections, 373
    joints, 393, 396–397
    mixing and placing of, 391–393
    parapet walls, 402
    permeability, 379
    precast concrete, 374
    prestressed concrete, 374
    properties of masonry units, 511
    quality of, 384–391
    ratios of mix, 384–385
    reinforced concrete, 374–375, 377, 397–401
    required average strength, 385–388
    seismic safety, 402–409, 404–409
    slab on ground, 401
    special exposure conditions, 382
    storing materials, 378
    strength of, 384–391
    substances added to concrete, 379–380
    sulfate exposure, 380
    testing, 378, 389–391
    water for mixing, 375–377, 378–379, 380, 388–389
Concrete block, 107
Condominums, 38, 117
    (*See also* Townhouses)
Congregate residences, 38

Construction documents, 38
Construction types (*see* Types of construction), 38
Contractors and permit responsibility, 17–18
Control area, 39
Cooling towers on roof, 264
Corridors, 39
Corrosive materials, 39
Courts, 39
Crane loads, 296
Crawl spaces, 56, 103, 362, 439–440
Cripple wall seismic tie, 353
Cupolas on roof, 265
Curtain walls, 57

Dampers, 103, 118–133
Dead loads, 39, 295, 296–297
Dead-end hallways, 39, 181
Decay protection for wood, 437–441
Deflection, seismic design, 330
Deformation, seismic design, 324
Design professionals (*see* Architects; Engineers)
Diaphragms:
 gypsum board diaphragms, 481–482
 particleboard diaphragm, 464–466
 seismic safety, 316, 328, 467, 468, 470
 wood structural panel diaphragms, 460–464
Dispensing materials, 40
Domes on roof, 265
Doors, 172, 199–204
 assembly occupancy, 212
 burglar bars, 187
 exterior doors, 200–201
 fire doors, 103, 155, 199–204
 fire resistance, 103, 118–133
 locked doors, 204
 panic hardware, 172, 226–227
 power-operated doors, 201
 revolving doors, 201–203
 self-closing, 54, 104
 sliding doors, 203–204
 special doors and egress, 171
 (*See also* Egress)
Dormitories, 40, 75
Doubled seismic ties, 355, 356
Draftstopping, 40, 103, 109, 129–130, 442–443
Drawings, 12–15
 additional data required, 14

Drawings (*Cont.*):
 design professional requirements, 14
 fire-resistance integrity, 14
 hazardous occupancies, 15
 plans, 18–19
 requirements, 13
 site drawings, 15
 structural integrity proven, 14
 (*See also* Specifications)
Drift, seismic design, 330
Ductwork, fire resistance, 118–133
Dwelling units, 89
Dwellings, 40, 89

Earthquake loads, 295, 313–338
 (*See also* Seismic safety)
Educational occupancy, 40, 67–68
 egress requirements, 226–227
 fire safety separations, 111–112
 panic hardware, 226–227
Efficiency dwelling, 40
Egress and exits, 49, 169–237
 access corridors to egress, 181
 accessibility to egress, 178–179
 arrangement of exits, 176–181
 assembly occupancy requirements, 210–225
 balconies, porches, galleries, 206
 boiler, incinerator, furnace rooms, 181
 burglar bars, 187
 capacity requirements, 175–176
 courts, passageways, vestibules from egress point, 197
 dead-ends, hallways, 181
 doors, 171, 172, 199–204
 educational occupancies, 226–227
 elevators, escalators, moving walks, 176
 emergency egress openings, 181–182
 factory-industrial occupancy, 227
 fire escapes, 198–199
 guardrails, 206–207
 hazardous occupancy, 227–228
 height requirements, 81
 illumination requirements, 172, 207–210
 institutional occupancy, 228–234
 measuring means of egress, 175
 mercantile occupancies, 234
 mezzanines, 185–187
 minimum number of exits, 178
 minimum requirements table, 517

Egress and exits (*Cont.*):
  number of exits required, 176–181
  occupant load vs. number of exits, 169, 173–176, 210
  panic hardware, 172, 226–227
  position of egress, 170–171
  public ways, 171
  ramps, 204–206
  refuge areas, 179–181, 196
  residential occupancy, 235
  roof access, 195
  signs for exits, 207–210
  smoke/heat vents, 177
  smokeproof enclosures, 182–185
  smoke-protected assembly seating, 218–225
  stairways, 172, 188–195
  storage occupancy, 236–237
  various terms for egress, 170–171
  width requirements, 171, 175–176, 517
  windows as egress, 169–170
Electrical, 4, 40, 411–425
  fire protection, 133
  ground fault circuit interrupters (GFCI), 422
  ground fault protection, equipment, 419, 421
  ground fault protection, personnel, 420
  grounding, 422–423
  hazardous locations, 424–424
  inspection, 23
  live parts guarding, 419–420
  National Electric Code (NEC), 411–425
  panel box, 421, 421
  testing electrical systems, 424–425
Elevators, 40, 102, 176
Enforcement of codes, 1
Engineered lumber, 145–146
Engineers, 14, 40, 53
Escalators, 176
Evaluation reports, 40–41
Excavations, 340–342
Existing buildings, 40
Exits (*see* Egress)
Explosives materials, 488
Exterior walls, 57, 239–257
  adhered veneer, 246–248
  architectural trim, 255
  asbestos shingles, 252
  balconies, 255
  bay windows, 255
  fire department access in exterior walls, 255–256

Exterior walls (*Cont.*):
  framing, 450–457
  glass veneers, 249
  housewraps, 256–257
  joints in veneered walls, 249–250
  masonry veneer, 241–248
  mechanical fastenings for veneers, 250
  metal veneers, 248–249
  slab-type veneers, 245–246
  stone veneer, 244–245
  stucco, 252, 253, 254
  tile veneers, 246
  time assigned to, 515
  veneered walls, 239–255
  vinyl siding, 255
  wood siding, 250–252
  (*See also* Framing; Walls)

Factory/industrial occupancy, 27, 47, 68, 227
Families defined, 41
Farm buildings, 41
Fastenings, nails, screws, etc., 443–444
Federal authority vs. National codes, 5
Fiberboard, 433
Fire codes, 41
Fire cuts in framing, 443
Fire escapes, 198–199
Fire protection systems (*see* Fire safety)
Fire resistant materials, 14, 91, 99–143, 496–500, 502–504
  acoustical ceilings, 153
  annular space, 103, 132
  apartments, 117–118
  application of finishes, 152–153
  boilers, 112
  brick, 105
  calculated fire resistance, 136
  carpeting, 150–151
  columns, 102
  combustibles in fire-rated assemblies, 133
  component systems, fire resistant, 142–143
  concealed space combustibles, 134
  concrete block, 107
  concrete, 137–138, 140–142
  dampers, 103, 118–133
  design matrix, 153–154
  doors, 103, 118–133
  draftstopping, 103, 109, 129–130
  ductwork, 118–133

Fire resistant materials (*Cont.*):
  electrical, 133
  elevators, 102
  engineered lumber, 145–146
  fire area, 103
  fire walls, 105, 118
  fireblocking, 103, 109, 129
  floors, 151–152
  foam plastics, 150
  garages, 112
  glass block, 107
  gypsum board, 106–107, 109, 139–140
  heating/cooling, 133
  incinerator rooms, 129
  insulation, 109–110, 134–136
  interior finish classificiations, 149–150
  joint systems, 104
  lintels, 102
  metal lath, 107
  minimum finish classifications, 149–150
  multi-wythe walls, 139
  noncombustible materials, 99–101
  occupancy separation requirements, 110–114
  occupancy-based restrictions, 148–149, 148
  openings in construction, 118–133
  partitions, 103, 114–118
  party walls, 105
  perlite, 107
  plenums, 134
  plumbing, 133
  protected construction, 104
  ratings, fire protection ratings, 103, 104
  refrigerant systems, 112
  refuse and laundry chutes, 128–129
  restrictions on interior finishes, 148–153
  roofing materials, 268
  self-closing fire doors, 104
  separation requirements, 103, 108–118
  shafts, 104, 118–133
  shutters, 118–133
  signs, plastic signs in interior, 153
  single membrane penetration, 104
  smoke barriers, compartments, 104
  sprinkler systems, 129
  stairways, 123–125
  through-penetration, 105, 130–133
  tile, 106
  townhouses, 117
  vermiculite, 107

Fire resistant materials (*Cont.*):
  vertical openings, 105
  walls, 105, 114–118
  windows, 103, 118–133
  (*See also* Fire safety)
Fire retardant, 42
Fire safety, 38, 41–42, 49, 50, 54, 56, 90, 441–443
  alarms, 156, 164–167
  area limitations, 86–87
  automatic systems, 36
  burglar bars, 187
  combustible materials, 38
  crawl spaces, 103
  doors, 199–204
  draftstopping, 442–443
  emergency egress openings, 182
  fire cuts in framing, 443
  fire department access in exterior, 255–256
  fire districts, 87
  fire doors, 155
  fire escapes, 198–199
  fireblocking, 42, 103, 109, 129, 441–442
  fire-retardant treated wood, 435
  height limitations, 78–80, 86–87
  ramps, 204–206
  refuge areas, 179–181, 196
  roof access, 195
  smoke detectors, 156
  smoke/heat vents, 177
  smokeproof enclosures, 182–185
  smoke-protected assembly seating, 218–225
  sprinkler systems, 155, 156–161
  stairways, 187–195
  standpipe systems, 155, 156–157, 161–164
  Type I construction, 91–92
  Type II construction, 92–93
  Type III construction, 93–96
  Type IV construction, 96
  Type V construction, 96–97
  Type VI construction, 97
  types of construction, 90–91
  (*See also* Egress; Fire-resistant materials)
Fire walls, 42, 118
Fireblocking, 42, 103, 109, 129, 441–442
Fire-retardant treated wood, 435
Firestops, 56
Flammable materials, 42, 489, 490
Flashing, 267, 276–277, 282–283, 286, 287–288

Floors, 434
  access floor system, 34
  area calculations, 42
  beams and girders, 446, 447
  decking, 95
  framing, 94, 444–450
  interior finishes, 151–152
  joists, 446, 448, 449
  live loads, 300–301
  masonry wall to floor framing, 450
  openings, 449
  partition supports, 449
  plank-and-beam framing, 450
  sill on foundation, 445–446
  stairways, 450
  subflooring, 449–450
  under-floor inspection, 21
  (*See also* Framing; Loads)
Foam plastics, 43, 150
Footboards, 43
Footings, 43, 342–343, 346–347
  footing bolted to seismic tie, 348
  timber, 362
Formwork for concrete, 393–395
Foundations, 57, 339–371, 436–437
  access, 362
  concrete and masonry, 347
  cripple wall seismic tie, 353
  doubled seismic ties, 355, 356
  excavation, 340–342
  footing bolted to seismic tie, 348
  footings, 342–343, 346–347
  groundwater table, 344
  hollow masonry Type VI construction, 362
  inspection, 21, 23
  insulating concrete forms (ICFs), 367–371
  ledger seismic tie, 349
  pier foundations, 347
  piles, 363–367
  seismic safety design, 362–363
  seismic strap tie, 351
  soil testing, 343–346
  sole plate seismic tie, 357
  special permits, 17
  three-way seismic tie, 350, 358
  timber footings, 362
  truss seismic tie, 352
  two-way seismic tie, 354
  ventilation, 360, 362
  wall design, 347
  wood foundation systems, 362

Foundations (*Cont.*):
  (*See also* Concrete)
Framing, 427–477
  area separation walls, 428
  ceiling framing, 457–460
  chord truss to top plate load factors, 472
  clearing site and preparation, 436
  concentrically braced frame (CBF), 316
  decay protection, 437–441
  design criteria, 429–430
  draftstopping, 442–443
  end-jointed lumber, 435
  fastenings, nails, screws, etc., 443–444
  fiberboard, 433
  fire cuts, 443
  fire protection, 441–443
  fire resistance ratings, 443
  fireblocking, 427, 441–442
  fire-retardant treated wood, 435
  floors, 94, 434, 444–450
  foundations, 436–437
  holes in framing, 427, 457
  inspection, 23, 25, 427
  link beams, 317
  minimum lumber grades, 434–435
  moisture content, 435
  moisture protection, 437
  moment frames, 317
  particleboard, 433
  particleboard diaphragm, 464–466
  piles, 434
  plan structural irregularities, 529
  plumbing, 429
  pressure-treated wood, 436
  quality of material, 432–435
  roof framing, 94, 457–460
  seismic design safety, 316–319, 321–338, 466–473, 525, 527–528
  shear panels, 470–473
  siding, 434
  size of lumber required, 432
  standards, 430
  steel framing, 473–477
  stiffeners, 317
  structual wood wall panel, 460–464
  structural members, 431, 525
  termite/insect protection, 437–441
  three-way seismic connector, 469
  time assigned to, 514
  treated lumber, 435–436
  trusses, 434
  vertical structure irregularities, 526
  wall framing, 450–457

Framing (*Cont.*):
  wood structural panel diaphragms, 460–464
  (*See also* Exterior walls; Floors; Loads; Roofs; Walls)
Front of lot, 43
Furnace rooms, egress, 181

Galleries, 43, 206
Garages, 43, 74
  fire safety separations, 112
  height limitations, 80
  sprinkler systems, 158
Gas utilities, 4, 24
GCp coefficients for MWFRS, 523–524
General building limitations (*see* Limitations to building)
Girders, 94–95, 446, 447, 512–513
Glass block, 107
Glass veneer walls, 249
Glass-fiber reinforced concrete (GFRC) wall panels, 401–402
Grading, 43, 340–342
Grandstands (*see* Bleachers and grandstands)
Ground fault circuit interrupters (GFCI), 422
Grounding, 422–423
Groundwater table, 344
Grout, 44
Guardrails, 44, 206–207, 214, 225
Guest occupancy, 44
Gutters and downspouts, 266
Gypsum board, lath, plaster, 44, 479–485
  application methods, 479–480, 479
  coefficients for plaster, 505
  diaphragms, 481–482, 481
  fire resistance, 106–107, 109, 139–140
  heights of partitions, 481
  inspection, 21, 25, 147
  materials, 479
  structural wall panels, 482–485
  time assigned to, 514

H1/4 materials, 488
Habitable space, 44
Handrails, 44, 205, 222
Hazardous materials, 44–45, 69–72, 113–114
Hazardous occupancies, 15, 159–160, 227–228

Heating/cooling, 38, 45
  fire protection, 112, 133
  plenums, 51–52, 52
Height/height limitations, 45, 77–85, 98, 492–495
  basements, 81–82
  egress areas, 81
  fire resistance and height, 86–87
  garages, 80
  gypsum wallboard walls, 481
  mezzanines, 80–81
  modifications to change, 79–80
  rooftop structures, 80
  seismic design, 324
  special unlimited height structures, 82
Heliports, 45–46, 236
Hood vents, 283
Horizontal separation, 46
Hospitals (*see* Institutional occupancy)
Hotels, 46–47, 75
Housewraps, 256–257
Hydrostatic uplift loads, 295

Illumination in egress areas, 172, 207–210
Impact loads, 303–304
Incinerator rooms, 129, 181
Industrial (*see* Factory/industrial), 27
Inspections, 3, 20–25
  concrete, 373
  electrical, 23
  existing buildings, 21–22
  final, 21, 23, 24
  foundation, 21, 23
  frame, 21, 23, 427
  gas, 24
  gypsum board, lath, plaster, 21, 147
  interior finishes, 147
  manufacturers and fabricators, 22
  mechanical, 24
  plaster fire protection, 25
  plumbing, 23
  prior to permit, 22
  reinforcing steel/structural framing, 25
  rough-in, 23, 24
  slab and under-floor, 21
  underground utilities, 23, 24
  written release, 24
Inspectors, 6–7, 8–10
Institutional occupancy, 47, 72–73, 228–234
Insulating concrete forms (ICFs), 367–371

Insulation:
  fire resistance, 109–110, 134–136
  foam, 43
  housewraps, 256–257
  insulating concrete forms (ICFs), 367–371
  roofs, 266–267
Interior finishes, 145–154
Inverted pendulum-type structures, 318, 329
Irritants, 47

Joists, 446, 448, 449, 457–458
Jurisdictions, 47

Labeled materials, 33, 47
Laundry chutes, 128–129
Ledger seismic tie, 349
Liability for damages, 7
Lighting, self-luminous, 54
Limitations to building, 77–87
  area limitations, 77–85
  buildings located on same lot, 85–87
  fire protection vs. limitations, 78–80
  height, 77–85
Link beams, 317
Lintels, 47, 102
Listed materials, 33, 47
Live loads, 48, 295, 297–304, 519–520
Loads:
  chord truss to top plate load factors, 472
  crane loads, 296
  dead loads, 39, 295, 296–297
  duration, 48, 295
  earthquake loads, 295, 313–338
    (*See also* Seismic safety)
  floor load certificates, 27
  GCp coefficients for MWFRS, 523–524
  hydrostatic uplift loads, 295
  impact loads, 303–304
  live loads, 48, 295, 297–304, 519–520
  occupancy vs. loading, 50, 296, 516
  orthogonal effects, 327
  rain loads, 303
  restrictions on loading, 296
  seismic, 54
  sidewalks, 304
  snow loads, 54, 295, 303, 304
  structural loads, 293–338
  structural safety standards, 295–296
  unit live load, 295

Loads (*Cont.*):
  uplift loads, 295
  velocity pressure chart, 312, 522
  walkways, 304
  wind load, 58, 268–269, 294, 303, 304–313, 522
  (*See also* Floors; Framing; Roofs)
Local vs. National codes, 2
Lodging houses, 48
Lots, 48
  number of buildings on, 85–87
Lumber (*see* Framing)

Manufacturers/fabricators, 35
Marquees, 48
Masonry veneer walls, 241–248
Mechanical systems, 4, 37
  closed, 56
  inspection, 24
  open systems, 57
Mercantile occupancy, 73, 234
Metal lath, 107
Metal veneer walls, 248–249
Mezzanines, 49, 80–81, 185–187
Mineral fiber shingles, 272
Minor repairs, 12
Modes/modal base shear, seismic design, 336–338
Moment frames, 317, 324
Moment resistance, 318, 322–323
Motels (*see* Hotels), 49, 75
Moving walks, 176
Multi-storied construction, 294

National Electric Code (NEC), 411–425
  (*See also* Electrical)
National Forest Products Association, 430
National Particleboard Association, 432
Nationally recognized, 33
Nonbearing walls, 57
Noncombustible materials, 49, 99–101
Number of buildings on lot, 85–87
Nursing homes (*see* Institutional occupancy)

Occupancy, 25–26, 49–50, 59–76
  assembly occupancy, 111–112
  business occupancy, 37, 65–67
  categories of occupancy, 60–61
  certificates of, 25–26, 27

Occupancy (*Cont.*):
  education occupancy, 40, 67–68, 111–112
  egress requirements, 169, 173–176, 210
  factory/industrial occupancy, 47, 68
  fire safety separation requirements, 110–114
  garages, 112
  guest occupancy, 44
  hazardous occupancies, 15, 69–72, 113–114
  institutional occupancy, 47, 72–73
  interior finish restrictions, 148–149
  load limits, 296
  mercantile occupancy, 73
  minimum occupant load, 516
  mixed occupancies, 61–62
  residential occupancy, 53, 73–75
  rooming houses, 61
  signs required, 27
  storage occupancy, 27, 75–76
  temporary/partial, 26
  uncertain classifications, 61
  use factors, 521
Open systems, 57
Open-air/open-space areas, 83–85
Oriel windows, 50
Orthogonal effects, seismic design, 327, 329
Overturning, seismic design, 334
Owner, 50

Panel box, 421
Panels, 57
Panic hardware, 172, 226–227
Parapet walls, 57, 265–266, 402
Parking areas, 98
Particleboard, 433
Particleboard diaphragm, 464–466
Partitions, 50, 103, 114–118
Party walls, 57
P-Delta effect, 318
Pedestrian walkways, 50
Penalties for violation, 31
Penthouses, 80, 259, 261–263
Perlite, 107
Permanent seating, 50
Permits, 3, 9–10, 38, 50
  affidavits, 16, 18
  alternate materials/methods, 10
  appeals process, 27–31
  application process, 11–12
  authorized work, 12

Permits (*Cont.*):
  contractor's responsibility, 17–18
  drawings and specifications, 12–15
  examination process, 15–17
  fees, 10–11, 19–20
  information required, 12
  inspections prior to permit, 22
  issuance, 16–17
  minor repairs, 12
  misrepresentation of application, 9
  plan reviews, 15
  plans, 18–19
  posting, 22–23
  public right-of-way, 17
  refusal, 17
  requirements not covered by code, 10
  revocation, 9
  special foundation permit, 17
  temporary structures, 11
  time limit, 12, 18
  unsafe buildings/systems, 9
  valuations to set fees, 20
  variances, 29–30
  violations of code, 9
  work commenced before permit, 20
Photoluminescent materials, 51
Pier foundations, 347
Piles, 363–367, 434
Pipes protruding through roof, 288
Plan structural irregularities, 529
Plans, 18–19
  (*See also* Drawings)
Plastics, 51, 150
Platforms, 51
Plenums, 51–52, 52, 134
Plumbing, 4, 429
  fire protection, 133
  inspection, 23
Porches, 206
Pressure treated wood, 436
Prisons (*see* Institutional occupancy)
Property lines, 53
Public spaces, 53

Quality control, 3

Rafters, 457–458
Rain loads, 303
Ramps, 204–206
Recordkeeping by building dept., 7
Refrigerant systems, 112

542 Index

Refuge areas, 179–181, 196
Refuse chutes, 128–129
Reinforcement (*see* Concrete)
Repairs, 53
Residential occupancy, 53, 73–75, 89
  egress requirements, 235
  sprinkler systems, 160–161
Resilient stable mounting systems, 318
Restraining devices, seismic designs, 318
Retaining wall, 57, 361
Revolving doors, 201–203
Right-of-way, 17
Roofs, 259–291, 457–460
  access/egress, 195
  adhesives, sealants, 275
  anchoring roof to masonry walls, 459
  asphalt shingles, 269–272
  attic space access, 459
  block bulkhead-to-roof connection, 262
  cement board roofing, 288–291
  chimneys, 283–284
  cooling towers on roof, 264
  covering systems, 259–260, 266–288
  decking, 95
  fasteners, 267
  fire resistant materials, 268
  flashing, 267, 276–277, 282–283, 286–288
  framing, 94
  GCp coefficients for MWFRS, 523–524
  gutters and leaders, 266
  hood vents, 283
  insulation, 266–267
  joists, 457–458
  live loads, 302–304
  mineral fiber shingles, 272
  overhangs, 524
  parapet walls, 265–266
  particleboard diaphragm, 464–466
  penthouses, 259, 261, 263
  pipes protruding through roof, 288
  plank-and-beam roofs, 459
  rafters, 457–458
  rake treatment, 285
  scuppers, 261
  sheathing, 458–459
  skylights, 283
  slate shingles, 272
  spires, domes, cupolas, towers on roof, 265
  structual wood wall panel, 460–464
  tanks mounted on roof, 263–264
  tile roofs, 273–288

Roofs (*Cont.*):
  time assigned to, 515
  truss torsion factors, 260
  trusses, 271, 271, 458
  turbines, 283, 288
  underlayments, 274, 281–282, 285, 286–288
  valleys, 287
  ventilation, 459–460
  vents, 283, 288, 518
  wall abutments, 283–284
  wind loads, wind resistance, 268–269
  (*See also* Loads)
Rooftop structures, 80
Rooming houses, 75
Rough-in inspection, 23, 24

Safe dispersal areas, 53–54
Scaffolding, 242
Scuppers, 261
Seating (*see* Assembly occupancy)
Seismic safety, 313–338
  acceleration, 316
  additions to existing buildings, 314–315
  analysis procedures, 325
  base/base shear, 316, 330–331, 336–338
  bearing wall systems, 316, 328
  Category A buildings, 326
  Category B buildings, 326
  Category C buildings, 326, 329
  Category D buildings, 326, 329
  Category E buildings, 325, 326, 329
  chord truss to top plate load factors, 472
  collector elements, 328
  concrete, 403–409, 404–409
  concrete/masonry wall anchors, 327
  cripple wall seismic tie, 353
  deflection, 330
  deformation, 324
  design earthquake, 316
  design requirements, 319
  designated seismic systems, 316
  diaphragms, 316, 328, 467, 468, 470
  discontinuities, 327
  doubled seismic ties, 355, 356
  drift, 330, 334–335
  effective peak/peak velocity, 316
  equivalent lateral force procedures, 330
  excepted buildings, 313–314
  footing bolted to seismic tie, 348
  foundation design, 362–363

Seismic safety (*Cont.*):
  framing guidelines, 316–319, 321–338, 466–473
  height limitations, 324
  horizontal shear, 334
  interaction effects, 324
  inverted pendulum-type structures, 318, 329
  irregularities, 325, 329
  ledger seismic tie, 349
  link beams, 317
  loads, 54
  materials, 327
  mixed use buildings, 315
  modeling, 335
  modes/modal base shear, 336–338
  moment frames, 317, 324
  moment resistance, 318, 322–323
  occupancy changes, 315
  openings, 327
  operational access to hazard group III, 315
  orthogonal effects, 327, 329
  overturning, 334
  P-Delta effect, 318, 334–335
  period calculations, 332–333, 336
  ratings, 525, 527–528, 530
  required design data, 314
  resilient stable mounting systems, 318
  restraining devices, 318
  seismic coefficients, 331–332
  seismic ground acceleration maps, 315
  seismic hazard exposure groups, 315
  seismic performance categories, 316, 323–338
  shear panels/walls, 319, 470–473
  site coefficients, 320–321
  site limitations, 316
  soil and structure interaction, 321
  sole plate seismic tie, 357
  stiffeners, 317
  story drift ratio, 319, 334–335
  story shear, 319
  strap tie, 351
  three-way seismic connector, 469
  three-way truss connectors, 453
  three-way seismic tie, 350, 358
  through-floor seismic connectors, 452
  ties and continuity, 326
  torsion, 334
  truss seismic tie, 352
  two-way seismic tie, 354
  vertical distribution of forces, 333–334

Seismic safety (*Cont.*):
  vertical seismic loads, 330
Self-closing doors, 54, 104
Self-luminous lighting, 54
Sensitizer materials, 54
Separation, 46
Service corridors, 54
Shafts, 54, 104, 118–133
Shear panels/walls, 319, 470–473
Shear, seismic design, 316, 334
Shingles, 252
Shutters, 118–133
Sidewalks, 304
Siding, 250–252, 434
Signs, 27
  exit signs, 207–210
  plastic signs in interior, 153
Single membrane penetration, 104
Site, 37
Site drawings, 15
Skylights, 283
Slab, 21, 440
  (*See also* Concrete; Foundations)
Slate shingles, 272
Sliding doors, 203–204
Smoke barriers, compartments, 104
Smoke detectors, 54, 156
Smokeproof enclosures, 182–185
Smoke-protected assembly seating, 218–225
Snow loads, 54, 295, 303, 304
Soil testing, 343–346
Sole plate seismic tie, 357
Specifications, 12–15
  additional data required, 13–14
  design professional requirements, 14
  fire resistance, 14
  requirements, 13
  structural integrity, 14
  (*See also* Drawings)
Spires on roof, 265
Sprinkler systems, 54, 82–83, 129, 155, 156–161
Stairways, 54–55, 172, 187–195
  assembly occupancy, 213–214
  fire protection, 123–125
  framing, 450
  signs identifying stairs, 209–210
  smokeproof enclosures, 182–185
Standard Unsafe Building Abatement Code, 9
Standards, 5

Standpipe systems, 155, 156–157, 161–164
State authority vs. National codes, 5
Steel framing, 473–477
Stiffeners, 317
Stone walls, 244–245
Stop work orders, 9
Storage occupancies, 27, 75–76, 159, 236–237
Stories, 55
Story drift ratio, 319
Story shear, 319
Streets, 55
Structural integrity, 14
Structural loads (*see* Loads)
Structural observation, 55
Structural systems, 525
Stucco, 252, 253, 254
Subflooring, 449–450
Surgical areas, 56
Synagogues and temples (*see* Assembly occupancy)

Tanks mounted on roof, 263–264
Technical codes, 2
Temporary structures, 11
Tenants, 56
Termite/insect protection for wood, 437–441
Terms and definitions, 33–58
Testing, 27
Theaters (*see* Assembly occupancy)
Three-way seismic tie, 350, 358
Through-penetration, 105, 130–133
Tile, 106, 246, 273–288
Time limit of permits, 12
Torsion, seismic design, 334
Towers on roof, 265
Townhouses, 56, 117–118
  (*See also* Apartments; Condominiums)
Toxic materials, 46, 491
Treated lumber, 435–436
Trim, 35, 255
Truss Plate Institute Inc., 432
Truss seismic tie, 352
Trusses, 271, 434
Turbines, 283, 288
Two-way seismic tie, 354
Type I through VI construction, 91–97
Types of construction, 38–39, 89–98
  classifications, 89–90
  dwellings, 89

Types of construction (*Cont.*):
  fire-resistance requirements, 90–91
  mixed types of construction, 98
  residential construction, 89
  Type I through VI construction, 91–97

Underground utilities, 23, 24
Underlayment for roofing, 274, 281–282, 285, 286–287, 288
Underwriters Laboratories (UL), 56
Unit live load, 295
Unsafe buildings, 30
Uplift loads, 295
Use, 56, 57
Use factors, 521
Utilities, certificate of connection, 26

Valuation, 57
Variances, 29–30
Velocity pressure chart, 312, 522
Veneer, 57
Vents, 283, 288
Vermiculite, 107
Vertical opening, 57
Vertical structure irregularities, 526
Vinyl siding, 255
Violations of code, 31

Waiver permission, 1–2
Walkways, 304
Walls, 57, 94–96, 440
  bearing walls, 456
  bracing, 454–455
  cement board, 288–291
  corner bracing, 454–455
  fire resistance periods, 502–504
  fire safety separation requirements, 108–118
  fire walls, 105
  foundation walls, 347
  framing, 450–457
  glass-fiber reinforced concrete (GFRC) wall panels, 401–402
  gypsum board, 482–485
  holes in framing, 457
  loads, 302
  minimum thicknesses, 501, 505
  nailing, 455
  nonbearing walls, 456
  openings in walls, 455–456

Walls (*Cont.*):
 particleboard diaphragm, 464–466
 partitions, 456
 party walls, 105
 plan structural irregularities, 529
 post-and-beam framing, 456
 seismic ratings, 527–528
 sheathing, 455
 structual wood wall panel, 460–464
 three-way truss connectors, 453, 453
 through-floor seismic connectors, 452
 vertical structure irregularities, 526
 (*See also* Exterior walls; Framing)
Weather-exposed surfaces, 57–58

Wind load, 58, 268–269, 295, 303–313
Windows, 50, 58
 bay windows, 255
 burglar bars, 187
 egress use, 169–170
 fire resistance, 103, 118–133
Wood siding, 250–252
Written notices, 58

Yards, 58

Zoning, 58

ABOUT THE AUTHOR

Jonathan F. Hutchings is a licensed building contractor and a registered building inspector. Executive Director of the Construction Management Institute, he has compiled one of the most comprehensive databases of national building codes in the country. Mr. Hutchings is also the author of McGraw-Hill's *CPM Construction Scheduler's Manual, Builder's Guide to Landscaping,* and *Builder's Guide to Modular Construction.*